Matthew Moncrieff Pattison Muir, Charles Slater

Elementary Chemistry

A Companion Volume to Pattison Muir and Carnegie's Practical Chemistry

Matthew Moncrieff Pattison Muir, Charles Slater

Elementary Chemistry
A Companion Volume to Pattison Muir and Carnegie's Practical Chemistry

ISBN/EAN: 9783337059989

Printed in Europe, USA, Canada, Australia, Japan

Cover: Foto ©berggeist007 / pixelio.de

More available books at **www.hansebooks.com**

ELEMENTARY CHEMISTRY

BY

M. M. PATTISON MUIR, M.A.

FELLOW AND PRÆLECTOR IN CHEMISTRY OF GONVILLE AND
CAIUS COLLEGE,

AND

CHARLES SLATER, M.A. M.B.

FORMERLY SCHOLAR OF ST JOHN'S COLLEGE,
CAMBRIDGE.

A COMPANION-VOLUME TO PATTISON MUIR AND CARNEGIE'S
PRACTICAL CHEMISTRY.

CAMBRIDGE:
AT THE UNIVERSITY PRESS.
1887

PREFACE.

THIS book forms one part of a course of elementary chemistry; the other part is contained in a companion-volume on *Practical Chemistry*. The two books are intended to be used together, the one being complementary to the other; their object is to teach the elements of chemical science.

A real knowledge of chemistry can be gained only by the intimate blending of a properly graduated course of laboratory-work with the lecture-work and reading of the student.

An attempt is made in this volume to present the principles of chemistry as rising out of and resting upon chemical facts, and chemical facts as furnishing the data from which principles are deduced. The book is intended to be used in conjunction with lectures on elementary chemistry, wherein more details will be given concerning important and typical bodies than are found in this volume. The book does not profess to be a descriptive catalogue of chemical facts regarding the properties of the individual elements and compounds.

The authors entertain views rather different from those which generally prevail regarding the relative importance of the various parts of chemistry; they have endeavoured to make the teaching given in this book sound so far as it goes; they have tried to bind together the facts and principles of the science, and at the same time to avoid speculation.

M. M. PATTISON MUIR.

CHARLES SLATER.

CAMBRIDGE, *October*, 1887.

TABLE OF CONTENTS.

CORRECTIONS AND ADDITIONS.

ELEMENTARY CHEMISTRY.

CHAPTER I.

CHEMICAL CHANGE.

CHEMISTRY is a branch of natural science. The aim of **1**
science is to "see things as they are." But *to see things as
they are* it is necessary to study the relations of things, because
in nature nothing is wholly cut off from other things, but
everything is either a cause or a consequence of many others,
and is related in manifold ways even to things which may
seem to be wholly unconnected with it.

For the purposes of exact study however some boundary
lines must be drawn between what we call the different parts
of each natural occurrence. Every natural occurrence, in
relation to our powers of comprehending it, is infinitely complex; in order to explain we must simplify; and to simplify
we must overlook some portions of the complete phenomenon.

Chemistry deals with certain portions of one class of
material phenomena. The mark of this class of phenomena
is, *change of properties accompanying change of composition.*
The object of chemistry is to classify the phenomena it studies
in order to discover general laws.

The object of this book is to place before the student an
outline of the methods by which chemistry proceeds; to teach
him some of the general laws of the science; and above all
things to shew him that the laws are gained by studying
natural occurrences, that the detailed study of these is the
foundation on which the science rests, but that, in so far as it
is a real branch of science, chemistry is much more than a
descriptive catalogue of interesting facts.

M. E. C. **1**

A little observation suffices to shew that all things are undergoing change. Physics and chemistry deal with the phenomena presented in material changes. Certain aspects of these changes we call physical; certain aspects of them we call chemical.

A fire burns on the hearth : when the fire was kindled the grate was filled with lumps of coal; as the fire sparkles and blazes up the black coal changes to a light-giving, glowing, mass, radiating heat on all sides; as the flames cease to play about the glowing coals the colour fades, the ashes accumulate, and the burning slackens; at last the change stops, there remain only ashes and some pieces of unburnt coal.

Many of the changes which pass before us as we watch the progress of a coal-fire are chemical changes. It is with such processes as this that we are concerned.

The burning and slow extinction of an ordinary fire is however an extremely complex event; we must turn to comparatively simple occurrences if we are to learn the characteristic features of chemical change.

2   When a piece of *platinum* wire is held in the flame of a Bunsen-lamp it becomes hot and gives out light; when the wire is removed from the source of heat it quickly cools, ceases to emit light, and returns to the same condition as before heating. When a piece of *magnesium* wire or ribbon is brought into the lamp-flame it also becomes hot and gives out light, but at the same time it rapidly burns away; when removed from the source of heat it continues to burn with emission of dazzling white light; after a little the burning ceases; the magnesium has now disappeared and in its place there is formed a white, soft, powder, called *magnesia*, very unlike the hard, lustrous, magnesium which was placed in the lamp-flame.

Some change was here produced in the properties both of the platinum and magnesium. In the case of the platinum, the properties of glowing and of communicating heat to colder bodies brought into contact with it were temporarily added to the other properties—hardness, lustre, tenacity, high specific gravity, infusibility, &c.—which distinguish platinum from other kinds of matter. When those properties which had been temporarily added were withdrawn, the platinum was found to exhibit the same properties which characterised it before it was brought into the lamp-flame. In the second case, the magnesium also temporarily acquired the properties of glowing and communicating heat to colder bodies brought into contact with it. But the withdrawal of these properties did not leave the magnesium as it was before heating; accompanying the exhibition of these properties there was a gradual change of the magnesium into a substance so unlike magnesium as to be at once recognised as a different kind of matter.

If a thin copper wire, covered with cotton or silk, is **3** wrapped many times round a piece of *soft-iron* and an electric current is then passed through the wire, the iron will acquire the property of attracting iron-filings; if the electric current is stopped the iron-filings cease to be attracted; if the current is again passed through the wire the soft-iron at once acquires the attractive power. Before, during, and after, the passage of the electric current, the soft-iron exhibits all those properties—colour, relative density, tenacity, malleability, texture, &c.—which mark it off from other kinds of matter; while the current is passing there is added to the iron the property of attracting iron-filings. If the same piece of soft-iron is exposed to damp air for a considerable time, a portion of it, or if sufficient time be given the whole of it, will be changed into *iron-rust*, which is a reddish powder unlike the iron in texture, colour, lustre, tenacity, malleability, and many other properties.

The change of iron to iron-rust resembles the change of magnesium to magnesia, in that in both cases there is produced a new kind of matter. The temporary addition to iron of the property of attracting iron-filings resembles the temporary addition to platinum of the property of glowing and communicating heat to colder bodies, in that in both cases the change consists in the addition of a property which does not destroy or mask the original properties, and which can be withdrawn by reverting to the conditions existent before this property was added.

Let a rod of *copper* and an electric bell be introduced into **4** the circuit of a galvanic battery (*s*. fig. 1); the ringing of the bell shews that the electric current is passing through the rod of copper. The moment the current is broken the bell ceases to ring. The characteristic properties of the copper are not modified by the passage through it of the electric current. Let the rod of copper be now removed; let a piece of sheet-platinum be attached to the end of the wire from the battery, and also to the end of the wire from the electric bell, and let these pieces of platinum be placed, side by side but not touching, in a dilute aqueous solution of *copper sulphate* to

1—2

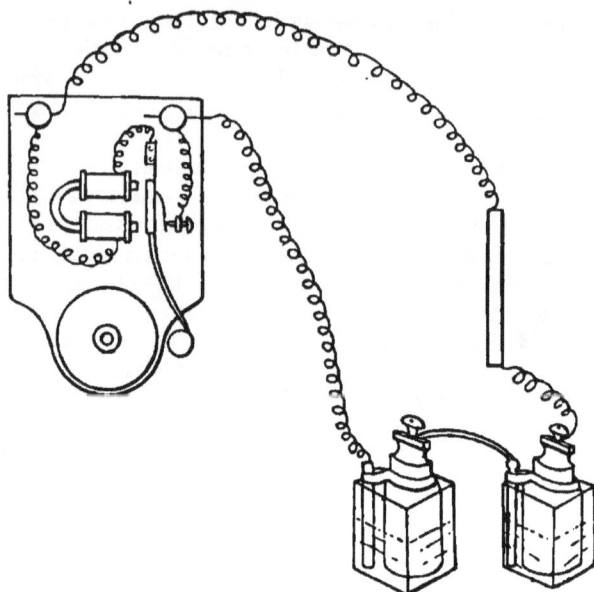

Fig. 1.

which some sulphuric acid has been added (*s.* fig. 2). The ringing of the bell shews that the electric current is passing through the solution of copper sulphate; but the passage of the current is accompanied by the deposition on one of the

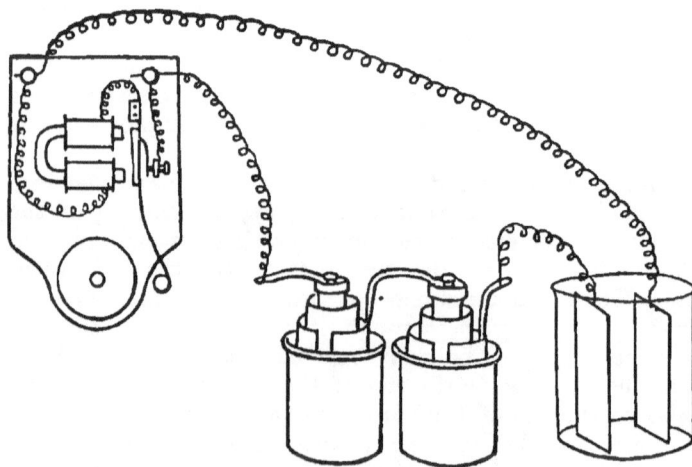

Fig. 2.

platinum plates of a reddish solid which may be proved to be
copper, and by the gradual disappearance of the copper
sulphate from the solution. When the current is stopped
there remains the new kind of matter, copper, which has been
formed by the action of the electric current on the kind of
matter originally present, copper sulphate; a certain amount
of one kind of matter has disappeared and a certain amount of
another kind of matter has been formed. The original matter
is not reproduced by reverting to the conditions which existed
before the change began; that is to say, by stopping the
passage of the electric current.

The three kinds of matter, platinum, soft-iron, and copper, **5**
have been changed by temporarily adding to each a property
which does not belong to it under ordinary conditions; this
property existed only so long as the special conditions which
caused its existence continued; the withdrawal of these
conditions was accompanied by the withdrawal of the special
property; when this property was withdrawn the platinum,
soft-iron, or copper, remained exactly as it was before the
change had been effected. On the other hand, the three kinds
of matter, magnesium, iron, and copper sulphate, have been
changed by each permanently losing certain properties which
characterise it, and at the same time permanently acquiring
new properties which characterise other kinds of matter.

Now we recognise different substances by their properties. **6**
One kind of matter is recognised, and distinguished from other
kinds of matter, by its colour, texture, brittleness, opacity,
relative density, hardness, &c.; also by its behaviour when
heated, when submitted to the action of electricity, placed in
direct sunlight, mixed with water, brought into contact with
other substances at high and low temperatures, &c. Substances
which have markedly different properties are said to differ in
kind, or to be different kinds of matter. Thus we say that
iron is a kind of matter different from glass; that sand is a
kind of matter different from wood, &c.

The prominent feature of the change undergone by the **7**
magnesium when heated, by the iron when allowed to remain
in damp air, and by the copper sulphate when the electric
current was passed through it, is, that in each case a kind of
matter has been produced different from, and in the place of,
that which existed before the change began. The prominent
feature of the change undergone by the platinum when heated,
by the iron when the current was passed round it, and by the

copper when the current was passed through it, is, that no
new kind of matter has been produced, but that the kind of
matter which existed before the change began existed also
while the change lasted and after the change ceased.

The first kind of change is called *chemical change;* the
second is called *physical change.* The differences and resemblances between these two kinds of change must be more fully
illustrated.

8   *Iodine* is a definite kind of matter, distinguished from other
kinds by its lustre, greyish purple colour, opacity, easy solubility in alcohol with production of a reddish brown liquid, and
by the fact that when a drop or two of this liquid is added to a
very little starch paste a substance is formed which colours the
liquid deep blue. *Lead nitrate* is a heavy, white, crystalline,
solid; it dissolves in a little hot water and separates from this
solution, as it cools, in white, lustrous, crystals.

Two retorts are arranged with the beaks passing into small
dry flasks, as shewn in fig. 3; a little *iodine* is placed in one retort,
and a little *lead nitrate* in the other; each retort is heated by

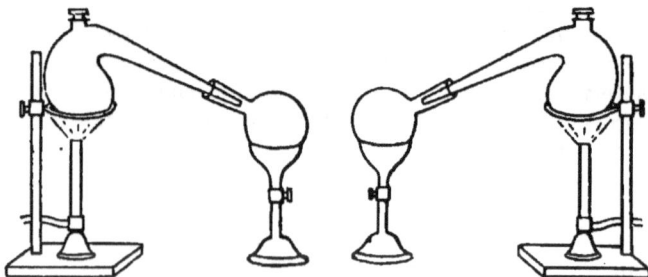

Fig. 3.

a Bunsen-lamp. The iodine soon changes to a dark purple,
almost opaque, gas; but this condenses on the cooler parts of
the retort and in the small flask, to a solid, which presents the
same appearance, and is possessed of the same properties, as
the iodine originally used. The lead nitrate is also changed;
a brownish red gas is produced which does not condense to a
liquid or solid; if the heating is continued so long as this gas
is produced, a yellowish coloured solid remains in the retort;
this solid is a different kind of matter from the lead nitrate
originally used. The change of solid iodine to gaseous iodine,
and of gaseous iodine to solid iodine, is a *physical change;* the
change of lead nitrate into two new kinds of matter—a brownish

red gas called *nitrogen oxide*, and a yellowish solid called *lead oxide*—is a *chemical change*.

The change which *water* undergoes when it is boiled is a **9** *physical change;* if the water is placed in a retort arranged as shewn in fig. 3, the water-gas (or steam) produced by heating the water is condensed to liquid water which is found in the small flask. But water may also be *chemically changed*. An electric current is passed through water to which some sulphuric acid has been added. The current passes from one plate of platinum to another; these plates are placed each within an inverted tube full of water and standing in a vessel of water (*s.* fig. 4). Bubbles of gas rise from each platinum plate and collect in the inverted tubes. If the process is continued the water will at last entirely disappear and in

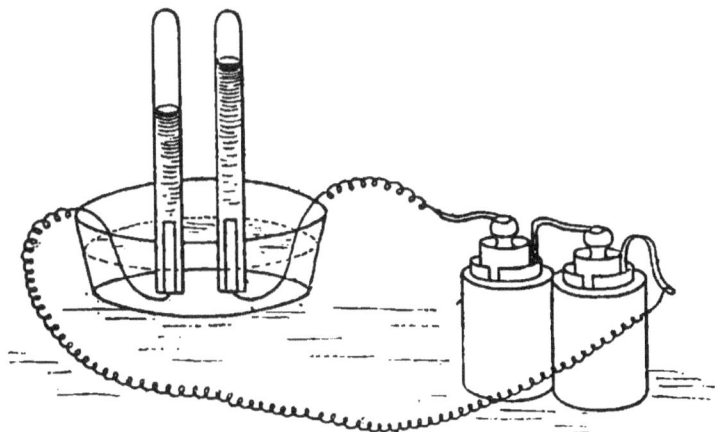

Fig. 4.

place of it we shall have two colourless gases. The gas in each tube is examined as regards its behaviour towards a.burning splint of wood: one of the gases takes fire, it is *hydrogen;* the other does not, but the splint of wood burns in the gas very rapidly and brilliantly, this gas is *oxygen*. These gases are definite kinds of matter; each is evidently very different from the water from which both have been produced. The change of water into the gases hydrogen and oxygen is a *chemical change*.

A few pieces of *loaf-sugar* are placed in a little water in a **10** porcelain dish; the sugar slowly disappears; some change has occurred. The water is now removed by boiling; the solid residue in the dish is sugar, it is characterised by all those properties which mark off sugar from other kinds of matter. The change of solid sugar into a solution of sugar, and of sugar-solution into solid sugar, is a *physical change.*

11   Into a small quantity of water are thrown one or two pieces of the metal *sodium;* the metal swims on the surface of the water; a gas is produced which may be collected and examined; when the sodium has disappeared the water is boiled off; there remains a white solid, which dissolves in water without formation of any gas, and which is evidently very different from either the sodium or the water by the mutual action of which it has been formed. This change of water and sodium into a solid body, *caustic soda,* and a gas, *hydrogen,* is a *chemical change.*

12   A little hot concentrated *sulphuric acid* is poured on to some pieces of *loaf-sugar;* much heat is produced, steam is given off, and a black, charcoal-like, solid remains. Sugar was only physically changed when it was brought into contact with water : but the mutual action of sugar and hot sulphuric acid is a *chemical change;* both the visible products of this change, *steam* and *carbon,* are different kinds of matter from the bodies, sugar and sulphuric acid, by the interaction of which they have been produced.

13   All those changes which we have classed together as chemical have this in common, that one, or more than one, kind of matter has disappeared, and another kind, or other kinds, of matter has been formed. So far as our experiments could tell us, the new matter formed did not exist as a part of the material system before the change began.

Those changes which we have classed together as physical have been characterised by the continued existence, during and after each process, of the same kind of matter which was present before the change began. This matter temporarily acquired a new property, or new properties; but the new property did not prevent us from recognising the other properties by which the special kind of matter was marked off from other kinds. In both classes of occurrences the matter experimented with was subjected to new conditions different from those which existed before the experiments began. When these conditions were removed, in one class of phenomena— the physical—we had a return to the state of things which prevailed at the beginning of the experiments; we had the

same kind or kinds of matter exhibiting the same properties :
in the other class of phenomena—the chemical—we had not a
return to the original state of things ; we had new kinds of
matter exhibiting new properties.

If the occurrences we have been considering were very **14**
closely and accurately examined it would be found that those
we have called chemical include changes which belong to the
physical class. The emission of light by the burning magnesium,
the conduction of heat through the mass of lead nitrate, the
heating and volatilisation of water in the interaction of sugar
and sulphuric acid ; these are physical rather than chemical
changes. The physical, and the chemical, are different aspects
of the complete phenomenon. We try to separate them as far
as we can that we may study each more accurately, and so find
·the general laws which hold good for each ; for the more we
understand natural events the more are we convinced that one
law of nature is never suspended or stopped by another law,
however the effects of one may be modified by the effects of
another.

But we cannot at present attempt minutely to analyse the
phenomena we have to study into chemical and physical parts;
we must be content to learn the broad features of the two
classes of occurrences.

In reasoning on the data obtained in the experiments **15**
already described, certain assumptions have been made, and
certain expressions have been used somewhat vaguely. It
was asserted that when magnesium was burnt in air, the
matter called magnesium disappeared and its place was taken
by a new kind of matter called magnesia ; that when lead
nitrate was heated the matter characterised by the properties
summed up in the name lead nitrate disappeared and in its
place there were formed two other kinds of matter,—nitrogen
oxide, and lead oxide ; that the passage of the electric current
through copper sulphate solution was accompanied by the
disappearance of one kind of matter,—copper sulphate, and
the formation of another kind of matter,—copper ; and
similarly with the other experiments. Now one may well
ask : how can it be proved that the magnesium, or the lead
nitrate, or the copper sulphate, really disappeared ? how can it
be proved that the place of the magnesium was taken by magnesia, or of lead nitrate by lead oxide and nitrogen oxide, or of
copper sulphate by copper ? And questions such as these
must also be asked : what exactly is meant by saying that the

magnesium, or the lead nitrate, *disappeared ?* : magnesia took
the place of magnesium ; copper, the place of the copper sulphate ; what is meant by saying *one substance takes the place of
another ?* An electric current was passed through water, some of
the water disappeared and in its place two gases were produced ;
but may not these gases have come from the platinum plates,
or from the glass of which the vessel was made ? Or, assuming
that the water was indeed changed in this experiment into the
two gases called hydrogen and oxygen, we ask ; what definite
meaning is to be put upon the statement *water can be changed
into hydrogen and oxygen ?*

16   The history of alchemy, and of the transition from alchemy
to chemistry, teaches the necessity of putting, and of answering,
such questions as these. The alchemists not only thought
that they could, but asserted that they did, change water into
earth or into fire, lead into silver, and copper into iron. Their
conception of nature led them to regard all things as undergoing continual change ; but they were not able so accurately
to study these changes as to discern the unchanging sequences
in which they occurred, and to grasp the unchangeable parts
of the phenomena they observed.

The assertion that water could be changed into earth, or
into fire, was based upon such experimental evidence as this.
A quantity of water was heated in an open glass vessel ; the
water slowly disappeared, and a little white earthy solid
matter remained in the vessel. The water had disappeared
and an earthy solid had been produced in its place. A piece
of red hot iron was plunged into water contained in a bell-
shaped glass vessel ; bubbles of gas rose through the water
and collected in the vessel ; this gas took fire when a lighted
taper was brought into it. The water had been changed into
'*the matter of fire*'. To prove that copper could be changed
into iron, the alchemist placed a piece of copper in *aqua fortis*
(nitric acid) ; the copper slowly disappeared ; in the blue-green
liquid thus formed he placed a piece of iron ; the iron disappeared, and copper was produced in its place. The conclusion
which the alchemist drew from such experiments as these was
that one kind of matter could be changed into other kinds of
matter. But if this were so, why should not any kind of
matter be changeable into any other ? Heat brought about
the change of water to earth ; hot iron, the change of water
into '*the matter of fire*'; *aqua fortis*, the change of iron into
copper. There must surely be some one thing which would

effect all transmutations. The pursuit of this *One Thing* became the central quest of alchemy. "There abides in nature" we read in an alchemical treatise "a certain pure matter, which, being discovered and brought by art to perfection, converts to itself proportionately all imperfect bodies that it touches."

Alchemy was a fascinating dream; but chemistry is a more satisfying reality.

Let us return to experiment. Let a small weighed quantity **17** of magnesium be burnt in air under conditions such that the whole of the magnesia produced in the burning remains in the vessel in which the burning proceeds. The apparatus shewn in fig. 5 is a simple one for the purpose. Some magnesium ribbon is placed on a piece of wire-gauze and covered with an inverted funnel, the stem of which is connected by caoutchouc tubing with another funnel; the upper funnel is covered with filter-paper. The whole apparatus is counterpoised; the funnels are removed; the magnesium is ignited by allowing a Bunsen-flame to play on to it from above; the funnels are then replaced. When the burning is complete the apparatus is allowed to cool and is then counterpoised. It is found that the magnesia produced weighs more than the magnesium before burning. We therefore conclude that the magnesia is produced by adding to, or combining with, the magnesium, some other kind of matter. As the change from magnesium to magnesia proceeded in air, it is probable that the new kind of matter, which, by our hypothesis, has combined with magnesium and so produced magnesia, is derived from the air. To find whether this conclusion is correct or not, it would be necessary, (1) to burn a known weight of magnesium in a known quantity of air; (2) to determine the weight of magnesia produced, and the diminution in the quantity of air which accompanied the production of this weight of magnesia; (3) to change the magnesia back to magnesium and air, and to determine the weight of each of these obtained. If the difference between the weight of the magnesia and that of

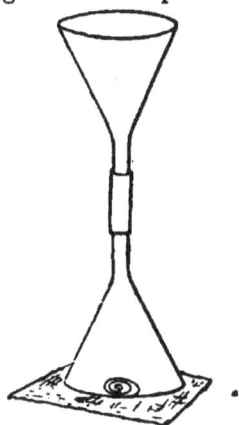

Fig. 5.

the magnesium, by burning which the magnesia was formed, was equal to the weight of air which disappeared during the burning; and if the magnesium obtained from the magnesia weighed the same as the magnesium originally burnt; and if the weight of air obtained from the magnesia was the same as the weight of air which disappeared during burning; then we should be justified in concluding that the chemical change which occurs when magnesium is burnt consists in the addition to, or combination with, magnesium, of a portion of the surrounding air, and that the new kind of matter produced is composed of two kinds of matter, viz. magnesium and air. We should further have learned that although the magnesium has disappeared it has not been destroyed; and that although magnesia has been produced it has not been produced from previously non-existent matter. We should have given a definite meaning to the terms '*produced*' and '*disappeared*', as regards the change of magnesium to magnesia at any rate; and we should, to some extent at least, understand what is meant by saying *the magnesia has taken the place of the magnesium which was burnt*.

It is not easy to arrange quantitative experiments by which the chemical change of magnesium to magnesia may thus be examined. But if we use mercury in place of magnesium, we can arrange an experiment which will enable us to find an answer to each of the three questions stated in par. 15.

18   The change of mercury to burnt mercury, or as we now call it oxide of mercury, was examined quantitatively by Lavoisier. A sketch of the essential parts of his apparatus is shewn in fig. 6. Lavoisier placed 4 oz. of mercury in a glass balloon, the neck of which, drawn out and bent, passed under mercury and then into the air contained in a bell-jar; the bell-jar contained 50 cub. inches of air. The mercury was heated nearly to its boiling point by means of a furnace; red specks appeared on the surface of the mercury and the volume of air in the bell-jar slowly decreased. After some days the production of red solid matter on the surface of the mercury seemed to have ceased; heating was continued for a few days more (12 days in all), and was then stopped. The air in the bell-jar now measured between 42 and 43 cub. inches, the diminution in volume was therefore between 7 and 8 cub. inches; the red solid was collected, and was found to weigh 45 grains. These 45 grains of the red solid produced by slowly burning mercury in air were placed in a glass tube

closed at one end and drawn out at the other ; the open end
passed, under mercury, a little way into a graduated glass
vessel filled with mercury (s. fig. 6). When the red solid was
heated mercury was formed and deposited on the colder parts
of the tube, and a gas collected in the graduated vessel. The
mercury thus formed weighed 41½ grains; the gas measured

Fig. 6.

between 7 and 8 cub. inches. The gas was proved to be, not
air, but oxygen : now 7½ cub. inches of oxygen measured at
the temperature and pressure of Lavoisier's experiment, weigh
3½ grains. Therefore the 45 grains of red solid formed by
slowly burning mercury in air consisted of, or were formed by
the chemical combination of, 41½ grains of mercury and 3½
grains of oxygen ; and these 3½ grains (or 7½ cub. inches)
of oxygen were originally present in the 50 cub. inches of air
contained in the bell-jar. When the mercury was burnt 41½
grains of it disappeared, and at the same time 3½ grains of one
of the constituents of air disappeared, and 45 grains of a new
kind of matter were produced ; but these 45 grains of this new
matter were composed of the 41½ grains of mercury and the
3½ grains of oxygen which had disappeared. No loss or
destruction of matter occurred during the burning. The hot
mercury so interacted with the oxygen contained in the air
that there was produced a kind of matter altogether different
from either of the interacting bodies.

19   Now if all the experiments already described were repeated so that the weights of the different kinds of matter taking part in each chemical change were determined, and the weights of each and every new kind of matter produced in each of these changes were also determined,—and this has actually been done,—we should find that the new kinds of matter formed were formed by the union or combination of the different kinds of matter which constituted the material system at the beginning of each experiment.

20   The terms *disappeared*, and *was produced*, do not then mean *were destroyed*, and *were created;* they rather mean, *ceased to exist under the conditions of experiment as a distinct kind of matter*, and, *was the product of the chemical interaction of two or more kinds of matter which previously existed each as a distinct kind of matter.*   Similarly the expression used in previous paragraphs, *has taken the place of*, is now seen to mean, *has been formed by the chemical interaction of;* the expression also implies that the weight of the new matter which has taken the place of that formerly present is equal to the weight of that which it has replaced.   In the case of the magnesium burnt to magnesia, it would be correct to say that the magnesia has taken the place of the magnesium and a certain weight of oxygen in the air.

We now see more clearly than before what is meant by saying that this or that body has been chemically changed into certain other bodies.   But a definite and accurate meaning can be given to this and similar expressions only when we have learned more about chemical occurrences.

21   In the preceding paragraphs the important truth has been assumed that the one fundamental property of matter is its mass or quantity.   Moreover it is assumed that the student has learned the proportionality of mass and weight*; that any portions of matter whose masses are equal, however different they may be in other properties, are of equal weights.   The mass of any portion of matter is the quantity of matter in that portion; the weight is the force with which that portion is attracted towards the earth's centre.   But in all practical problems with which we shall have to deal, the terms mass and weight may be taken as synonymous; because the relative

* If the student is not familiar with the connection between mass and weight he ought to consult a treatise on physics.

masses of substances are determined in chemistry by weighing them against a standard mass of brass, or other material, called 1 gram, or 1 decigram, or 1 milligram, and the weights of substances as thus determined are independent of variations in the force of gravity.

We have now some notion of what is implied in saying that in a chemical change some kinds of matter cease to exist as such, and some other kinds of matter are produced by the interactions of those originally present. To some extent we see that in chemical occurrences change of properties is connected with change of composition.

CHAPTER II.

22   WE must now more fully examine the conception expressed
in the phrase *composition of this or that kind of matter.*
It has been already proved by experiment that when a
specified mass of magnesium is burnt in air, a new body is
produced composed of the magnesium and oxygen taken from
the air, and that the mass of this product of the burning
of magnesium is greater than the mass of the magnesium
used.   The magnesium has been changed by adding to, or
combining with, it, another kind of matter, viz. oxygen.
If a little very finely divided *iron* is weighed, and then
heated to redness, the iron will glow brightly ; if the source of
heat is then withdrawn it will be seen that a reddish brown
substance, quite unlike the original iron, has been produced ;
if this substance is weighed, when cold, it will be found to
weigh more than the iron used.   Iron, like magnesium, has
been changed into a new substance, and this change has been
effected by causing the iron to combine with some kind of
matter different from itself.   As we know that the burning of
magnesium consists in combination with oxygen, and as the
conditions under which the iron has been chemically changed
are similar to those which prevailed during the burning of
magnesium, we conclude that the change which the iron has
suffered probably consists in combination with oxygen.   This
conclusion has been verified by experiments.

23   A weighed quantity of finely divided *copper* is dissolved in
moderately concentrated warm *sulphuric acid;* when solution
is complete, the greenish blue liquid is evaporated to dryness
by steam ; the blue solid is kept at 100° until it is perfectly
dry, when it is collected and weighed.   It weighs more than

the copper did. The blue solid is now dissolved in water, some sulphuric acid is added, and an electric current is passed through the liquid until it is perfectly colourless. Every particle of the red solid which has formed on one of the platinum plates is washed off the plate into a small basin, where it is repeatedly washed with water, then with alcohol, then dried over oil of vitriol, and weighed. It weighs the same as the copper did at the beginning of the experiment; moreover an examination of the properties of this red solid proves that it is copper. In this series of changes, copper has been converted into a blue solid—called copper sulphate—by causing it to react with warm sulphuric acid, and this copper sulphate has been re-converted into copper by electrolysing it. The copper sulphate was produced by combining some other kind of matter with the copper: the experimentally determined fact, that the mass of the copper obtained from the copper sulphate was equal to the mass of copper dissolved in sulphuric acid, proved that the copper sulphate was produced by the union of the copper with some other substance.

A little concentrated *nitric acid* is heated in a porcelain **24** dish; after a time the whole of the liquid has disappeared; it has been entirely gasified. A few scraps of thin *tin-foil* are now weighed into a porcelain dish and a little concentrated nitric acid is allowed to fall, drop by drop, on to the tin; the tin is changed to a loose white powdery solid; this is heated so long as any gas comes off, then allowed to cool, and weighed. The white powder weighs more than the tin did. But we know that nitric acid is entirely volatilised by heating in an open dish; hence we conclude that in the reaction between the two kinds of matter, tin and nitric acid, the tin has probably laid hold of some constituent or constituents of the acid, and that the white powder formed is the product of the union of the tin with this substance. This conclusion has been verified by carefully conducted experiments: it has been proved that the change of tin to the white powdery solid produced in the last experiment consists in the combination of the tin with one of the constituents of nitric acid, namely, oxygen.

The experiments now described have this feature in com **25** mon: in each a chemical change has occurred; one kind of matter has been changed into another, and the change has consisted in the combination of the original matter with another kind of matter unlike itself. Each change has been

43

an interaction between at least two definite kinds of matter ; the product of the change has consisted of the whole of one of the interacting substances, and either the whole, or a part, of the other. The mass of the product has therefore been greater than the mass of that one of the interacting substances which was weighed before the change began.

Magnesium and iron combined with oxygen contained in the air surrounding them ; copper combined with sulphur and oxygen obtained from the sulphuric acid with which it interacted ; tin combined with oxygen obtained from nitric acid : the óxide of magnesium, or iron, or tin, thus produced weighed more than the magnesium, or iron, or tin, used ; and the sulphate of copper produced weighed more than the copper used. But the whole of the magnesium, or iron, or tin, or copper, formed a part of the new kind of matter into which it was changed.

26   Let us now turn to some chemical changes which differ from those just considered in that in each of them a specified mass of one definite substance is converted into two or more different substances the mass of each of which is less than that of the original substance.

A flask with a good fitting cork and exit tube is arranged as shewn in fig. 7 ; the glass cylinder is graduated, filled

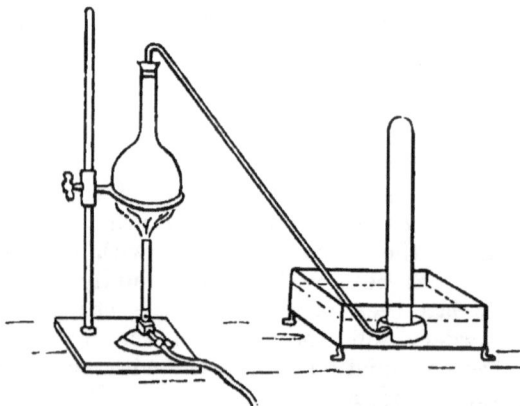

Fig. 7.

with mercury, and inverted in a vessel containing mercury. A weighed quantity of a white solid called *potassium chlorate* is placed in the flask : this solid is heated until it melts; the gas which comes off is collected in the graduated cylinder. The heating is continued as long as any gas is produced. The apparatus and its contents are allowed to cool, precautions being taken to prevent the mercury from rushing back into the flask. By methods which need not be described here, the whole of the gas which is in the flask and exit tube when the experiment is finished is driven into the graduated cylinder. The white solid in the flask is now weighed; the small quantity of air in the cylinder (which air was in the flask and exit tube at the beginning of the experiment) is removed by suitable methods, and the gas in the cylinder is measured. This gas is now proved to be *oxygen.* The properties of the white solid left in the flask are compared with those of the potassium chlorate, i.e. with those of the kind of matter used in the experiment: the two substances are easily proved to be very different. The white solid produced in the change is called *potassium chloride.* As the weight of a specified volume of oxygen has been accurately determined, the weight of oxygen produced in the process is easily calculated from the observed volume of the oxygen.

In this experiment, one kind of matter—potassium chlorate —has been changed, by the agency of heat, into two kinds of matter—oxygen, and potassium chloride—; the mass of each of these is less than that of the potassium chlorate, but the sum of the masses of the oxygen and the potassium chloride is equal to the mass of the potassium chlorate.

An electric current is passed through acidulated *water.* **27** The experiment is conducted as described in par. 9. (*s.* fig. 4). But the water used is weighed, and the water remaining at the close of the experiment is weighed; the volumes of *hydrogen* and *oxygen* produced are measured; and special precautions are taken that no water is spilt or lost, and that all the hydrogen and oxygen produced are collected in the tubes. When certain corrections have been made on account of the slight solubility in water of the gases hydrogen and oxygen, the result of this experiment is, that a specified mass of water has been changed into hydrogen and oxygen, that the mass of each of these is less than that of the water used, and that the sum of the masses of the two gases is equal to the mass of the water.

A small weighed quantity of a black solid called *copper* **28** *oxide* is placed in a bulb of hard glass arranged as shewn in fig. 8. The U tubes contain calcium chloride, a substance

which greedily absorbs water; these tubes, as well as the little
dry bulb *a*, are accurately weighed; the bulb containing the

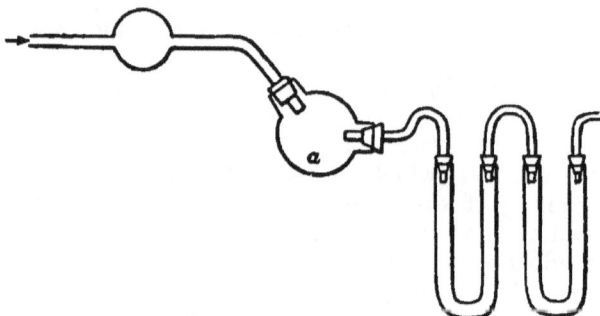

Fig. 8.

copper oxide is also weighed. Precautions are taken that the
entire apparatus is quite air-tight. Pure dry hydrogen is
passed slowly in as shewn by the arrow; after a few minutes
the copper oxide is heated; drops of a liquid resembling water
begin to trickle down into the bulb *a*; heating in a slow
stream of hydrogen is continued as long as any trace of what
seems to be water is produced. When the change is complete, the apparatus is allowed to cool; and the various parts
are then weighed. The liquid formed can be proved to be
water, and the red solid left can be proved to be *copper*.
Assuming that the proofs are conclusive; and assuming that
1 part by weight of hydrogen combines with 8 parts by weight
of oxygen to produce 9 parts by weight of water—the results
of the last experiment shew that this is so, and the statement
has been amply verified experimentally—; assuming these
points, the results of the present experiment teach, (1) that
the only products of the interaction of copper oxide and
hydrogen are water and copper; (2) that the water is formed
by the union of the hydrogen with oxygen previously combined
with copper; (3) that the mass of the oxygen thus taken away
from combination with copper is less than the mass of the
copper oxide, and the mass of the copper thus removed from
combination with oxygen is also less than the mass of the
copper oxide; (4) that the sum of the masses of the copper
and the oxygen is equal to the mass of the copper oxide.

The results of the experiments described in the preceding
paragraphs present certain points of similarity. In each, a
specified mass of one kind of matter was changed into two

(or more) kinds of matter, each different from, and each weighing less than, the original matter. Potassium chlorate was changed into potassium chloride and oxygen; water into hydrogen and oxygen; copper oxide into copper and oxygen: the potassium chloride, or the oxygen, weighed less than the potassium chlorate; the hydrogen, or the oxygen, weighed less than the water; the copper, or the oxygen, weighed less than the copper oxide[1]. But the sum of the masses of the products of each change was equal to the mass of the kind of matter which was changed into these products: the mass of the potassium chloride added to that of the oxygen was equal to the mass of the potassium chlorate changed; the mass of the hydrogen added to that of the oxygen was equal to that of the water; the mass of the oxygen added to that of the copper was equal to that of the copper oxide by the decomposition of which the oxygen and copper were produced.

We have now examined two classes of chemical changes. **30** One class presented to us interactions between two, or more, definite kinds of matter, resulting in the disappearance of the interacting substances, and the production in their place of one, or more, substances very unlike the original kinds of matter. We paid attention to the mass of one of the interacting substances, and to the mass of that product of the change which contained the whole of this substance (neglecting other products if other products there were); we found that the parts of the change to which we paid attention consisted in the combination of the whole of one of the interacting substances with either the whole, or a part, of the other substance. In each experiment a certain kind of matter was changed into a different kind of matter, by entering into combination with some substance different from itself.

The other class of chemical changes presented to us decompositions of one definite kind of matter into two, or more, different substances; the original kind of matter disappeared, and its place was taken by the new substances formed from it. We found that the mass of any one of the new kinds of matter was less than the mass of the matter from which it was derived, but that the sum of the masses of all the new kinds of matter was equal to the mass of the matter which had been changed into these new kinds of matter.

One of the changes considered presented features common to

[1] The student should carefully follow the reasoning on which this conclusion was based. (s. par. 28.)

both classes of changes. Copper oxide interacted with hydrogen; water and copper were produced : but the water was itself shewn to be produced by the union of the reacting hydrogen with oxygen separated from the copper oxide. The mass of the water added to that of the copper was greater than that of the copper oxide; this was because the water formed was composed of the oxygen at first combined with the copper and also the hydrogen which was one of the constituents of the whole changing system. The change of the copper oxide into copper and oxygen was a change belonging to our second class of chemical reactions; but this was accompanied by a change belonging to the other class of reactions, viz. the production of water by the combination of the oxygen separated from the copper with the hydrogen, the presence of which hydrogen in contact with heated copper oxide was the condition under which the separation of copper oxide into copper and oxygen was accomplished.

31   When various kinds of matter are examined chemically, it is found that they all belong to one or other of two classes, which we may at present call the *hydrogen-class* and the *water-class*.

Those kinds of matter which are placed in the *hydrogen-class* are characterised by this;—when any one of them is changed into a totally different kind of matter, the mass of this kind of matter is greater than the mass of the substance belonging to the hydrogen-class which was thus changed. These substances have never been changed by separation into unlike parts. They suffer chemical change by combining with another kind, or other kinds, of matter, the test of this combination being that the substance produced differs from, and weighs more than, the original substance.

Those kinds of matter which are placed in the *water-class* are characterised by this;—any one of them may be chemically changed by separating it into unlike parts, the test of this separation being the production from a specified mass of the original substance of at least two different kinds of matter the mass of each of which is less than that of the original substance, while the sum of their masses is equal to that of the original substance. Most, if not all, of the kinds of matter placed in this class may also be chemically changed by combining with some other kind, or kinds, of matter different from themselves, and so producing a new kind of matter weighing more than the original substance.

Those kinds of matter which belong to the *hydrogen-class* **32** are called Elements: those which belong to the *water-class* we shall at present call Not-Elements.

About seventy different kinds of matter belong to the class *Elements*. By no experiments hitherto tried have chemists succeeded in separating any one of these into unlike parts. When the elements are brought into contact with each other or with not-elements; or are subjected to the action of heat, light, electricity, or magnetism, either in the presence or absence of other kinds of matter; or are compressed, or hammered, or drawn into wires, or otherwise mechanically treated; they either remain unchanged into kinds of matter different from themselves, or they combine with other substances and produce new kinds of matter each weighing more than the element from which it has been produced.

In the present state of knowledge then we regard an element as a completely homogeneous kind of matter. We do not assert that an element is completely homogeneous; that every attempt to separate an element into unlike parts must necessarily fail; but we say that so far as experimental investigation has gone those kinds of matter which are called elements behave as if each were a distinct substance different in kind from all other substances, and were composed of one kind of matter only.

The properties of the elements differ much. The following well-known substances are elements:—*iron, lead, tin, silver, gold, copper*. These are all heavy, lustrous, malleable, solids. A few elements are gases at ordinary temperatures and pressures, viz. *oxygen, hydrogen, nitrogen, chlorine, fluorine* (?): two are liquids under ordinary conditions, viz. *bromine* and *mercury*: the others are solids. Some of the elements are found uncombined with other elements in rocks, e.g. *carbon, iron, tin, copper, gold, platinum, sulphur; oxygen* and *nitrogen* form the chief constituents of the atmosphere; *hydrogen* is sometimes found in volcanic gases. Most of the elements however have been separated from those combinations of them with other elements which are found in rocks, soils, waters, or parts of animal or vegetable organisms.

The colour of many elements is grey to white; a few are yellowish-white, or yellow; one or two are reddish-brown; three are colourless gases. Some of the elements are very malleable and very ductile; others are very brittle: some melt at very low temperatures; others only at the highest attainable temperatures, one or two have never been fused. Most of the elements are heavier, some as much as 20 or 22 times heavier, than an equal volume of water; a few are specifically lighter than water. Some are very good conductors of heat and electricity; others are practically nonconductors: most elements are opaque; a few are translucent. Some again very readily react chemically with most of the others to pro-duce new kinds of matter; e.g. compounds of *oxygen*, of *chlorine*, of *bromine*, or of *sulphur*, with most other elements, are known. On the other hand, some elements e.g. *boron* and *nitrogen*, combine directly only with a comparatively small number of other elements.

To state the name of an element is to state the composition of the element: the name is a short symbol for certain properties which characterise that kind of matter to which the name is given, and mark it off from other kinds of matter. So far as we know at present the element is composed of itself; i.e. any quantity of it is not made up of, or formed by the union of, two or more different kinds of matter, but is completely homogeneous.

CHAPTER III.

THE different kinds of matter classed together as *Not-* **33** *Elements* are composed each of two or more elements. But we must attempt a subdivision of the class *not-element* : we are not specially concerned with all the members of this class. If some finely divided iron is intimately mixed with some powdered sulphur, a heavy, greenish-grey, solid is formed. This solid cannot be an element; the method of its preparation precludes this. It is composed of the two distinct kinds of matter, iron and sulphur. It belongs to the class *Not-Elements*. But is it one of those *not-elements* whose properties and composition are studied in chemistry?

Make a very intimate mixture of finely divided *iron* and *sulphur*, in the ratio of 1 part sulphur to $1\frac{3}{4}$ parts iron by weight. Compare the colour and appearance of this mixture with the colour and appearance of each of its constituents, iron and sulphur : the mixture is neither brownish black like iron, nor yellow like sulphur; it is not lustrous like iron, nor is its texture that of sulphur. The colour and appearance of the mixture are approximately the mean of the colour and appearance of its constituents.

Place a little finely divided iron in water; the iron sinks : place a little powdered sulphur in water; part of the sulphur floats to the surface. Bring a magnet under a sheet of white paper on which is strewn a little finely divided iron, and blow (with the mouth) along the surface of the paper; a good deal of the iron remains held by the magnet, although the paper is between the iron and the magnet: examine the action of the magnet on powdered sulphur under similar conditions; the sulphur is entirely blown away, none is held by the magnet. Examine some of the iron, and some of the sulphur,

used, under the microscope, and note the differences in their
appearance. Pour a little carbon disulphide on to a very little
powdered sulphur, and on to a very little finely divided iron,
respectively, and very gently warm each; the sulphur slowly
dissolves in the carbon disulphide, the iron remains unchanged.
Pour a little hydrochloric acid on to portions of the iron, and
sulphur, used: the iron slowly dissolves and a gas is evolved
which can be proved to be hydrogen; in the case of the sulphur
no gas is evolved, nor is there any apparent change.

Now turn to the mixture of iron and sulphur, and determine whether the iron in it is characterised by those properties which we have found belong to iron as a definite kind of
matter, and whether the sulphur in it exhibits the properties
which belong to sulphur when it is unmixed with other kinds
of matter. Experiment proves that the mixture may be
separated into iron and sulphur, by shaking it with water,
or by dissolving out the sulphur by carbon disulphide, or by
holding the iron by a magnet and blowing away the sulphur.
Experiment also proves that hydrochloric acid reacts with the
mixture to dissolve the iron and leave the sulphur, and that
hydrogen is produced in this reaction. Examination of the
mixture under the microscope shews the particles of iron, and
the particles of sulphur.

Now heat another portion of the mixture of iron and
sulphur; it glows throughout: powder the black mass which
remains after cooling, and heat it again. Again powder the
heated substance, and determine whether iron or sulphur can
be detected in it by making use of those properties of iron
and sulphur, respectively, which we know characterise these
kinds of matter. The appearance and colour of the substance
are distinctly different from those either of iron or sulphur;
the substance is not separated into iron and sulphur by any one
of the three methods (water, magnet, carbon disulphide), each
of which separated the mixture of iron and sulphur into its
constituents; the substance appears under the microscope to
be homogeneous; interaction with hydrochloric acid results in
solution of the substance as a whole, and production of a gas
which is not hydrogen, but is sulphuretted hydrogen, a body
easily distinguished from hydrogen by many prominent properties.

The substance produced by heating a mixture of one part
sulphur with 1¾ parts iron is thus proved to be a kind of
matter quite different from either iron or sulphur; the sub-

stance produced by mixing sulphur and iron in the ratio $1:1\frac{3}{4}$
was proved to possess properties characteristic both of iron
and of sulphur. Mixing iron and sulphur has evidently not
produced a chemical change : heating the mixture of iron and
sulphur has produced a chemical change.

With a mixture of iron and sulphur, we are not especially **34**
concerned in chemistry ; with the new kind of matter, called.
iron sulphide, produced by heating the mixture of 1 part sulphur
with $1\frac{3}{4}$ parts iron, we are especially concerned. Both kinds of
matter are *Not-Elements* : the first is a *mixture* of different
kinds of matter, each of which can be recognised in the mixture by properties which characterise it when it is unmixed
with other kinds of matter ; the second is a *compound* formed
by the combination of different kinds of matter, none of which
can be recognised in the compound by properties which belong
to it when uncombined with other kinds of matter.

The class **Not-Elements** is divided into *compounds* and
mixtures. Chemistry deals with the changes of composition
and of properties of compounds. We are. at present endeavouring to understand what is meant by the composition
of compounds. In order to learn something about the composition of compounds we must gain as clear a notion of the
differences between compounds and mixtures as we can at this
stage of our progress.

Ammonia is a colourless, volatile, gas, with a very pungent **35**
and penetrating smell ; *charcoal* is a black, porous, light,
solid : these two kinds of matter may be recognised by these
properties. Let a quantity of ammonia be confined in a tube
over mercury ; and let a few pieces of charcoal (previously
heated to remove air from their pores) be passed into the tube.
The ammonia is rapidly absorbed by the charcoal, and the
mercury rises in the tube. The appearance of the charcoal
is not changed ; only it smells strongly of ammonia. The
ammonia is easily removed from the charcoal, with which
it is mixed, by warming the charcoal in a small dry
flask, and allowing the ammonia to collect in a tube filled
with mercury and placed mouth downwards in a vessel of
mercury.

Hydrogen chloride is a colourless, volatile, gas, with a
very pungent smell, and most irritating action on the skin.
Let a certain volume of ammonia be confined over mercury,
and let an equal volume of hydrogen chloride be passed into
the vessel ; instantly there is produced a white solid, utterly

unlike either the ammonia or the hydrogen chloride by the interaction of which it has been formed. If this solid is collected and examined it is found to be characterised by none of the properties which mark off ammonia and hydrogen chloride, respectively, from other kinds of matter.

Ammonia and charcoal when brought together form a *mixture*; both constituents are easily recognised in the mixture by the same properties as those by which they are recognised when unmixed. Ammonia and hydrogen chloride when brought together form a *compound*, called *ammonium chloride*; neither constituent can be recognised in the compound by the properties by which it is recognised when apart from other kinds of matter.

36   *Water* boils at 100° when the pressure on the surface of the water is equal to 760 mm. of mercury. At the same pressure *alcohol* boils at 78°·3. Each of these liquids may be recognised, and differentiated from other substances, by observing its boiling point. A mixture of water and alcohol in about equal parts is placed in a flask, fitted with a thermometer, and connected with a condenser and receiver, as shewn in fig. 9. When the liquid has been heated to boiling the thermometer registers a temperature higher than 78°·3 and lower than 100°; as boiling continues the temperature rises, but a fixed boiling point is not attained. The liquid in the receiver may be proved to be a mixture of water and alcohol, and the portions which distil over soon after boiling begins may be proved to be richer in alcohol than those which distil over after boiling has continued for some time. If the liquid which distils over (the distillate) is collected in a series of flasks, so that each contains that quantity which has come over for a temperature-interval of (say) 5° or 8°, and if each of these quantities is again distilled, and the distillate for every 3° or 4° is collected in separate vessels, it is possible to effect a rough separation of the original mixture of water and alcohol into two liquids, one of which consists for the most part of water and the other for the most part of alcohol. This separation of the mixture has been effected by making use of a property of each consti-tuent, which property is a characteristic physical property of that constituent when unmixed with other kinds of matter.

37   *Butylene* is a colourless liquid, boiling at (about) 3°. *Bromine* is a dark reddish brown, heavy, strongly smelling, liquid, boiling at (about) 60°. Each of these is a definite kind

of matter characterised by definite properties, of which the
boiling point is one. When butylene and bromine are mixed
in the ratio of 1 part butylene to 2·86 parts bromine, by
weight, a colourless liquid unlike either constituent is produced. The weight of the liquid formed is equal to the sum
of the weights of the butylene and bromine. If the liquid is
distilled (s. fig. 9) the thermometer registers 160⁰ from the
time when boiling begins until the last drops of the liquid

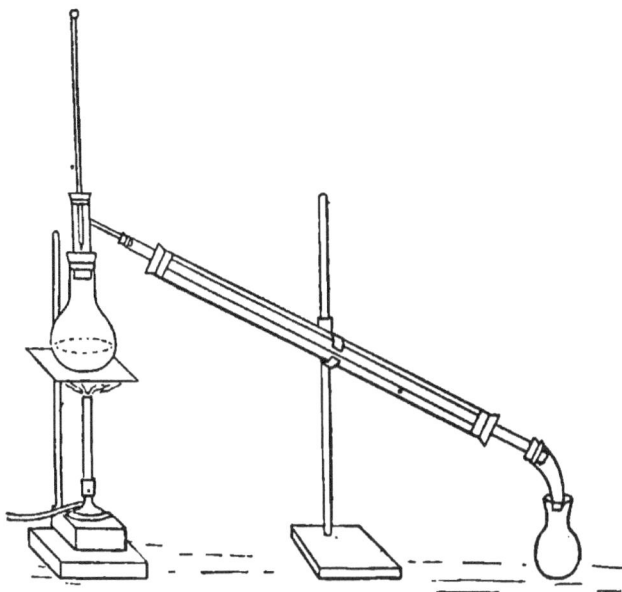

Fig. 9.

have passed over into the receiver ; moreover the distillate has
the same properties as the liquid before distillation. Butylene
and bromine have formed a *compound* (called *butylene bromide*),
whose properties are very different from those of either
constituent, and from which neither constituent can be
withdrawn by taking advantage of one of the physical properties, viz. boiling point, belonging to each constituent when
uncombined with other substances.

The constituents of a mixture of gases may frequently be **38**
separated by making use of the property which gases have of
passing through the fine pores of a mass of dry plaster of

Paris.   The passage of a gas through such a porous substance as plaster of Paris is called *diffusion*.

If two graduated glass tubes are each stopped at one end with a thin dry plate of plaster of Paris, if one is then filled with *hydrogen* and the other with *oxygen*, and if both are at once placed in water with the open ends under the water (*s.* fig. 10), the water will begin to rise in both tubes. As the gases cannot escape at the lower ends of the tubes, they must be passing outwards through the plates of plaster of Paris, and passing outwards more rapidly than air is passing inwards. If the tubes are of the same section and the

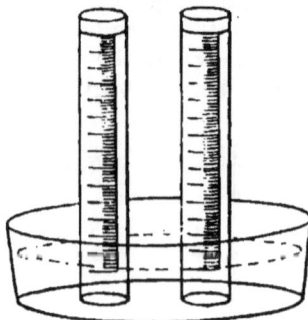

Fig. 10.

same length, and if the level of the water in each is observed after a little time, it will be found that the hydrogen has diffused through the porous plate about 4 times quicker than the oxygen.

If a similar experiment is made (with proper precautions) with chlorine,—a heavy, yellowish-green, very badly smelling, gas—it will be found that the rate of diffusion of hydrogen is about six times that of chlorine.

Now let there be prepared a mixture of two volumes hydrogen with one volume oxygen.   This mixture cannot be distinguished by the eye from its constituents; if a very little of it is placed in a strong glass tube and a flame is brought near a violent explosion occurs.   Let the mixture be collected in a gas-holder from which it may be forced at any desired rate by allowing water to enter the gas-holder from a reservoir above. The gas-holder communicates with an arrangement for drying the gases, and this is connected with a long, dry, clay-pipe placed inside a glass tube arranged as shewn in fig. 11.   The mixture of oxygen and hydrogen is caused to pass very slowly through the pipe; gas issues at *a* and *b*; a tube is filled with the gas issuing at *a* and another with that issuing at *b*.   The gas collected at *b* does not burn when a lighted wooden splint is brought near it, but the splint itself burns more.brightly ; the gas collected at *a* burns with a slight explosion.   The gas issuing at *b* consists chiefly of oxygen; that issuing at *a* consists chiefly of hydrogen.

The *mixture* of hydrogen and oxygen has been partially

separated into its constituents by making use of a physical property of these gases called rate of diffusion, which property

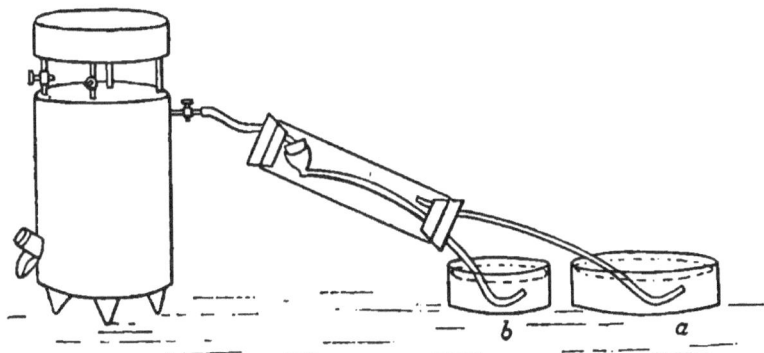

Fig. 11.

is a characteristic mark of each gas when unmixed with other substances.

If now a mixture is made of equal volumes of *hydrogen* **39** and *chlorine*, and this mixture is exposed to diffused sunlight for some time, a new gas will be formed; the new gas is colourless; chlorine is yellow, hydrogen is colourless—: the new gas fumes much in the air—neither chlorine nor hydrogen fumes in air—; it has an intensely acrid smell, quite different from the smell of chlorine. This gas is called *hydrogen chloride*. The weight of hydrogen chloride formed is equal to the sum of the weights of the hydrogen and chlorine which have combined to form it.

If hydrogen chloride is passed through a dry clay-pipe under conditions similar to those already described it can be proved that no separation into hydrogen and chlorine has occurred, but that the gas which issues at *a* is identical with that which issues at *b*, and that both are hydrogen chloride.

The *compound* of hydrogen and chlorine (*hydrogen chloride*) cannot be separated into its constituents by making use of a certain physical property—viz. rate of diffusion—which belongs to, and characterises, each of its constituents when these are uncombined with other kinds of matter.

The results of experiments such as those considered in **40** pars. 33 to 39 shew that the substances comprised in the class Not-Elements may be divided into two groups; *compounds* and *mixtures*.

Each constituent of a *mixture* retains in the mixture the properties which characterise it when unmixed with other substances: the properties of the mixture are, broadly, the sum of the properties of the constituents. No constituent of a *compound* retains in the compound the properties which characterise it when separated from other substances: the properties of a compound are not the sum of the properties of the constituents; the compound is a definite kind of matter, as distinct from each of its constituents as these are from one another, and yet formed by the combination of these constituents.

41   To say of a mixture, that it *contains* the bodies by mixing which it has been produced, is to use an expression which conveys a correct notion of the relations of the properties of the mixture to those of its constituents. But it is not so correct to say that a compound *contains* each of those kinds of matter by the interaction of which it has been formed. Thus, a mixture of iron and sulphur contains iron and contains sulphur; inasmuch as, not only is the mass of the mixture the sum of the masses of the mixed iron and sulphur, but the properties of the mixture are also the properties of iron added to those of sulphur. The mass of a compound of iron and sulphur is certainly the sum of the masses of the iron and sulphur which have combined to form it; but the properties of the compound are quite distinct from the properties by which iron or sulphur is marked off from other kinds of matter.

The *formation of a mixture is a physical process.*

The properties of every mixture probably differ slightly from the sum of the properties of its constituents; some change occurs in the formation of the mixture; nevertheless the properties of each kind of matter in the mixture are so slightly modified by the presence of the other kinds of matter that it is always possible, and generally easy, to recognise each of these kinds of matter by some, or all, of the properties which distinctly mark it off from other kinds of matter.

The *formation of a compound is a chemical process.*

The properties of each of those kinds of matter which combine are so largely modified by the presence of the other combining substances that it is impossible to recognise any of them by the properties which belong to it when uncombined. Iron sulphide is as distinct and definite a kind of matter as iron or sulphur; ammonium chloride is as distinct and definite a kind of matter as ammonia or hydrogen chloride; butylene bromide is marked off from other kinds of matter by properties as distinct and definite as those which characterise butylene or bromine; hydrogen chloride, so far as its physical properties indicate, is as homogeneous and as little formed of unlike parts as either hydrogen or chlorine.

Iron sulphide, or ammonium chloride, or butylene bromide, **42** or hydrogen chloride, can be separated into unlike parts; but this separation is accompanied by the disappearance of all the distinctive properties of the compound, and by the production, in each case, of two kinds of matter—iron and sulphur, ammonia and hydrogen chloride, butylene and bromine, hydrogen and chlorine—so unlike the compounds from which they have been produced that the only expression to be used regarding the occurrence is that each compound has ceased to exist and has been replaced by two new kinds of matter. Neither iron nor sulphur has yet been separated into unlike parts; the methods which succeed in separating iron sulphide into iron and sulphur fail to separate iron or sulphur into kinds of matter different from iron or sulphur. Bromine likewise refuses, at present, to reveal its composition, if composition it has in the sense in which it may be said that butylene bromide is composed of butylene and bromine. But ammonia and hydrogen chloride, which are produced by separating ammonium chloride into unlike parts, can, each, be further separated into two kinds of matter totally unlike either ammonia or hydrogen chloride. Ammonia is formed by the union of, and can be resolved into, two colourless, odourless, gases—nitrogen and hydrogen; hydrogen chloride is formed by the union of, and can be resolved into, hydrogen, and another, yellowish-green, badly smelling, gas, chlorine. All attempts to separate nitrogen, or hydrogen, or chlorine, into unlike parts, have hitherto failed.

A mixture is separated into its constituents by making **43** use of some property or properties of each constituent which belong to that substance when it exists apart from other kinds of matter. Thus the mixture of iron and sulphur was separated by making use of the fact that iron is attracted by a magnet while sulphur is not attracted; or of the fact that iron sinks in water, while sulphur floats, at least for a time; or of the fact that sulphur is soluble, while iron is not soluble, in carbon disulphide. The mixture of ammonia and charcoal was separated by taking advantage of one of the properties of ammonia viz.

that it is a very volatile gas. The property possessed by hydrogen of diffusing four times more rapidly than oxygen through a porous plate gave us a method for approximately separating a mixture of hydrogen and oxygen into its constituents.

But if a compound is to be separated into unlike parts it is necessary either to act upon it by some natural agency, or form of energy, such as heat, light, or electricity—or in some cases mechanical energy—or, and this is the more usual method, to cause it to interact under suitable conditions with some other kind, or kinds, of matter. Thus the compound water was separated into hydrogen and oxygen by passing an electric current through the water (s. experiment in par. 9). Similarly ammonia may be separated into nitrogen and hydrogen by passing electric sparks through it.

Copper oxide was separated into copper and oxygen (s. par. 28) by causing it to interact with hydrogen at a high temperature; the results of this interaction were copper and water; but the results of a previous experiment shewed that water is produced by the combination of hydrogen with oxygen.

As we proceed in our study we shall learn more of the methods employed for separating compounds into the different kinds of matter by the combination of which they are produced; meanwhile it is important to observe that the method does not consist in making use of the physical properties belonging to these different kinds of matter. The formation and decomposition of a compound are chemical processes.

44 We have already learned that the chemist puts in one class all those distinct kinds of matter which he has not been able to separate into unlike parts, and calls them Elements.

We now learn that certain Not-Elements are distinct kinds of matter, each marked by its own definite and characteristic properties, yet each capable of being separated into parts, totally unlike each other, and unlike the original. These not-elements the chemist puts in one class, and calls them Compounds. One marked characteristic, viz. the constancy of composition, of compounds will be dealt with later. (pars. 58 and 59.)

All other substances belonging to the group Not-Elements are classed together and called Mixtures. An infinite number of these exists, or may be formed, by mixing elements with elements, or compounds with compounds, or elements with compounds, or mixtures of any of these with other mixtures; they are all marked off from elements and compounds by the facts that their properties are, broadly, the sums of the properties of their constituents, and their constituents exist'in the mixtures each with its own properties scarcely, if at all, modified by the presence of the other constituent parts.

Chemistry deals with certain parts of the phenomena **45** presented in the.changes of elements into compounds, and of compounds into simpler compounds or into elements. Chemistry concerns itself but little with the formation of mixtures or the resolution of mixtures into their constituents.

By the *composition of an element* is meant the element **46** itself; so far as our knowledge goes at present, each kind of matter placed in the class element is entirely homogeneous.

By the *composition of a compound* is meant, primarily, a statement of the elements by the combination of which the compound is formed and into which it can be resolved, and also a statement of the mass of each element which goes to form, or can be obtained from, a specified mass of the compound. By the composition of a compound is frequently meant a statement of certain less complex compounds, and of the masses of these, which interact to produce a specified mass of the compound in question, or which can be obtained from a specified mass of this compound. Thus, experiment has shewn us (par. 35) that the compound ammonium chloride is produced by the interaction of ammonia and hydrogen chloride; experiment (par. 39) has also told us that hydrogen chloride is itself a compound of hydrogen and chlorine. It may also be proved that ammonia is a compound of nitrogen and hydrogen. The composition of ammonium chloride may be expressed by either of the following statements :—

(1)   100 parts by weight of ammonium chloride are formed by the combination of 31·77 parts by weight of ammonia and 68·23 parts by weight of hydrogen chloride;

(2)   100 parts by weight of ammonium chloride are formed by the combination of 26·17 parts by weight of nitrogen, 7·48 parts by weight of hydrogen, and 66·35 parts by weight of chlorine.

We have now gained a clearer conception of chemical **47** change. We now regard such a change as, either the change of a specified mass of a compound into fixed masses of two or more compounds or elements, or the interaction of fixed masses of two or more elements or compounds to produce

3—2

definite masses of new elements or compounds. We know
that in the first case the mass of each element or compound
produced is less than that of the original compound before the
change began. In both cases, we know that the sum of the
masses of the different kinds of matter produced in the change
is equal to the sum of the masses of the different kinds of
matter which suffered change.

Our present conception of chemical change requires us to
have clear notions of the classes of things called elements and
compounds, respectively; and this, in turn, demands that we
have grasped, as far as we can at this stage, the essential
points of difference between chemical and physical change,
and between elements and compounds, on one hand, and
mixtures, on the other.

We have also learned something of the meaning of the
term chemical properties of this or that kind of matter, as
contrasted with the term physical properties of the same
kind of matter. Sulphur, for instance, is a yellow, brittle,
solid, twice as heavy as water bulk for bulk; it crystallises
in rhombic octahedra, melts at about 115°, boils at about 440°,
and is a bad conductor of heat and electricity : these are some
of the physical properties of sulphur, that is, the properties
which are recognised as belonging to this kind of matter when
it is examined apart from other kinds of matter. But when
we examine the relations of sulphur to other kinds of matter,
we enter on the study of its chemical properties. We find
that one of the chemical properties of sulphur is its power of
combining with iron; we find that when one part by weight
of sulphur is heated with $1\frac{3}{4}$ parts of iron, $2\frac{3}{4}$ parts by
weight of a compound of iron and sulphur (iron sulphide) are
formed; we find that this compound is totally unlike either
iron or sulphur, but that the whole of the iron and the whole
of the sulphur have been used in its production. Further investigation would shew us that sulphur combines with oxygen
to form two distinct compounds, unlike each other, and both
unlike either sulphur or oxygen ; we should find that 2 parts
by weight of one of these compounds are produced by the combination of one part of sulphur with one part of oxygen, and
that $2\frac{1}{2}$ parts by weight of the other compound are produced
by the union of one part of sulphur with $1\frac{1}{2}$ parts of oxygen.

CHAPTER IV.

CONSERVATION OF MATTER.

CHEMICAL changes are evidently complex occurrences. **49** What we have learned regarding them has been learned by making quantitative experiments and by reasoning on the results of these experiments. So long as our experiments are merely qualitative we can attain to no just conceptions of those changes which it is our business, as chemists, to investigate.

Chemistry began to be a science, that is a department of exact and systematised knowledge of natural events, when quantitative investigation had superseded qualitative experiments.

Before the time of Lavoisier there was much vague **50** speculation about *elementary principles*. At one time the commonly accepted view was that all things were composed of the four principles, *earth, air, fire,* and *water*. A piece of green wood is burnt: smoke ascends, therefore, it was said, wood contains the element air; the flame which plays round the wood proves the presence of the element fire; the hissing sound proves that the element water is present in the wood; and the ashes which remain demonstrate that the element earth is one of the four constituents of the wood. Such reasoning, and such experiments, were possible only so long as chemists did not measure the quantities of the materials taking part in the changes which they observed.

About the middle of the eighteenth century, Black firmly established the fact that chalk and burnt lime have a definite and unalterable composition. By quantitative experiments he proved that when chalk is burnt it is changed into lime and carbon dioxide; and that when burnt lime is exposed to air, it slowly combines with carbon dioxide and chalk is re-formed.

Lavoisier carried on the work begun by Black. He gave the true interpretation of very many chemical changes, on the superficial qualitative examination of which the structure of alchemy had been raised.

51 That water had been repeatedly changed into earth was granted by all the alchemists. Water was boiled for a long time in a glass vessel; the water disappeared, and a considerable quantity of a white earth-like solid remained in the vessel. Lavoisier placed some water in a weighed glass vessel; he closed the vessel and weighed it with its contents; he kept the water hot for 101 days, and then poured out the water into another vessel and boiled it until the whole of it had disappeared; there remained $20\frac{1}{2}$ grains of solid earthy matter; he then dried and weighed the glass vessel in which the water had been heated, it weighed $17\frac{1}{2}$ grains less than it had weighed before the water was heated in it. Lavoisier concluded that the earthy matter was produced by the action of the water on the glass; that is to say, that the alleged transmutation of water into earth did not occur, but that the earth was a part of the material of the vessel in which the water was heated. The small difference between $20\frac{1}{2}$ and $17\frac{1}{2}$ grains was due, according to Lavoisier, to experimental errors: this conclusion was fully confirmed when more accurate methods of weighing became possible. From quantitative experiments such as these, Lavoisier drew the all-important conclusion, that the total quantity of matter which is concerned in any chemical change is the same at the end of the change as at the beginning.

52 Every accurate investigation conducted since the time of Lavoisier has confirmed this generalisation. Under the name of *the principle of the conservation of matter*, or sometimes *conservation of mass*, it is now one of the foundations of all modern science. Experimental proofs of this generalisation have been given in preceding paragraphs.

However we may change the form of matter, whatever transmutation we may succeed in accomplishing, there is one thing we cannot change, and that is the quantity, or mass, of matter taking part in each of these transmutations.

The statement of this principle, or law, sometimes takes such a form as this; *we cannot create or destroy a single particle of matter, we can only change its form.* It is important to notice that the test of creation, or destruction, is here, increase, or decrease, of the total mass of matter.

53 In place of the indefinite and indefinable *elementary* *principles* of the alchemists we have the 70, or so, elements of chemistry. Each element is a definite kind of matter characterised by its own properties which can be accurately stated frequently in terms having a quantitative signification. By bringing these elements into contact with each other under various conditions, we can accomplish stranger changes than those which alchemists dreamt of ; but we know that the new kinds of matter thus produced are formed by the combinations of the elements ; we have learned that no particle of any of the interacting elements is destroyed, but that the quantity of matter in the products is always exactly equal to the quantity of matter in the interacting elements.

CHAPTER V.

54 WE have seen that a mixture may be made of two elements or compounds in different proportions, that the properties of the resulting mixture are the sum, or nearly the sum, of the properties of its constituents, and that the greater the proportion of one of the constituents the more nearly do the properties of the mixture resemble those of that constituent. A chemical compound, on the other hand, is wholly unlike the elements or simpler compounds from which it is formed ; its properties are perfectly definite and fixed, and are different from those of any of its constituents. Does the compound differ also from the mixture in having a fixed composition ? Do the constituents of the compound combine in definite quantities ? It is evident that we must quantitatively examine the composition of compounds if we desire to discover the laws of their formation.

55 Let us return to the first experiment by which we gained a rough notion of the difference between chemical and physical change. Let us again burn the element magnesium in air; but let the magnesium be weighed before it is burnt, and let the magnesia which is produced be collected and weighed. The result of this experiment is ;—

1 gram of magnesium when completely burnt in air, or in oxygen, produces 1·66 grams of magnesium oxide or magnesia : we already know that the substance produced is a compound of magnesium and oxygen.

This result may also be stated thus ;—

100 parts by weight of magnesium oxide are formed by the combination of

60 parts by weight of magnesium, and
40 parts by weight of oxygen.

There are other ways of preparing magnesia, but 100 parts by weight of this compound, however it has been prepared, can always be resolved into 60 parts of magnesium and 40 parts of oxygen.

If the compound of iron and oxygen produced by burning iron in oxygen is analysed it is found that its composition per 100 parts is ;—

$$\text{iron} = 72\cdot41$$
$$\text{oxygen} = 27\cdot59.$$

By *composition per* 100 *parts* is meant a statement of the mass of each of the elements which by their combination produce 100 parts by weight of the compound (*s. par.* 46).

An experiment was already described by which the substance potassium chlorate was proved to be a compound of the element oxygen and the less complex compound potassium chloride. 100 parts by weight of potassium chlorate are resolved by heating into 39·13 parts by weight of oxygen and 60·87 parts by weight of potassium chloride; if 200 parts of the chlorate are used, 78·26 parts of oxygen and 121·74 parts of potassium chloride are obtained. Potassium chloride is itself produced by the combination of the two elements potassium and chlorine in the ratio 52·41 to 47·59; i.e. 100 parts of the compound are composed of 52·41 parts of potassium and 47·59 parts of chlorine. The composition of either potassium chlorate or chloride is definite and unchangeable. By whatever method either of these compounds is prepared, it is always composed of the same elements combined in the same proportions.

The composition per 100 parts of the iron sulphide produced **56** by heating together iron and sulphur is ;—

$$\text{iron} = 63\cdot63$$
$$\text{sulphur} = 36\cdot37.$$

In other words the ratio of sulphur to iron is 1 : 1·75.

Now if a mixture is made of very finely divided sulphur and iron in the ratio 1 : 2, and this mixture is heated, a black solid will be formed characterised by the properties of iron sulphide; but it can be experimentally proved that the substance thus produced is not iron sulphide only, but is a mixture of iron sulphide and iron; and further it can be proved that 2·75 parts by weight of iron sulphide have been formed and that ·25 parts of iron remain uncombined with sulphur. Again, if a mixture of 1·25 parts of sulphur with

1·75 parts of iron is heated, 2·75 parts of iron sulphide are formed and ·25 parts by weight of sulphur remain uncombined with iron. The compound known as iron sulphide is thus shewn to have a definite and fixed composition: a certain mass of sulphur combines with a fixed mass of iron; if there is more iron than this fixed mass, the iron· over and above the fixed mass—generally called the *excess* of iron—does not combine with the sulphur; if there is an *excess* of sulphur, some of the sulphur does not combine with the iron.

57   Experiments have been described by which water has been shewn to be a compound of the elements hydrogen and oxygen. If water is a compound, the composition of water must be definite and unchangeable.

A tube of stout glass is divided into a number of equal parts, preferably into cubic centimetres; the divisions are marked on the outside; the tube is closed at one end; two platinum wires pass through the walls of the tube near the closed end, and are bent so that the ends of the wires nearly, but not quite, touch inside the tube (*s.* fig. 12). The tube is filled with mercury, with proper precautions, and is inverted in a trough of mercury. A small quantity of oxygen is passed into the tube, and the volume of the oxygen is determined; let it be 10 c.c. 20 c.c. of hydrogen are now passed into the tube. The tube is then pressed down on a pad of caoutchouc, and firmly clamped (*s.* fig. 13). An electric spark from an induction-coil is passed from one platinum wire to the other; combination of the hydrogen and oxygen occurs instantly, and the inside of the tube is slightly dimmed by the minute quantity of water produced. The tube is now raised slightly from the caoutchouc pad; mercury rushes in and practically fills the tube. It may be proved conclusively that water, and nothing but water, is formed in this experiment.

Fig. 12.

The result of this experiment shews that 2 volumes of hydrogen combine with 1 volume of oxygen to produce water.

Let the experiment be repeated, but with different volumes of hydrogen and oxygen.

(1) Let there be 20 c.c. hydrogen and 20 c.c. oxygen: when the mercury is allowed to rush into the tube 10 c.c. of gas will remain; this gas may be proved to be oxygen.

(2) Let there be 30 c.c. hydrogen and 10 c.c. oxygen:

10 c.c. of gas will remain, which may be proved to be hydrogen.

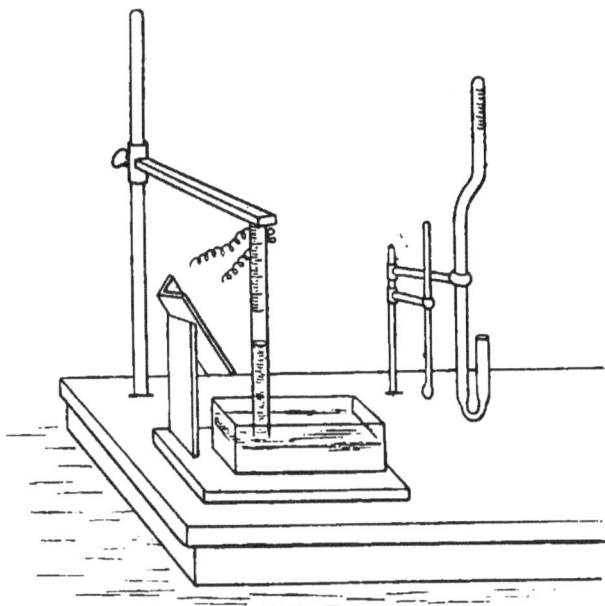

Fig. 13.

(3)   Let there be 50 c.c. hydrogen and 25 c.c. oxygen : no gas will remain.

[It is assumed that every precaution has been taken in measuring the gases, and that all necessary corrections for changes in temperature and pressure have been made.]

The result of these experiments is that hydrogen and oxygen combine to form water in the ratio 2 : 1 *by volume*, and in this ratio only.   Oxygen is 16 times heavier than hydrogen, bulk for bulk ; hence 1 volume of oxygen weighs 8 times as much as 2 volumes of hydrogen, measured at the same temperature and pressure ; hence the results of these experiments shew that hydrogen and oxygen combine to form water in the ratio 1 : 8 *by weight*, and in this ratio only.

The composition of several compounds has now been ex-   **58** amined quantitatively ; in every case it has been found that a specified mass of the compound has been produced by the combination of fixed and invariable masses of two, or more than two, elements.   What is stated regarding the quantita- .

tive composition of these compounds has been found to hold good for all compounds.

Every compound is a definite kind of matter, characterised by certain properties which mark it off from other kinds of matter; every compound is produced by the combination of two or more simpler compounds, or two or more elements; and these simpler compounds, or these elements, always combine in the same proportion to form the specified compound.

This result of the examination of the quantitative composition of compounds is of fundamental importance in chemistry. It at once enables us to draw a marked distinction between mixtures and compounds. The composition of a mixture is not unalterable; that of a compound is fixed and definite.

This fact regarding the composition of compounds is usually called

The law of constant, or definite, proportions: or the law of fixity of composition.

It may be stated in various ways; thus,

· *The proportions in which bodies unite together chemically are definite and constant.*

A given chemical compound is always formed by the union of the same elements in the same proportions.

The masses of the constituents of every compound stand in an unalterable proportion to each other, and also to the mass of the compound formed.

The evidence in support of this statement is really the whole body of chemical facts which are at present known. But special experiments have been conducted with the view of testing the law of fixity of composition.

The experiments made by Stas were characterised by the most rigorous and scrupulous accuracy. Stas prepared the compound *salammoniac*, or *ammonium chloride*, by four distinct methods; he purified each preparation with the utmost care, and then determined its composition. Ammonium chloride is a compound of the three elements nitrogen, hydrogen, and chlorine. When an aqueous solution of this compound is mixed with a solution of silver in nitric acid, the ammonium chloride is decomposed and the whole of the chlorine formerly combined with nitrogen and hydrogen enters into combination with the silver to form silver chloride. Silver chloride is a heavy white solid; when it is formed as described from ammonium chloride it settles down to the bottom of the vessel in which the experiment is conducted, and may be collected, washed, and accurately weighed.   In each experiment Stas added 100 parts by weight of pure silver, prepared with the greatest care and weighed with the greatest accuracy, to a solution of ammonium chloride prepared by one or other of four distinct methods; he collected, and most carefully weighed, the silver chloride produced; thus he determined the mass of ammonium chloride which was wholly decomposed by 100 parts of silver. The following numbers are selected from the results obtained by Stas.

100 parts by weight of silver were required to remove, and enter into combination with, all the chlorine from x parts by weight of ammonium chloride :—

$$x = 49\cdot600\,;\ 49\cdot599\,;\ 49\cdot597\,;\ 49\cdot598\,;\ 49\cdot593\,;\ 49\cdot5974\,;$$
$$49\cdot602\,;\ 49\cdot597\,;\ 49\cdot592.$$

Every experiment is attended with certain unavoidable errors. The results obtained by Stas prove beyond doubt that the quantitative composition of the ammonium chloride examined by him was the same, by whatever method that ammonium chloride had been prepared.

We must more fully examine the composition of compounds **61** with the view of learning more of the laws of combination. We found that the composition of magnesia is defined by the statement

$$\text{magnesium} = \ \ 60$$
$$\text{oxygen} = \ \ \underline{40}$$
$$magnesia = 100$$

The analytical results thus expressed tell that masses of magnesium and oxygen combine to form magnesia in the ratio $60 : 40 = 6 : 4 = 3 : 2 = 120 : 80$, &c.

But two elements often combine to produce two, or more **62** than two, distinct compounds. For instance carbon and oxygen combine to form two compounds. The composition of these oxides of carbon is represented, in parts per 100, thus;—

	I.	II.
carbon =	42·85	27·27
oxygen =	57·15	72·73
carbon oxide =	100·00 ·	carbon oxide = 100·00

But these analytical results may be stated in another form. We may ask, how many parts by weight of oxygen are combined with one part by weight of carbon in each compound?

The answers are easily found

I. $42 \cdot 85 : 1 = 57 \cdot 15 : x.$ II. $27 \cdot 27 : 1 = 72 \cdot 73 : x.$
            $x = 1 \cdot 33.$ $x = 2 \cdot 66.$

Here we see that the mass of oxygen which has combined with unit mass of carbon to form compound II. is exactly double that which has combined with unit mass of carbon to form compound I.

Carbon and hydrogen combine to form many compounds; let us select four of these and state their compositions in parts of carbon and hydrogen per 100 parts of each compound. The results are as follows;—

Compound I. is called *acetylene*, II. is called *ethylene*, III. *ethane*, and IV. *methane*.

	I.	II.	III.	IV.
carbon =	92·3	85·7	80·0	75·0
hydrogen =	7·7	14·3	20·0	25·0
	100·0	100·0	100·0	100·0

If these results are treated as we treated the analyses of the oxides of carbon, we find that

1 part by weight of carbon is combined with
            ·083 parts by weight of hydrogen in compound I.
1 with 2 × ·083   ,,   ,,   hydrogen   ,,   ,,   II.
1   ,,   3 × ·083   ,,   ,,   hydrogen   ,,   ,,   III.
1   ,,   4 × ·083   ,,   ,,   hydrogen   ,,   ,,   IV.

Five compounds of the two elements nitrogen and oxygen are known; if the composition of each is determined and is stated as parts by weight of oxygen combined with 1 part by weight of nitrogen, we have this result;

Compound I.   ·57 parts by weight of oxygen combined
                            with 1 part by weight of nitrogen.
      ,,   II.   2 × ·57 oxygen with 1 of nitrogen.
      ,,   III. 3 × ·57   ,,   ,,   ,,   ,,
      ,,   IV. 4 × ·57   ,,   ,,   ,,   ,,
      ,,   V. 5 × ·57   ,,   ,,   ,,   ,,

63   By examining the composition of series of compounds of the same two elements and tabulating the results as we have done for the compounds of carbon and oxygen, carbon and hydrogen, and nitrogen and oxygen, we arrive at the second law of chemical combination, which is generally known as

The law of multiple proportions. This law may be stated thus;

When one element combines with another in several proportions, these proportions bear a simple relation to one another. Or, better, thus;

When two elements combine to form more than one compound, the masses of one of the elements which combine with a constant mass of the other element bear a simple relation to each other.

In the cases considered we have kept the mass of one **64** of the combining elements constant and have taken this mass as unity, and we have found that the masses of the other element which combine with this constant mass are all whole multiples of one quantity, viz. the smallest mass of the other element which combines with the constant mass of the standard element.

• The relation between the combining masses of the second element is evidently in these cases a very simple one. But this relation is not always so simple. Thus iron and oxygen combine to form three distinct oxides of iron; the ratio of the quantities by weight of oxygen which severally combine with one part by weight of iron is

$$1 : 1·33 : 1·5.$$

Again lead and oxygen combine to form four distinct compounds; the masses of oxygen which combine with 1 part by weight of lead are, severally,

$$·077, \ ·103, \ ·116, \ \text{and} \ ·154.$$

The ratio of these is

$$1 : 1·33 : 1·5 : 2.$$

The law of multiple proportions confirms, and also goes **65** beyond, the law of constant proportions. The latter law is the statement of the fundamental fact that the composition of every compound is definite and unchangeable; the former law generalises the composition of series of compounds of pairs of elements, and states that such compounds are produced by the combination, with a constant mass of one element, of masses of the other element which are simple multiples of the smallest of these masses.

We are now ready to advance a step further, and to con- **66** sider the composition of compounds of one element with various other elements, and compounds of these other elements with each other.

We shall begin by considering compounds of the element potassium with (1) chlorine, (2) iodine; and then (3) the compound of chlorine with iodine.

The percentage compositions of these compounds are;—

I.	II.	III.
Potassium chloride.	*Potassium iodide.*	*Iodine chloride.*
potassium = 52·4	potassium = 23·6	iodine = 78·1
chlorine = 47·6	iodine = 76·4	chlorine = 21·9
100·0	100·0	100·0

Let us find the masses of chlorine and iodine which severally combine with the same mass of potassium: this may be done by finding, (1) the mass of iodine combined with 52·4 of potassium, or (2) the mass of chlorine combined with 23·6 of potassium;

$$(1) \quad 23\cdot6 : 52\cdot4 = 76\cdot4 : x.$$

$$(2) \quad 52\cdot4 : 23\cdot6 = 47\cdot6 : x'.$$

We shall choose (2); $x' = 21\cdot4$.

That is, with 23·6 parts by weight of potassium there combine, (1) 76·4 parts of iodine to form potassium iodide, (2) 21·4 parts of chlorine to form potassium chloride.

Now let us turn to the compound of chlorine and iodine. Let us ask; what is the mass of chlorine which is combined with 76·4 parts of iodine?

$$78\cdot1 : 76\cdot4 = 21\cdot9 : x.$$
$$x = 21\cdot4.$$

We have now this result;—

23·6 parts by weight of potassium combine with

76·4 parts by weight of iodine.
21·4   „   „   chlorine.

76·4 parts by weight of iodine combine with

21·4 parts by weight of chlorine.

Or, stated more generally, the masses of chlorine and iodine which severally combine with a constant mass of potassium are also the masses of chlorine and iodine which combine with each other.

67   If another element is used instead of potassium, will a similar result be obtained? Chlorine combines with hydrogen to form hydrogen chloride; iodine also combines with hydrogen to form hydrogen iodide. If the compositions of these compounds are tabulated we have the following results;—

Hydrogen iodide.		*Hydrogen chloride.*	
hydrogen =	0·79	hydrogen =	2·77
iodine =	99·21	chlorine =	97·23
	100·00		100·00

Treating these results as before, we find that
·79 parts by weight of hydrogen combine with
99·21 parts by weight of iodine.
27·8 „ „ chlorine.

Then we inquire; how much iodine combines with 27·8 chlorine? The answer to this is found from the composition of iodine chloride; it is 99·21. So that we complete the foregoing statement by adding
99·21 parts by weight of iodine combine with
27·8 parts by weight of chlorine.

Or, stated more generally, the masses of chlorine and iodine which severally combine with a constant mass of hydrogen are also the masses of chlorine and iodine which combine with each other.

Hydrogen combines with oxygen to form water; hydrogen **68** also combines with sulphur to form hydrogen sulphide; oxygen combines with sulphur to form oxide of sulphur. Let us examine the compositions of these compounds. We need not state the composition of each in parts per 100; let it suffice to state the results thus

1 part by weight of hydrogen combines with
8 parts by weight of oxygen.
16 „ „ sulphur.

Then we inquire; how many parts by weight of sulphur combine with 8 of oxygen? Experiment tells that 8 parts by weight of sulphur combine with 8 parts by weight of oxygen. We have then this result

1 part by weight of hydrogen combines with
8 parts by weight of oxygen
16 „ „ sulphur.
8 parts by weight of oxygen combine with
8 parts by weight of sulph

Phosphorus combines with hydrogen to form phosphorus **69**

hydride; we know that chlorine combines with hydrogen to form hydrogen chloride; phosphorus also combines with chlorine to form phosphorus chloride.

We know that phosphorus combines with hydrogen, and that oxygen combines with hydrogen; phosphorus also combines with oxygen to form phosphorus oxide.

If we determine the compositions of these various compounds, and treat the results as before, always in the case of a. hydrogen compound determining the mass of the other element combined with 1 part by weight of hydrogen, we have these results :—

1 part by weight of hydrogen combines with
$$10\cdot3 \text{ parts by weight of phosphorus.}$$
$$35\cdot5 \quad \text{,,} \quad \text{,,} \quad \text{chlorine.}$$

10·3 parts by weight of phosphorus combine with
$$35\cdot5 \text{ parts by weight of chlorine.}$$

1 part by weight of hydrogen combines with
$$10\cdot3 \text{ parts by weight of phosphorus.}$$
$$8 \quad \text{,,} \quad \text{,,} \quad \text{oxygen.}$$

10·3 parts by weight of phosphorus combine with
$$8 \text{ parts by weight of oxygen.}$$

Stating these results generally, we find that the masses of phosphorus and chlorine, or the masses of phosphorus and oxygen, which severally combine with a constant mass of hydrogen, are also the masses of. phosphorus and chlorine, or of phosphorus and oxygen, which combine with each other.

70   We·also find that the masses of oxygen and sulphur which combine with each other bear a simple relation to the masses of these elements which severally combine with a constant mass of hydrogen.

We have learned that

1 part by weight of hydrogen combines with

$$\begin{cases} 10\cdot3 \text{ parts by weight of phosphorus.} \\ 8 \quad \text{,,} \quad \text{,,} \quad \text{,,} \quad \text{,,} \text{ oxygen.} \\ 16 \quad \text{,,} \quad \text{,,} \quad \text{,,} \quad \text{,,} \text{ sulphur.} \\ 35\cdot5 \quad \text{,,} \quad \text{,,} \quad \text{,,} \quad \text{,,} \text{ chlorine.} \end{cases}$$

Also that

10·3 parts by weight of phosphorus combine with
$$8 \quad \text{parts by weight of oxygen.}$$
$$35\cdot5 \quad \text{,,} \quad \text{,,} \quad \text{,,} \quad \text{chlorine.}$$

Phosphorus forms two compounds with sulphur; when the

composition of that mass of each of these which is produced by combining 10·3 parts of phosphorus with sulphur is stated, we have

10·3 parts by weight of phosphorus combine with
$\begin{cases} \text{(1) } 8 \times 2 (=16) \text{ parts by weight of} \\ \quad \text{sulphur.} \\ \text{(2) } 8 \times 3\frac{1}{3} (=26·6) \text{ ,, ,, ,,} \\ \quad \text{sulphur.} \end{cases}$

Chlorine and oxygen form two compounds, the compositions of which are;

35·5 parts by weight of chlorine
combine with
$\begin{cases} \text{(1) } 8 \text{ parts by weight of} \\ \quad \text{oxygen.} \\ \text{(2) } 8 \times 4 (=32) \text{ parts by weight} \\ \quad \text{of oxygen.} \end{cases}$

Chlorine and sulphur form a compound the composition of which is;

35·5 parts by weight of chlorine combine with $8 \times 4 (=32)$ parts by weight of sulphur.

These results may be stated in more general terms thus :—

The masses of phosphorus, oxygen, sulphur, and chlorine, which severally combine with a constant mass of hydrogen are also the masses of those elements which combine with each other, or they bear a simple relation to these masses.

This statement, or a statement equivalent to this, holds good **71** for all the elements. The statement is known as *The law of reciprocal proportions.*

This law may be expressed in various forms; thus

When two elements, A and B, severally combine with a third element, C, then the proportions in which masses of A and B severally combine with C are also the proportions in which A and B combine with each other, or they bear a simple relation to these proportions. Or, better, thus

The masses of different elements which severally combine with one and the same mass of another element are also the masses of those different elements which combine with each other, or they bear a simple relation to these masses.

The student should particularly observe that the laws of **72** multiple, and reciprocal, proportions, are generalised statements of facts. He should also familiarise himself with the method by which these laws are deduced from the composition of compounds. Statements of the percentage composition of a series of compounds do not suggest the laws in question, although they contain the data from which the laws are deduced. It is necessary to compare the compositions of compounds of each of two, or more, elements with one and the same element; it is also necessary to state these compositions so that the mass of the element with which the others combine is kept constant throughout all the compounds.

Any element may be chosen as the standard element; and any mass of the standard element may be chosen as the fixed mass with which other elements are to be combined. It is found that the relations between the masses of elements which mutually combine are very clearly and simply exhibited by choosing hydrogen as the standard element, and one part by weight (say 1 gram) of hydrogen as the fixed mass.

73   The following table illustrates this way of looking at the composition of several compounds. Column I. exhibits the composition of three compounds of hydrogen, stated, (a) as parts of each element per 100 parts of the compound, (b) so as to shew the weight of the second element combined with 1 part by weight of hydrogen. Columns II., III., and IV., exhibit the composition of compounds of two elements neither of which is hydrogen, stated, (a) as parts per 100, and (b) so as to shew the weights of those elements which severally combine with that weight of oxygen, sulphur, or chlorine, which has been shewn in I. to unite with 1 part by weight of hydrogen.

74   The masses of oxygen, sulphur, and chlorine, which severally combine with 1 part by weight—i.e. with unit mass —of hydrogen are 8, 16, and 35·5, respectively. The masses of copper, lead, and thallium, which severally combine with 8 parts by weight of oxygen are 31·7, 103·5, and 204, respectively; and these are also the masses of those elements which severally combine with 16 parts by weight of sulphur, and with 35·5 parts by weight of chlorine, respectively.

Let us call those masses of oxygen, sulphur, &c. the *combining weights* of oxygen, sulphur, &c. We have then :—

Combining weights, deduced from composition of compounds with hydrogen.

Oxygen = 8 ; Sulphur = 16 ; Chlorine = 35·5.

These numbers represent parts by weight of each element which combine with one part by weight of hydrogen.

Combining weights, deduced from composition of compounds with oxygen.

Copper = 31·7 ; Lead = 103·5 ; Thallium = 204.

I.

Water
(a) Hydrogen = 11·1
    Oxygen = 88·9
              ————
              100·0
(b) Hydrogen = 1
    Oxygen = 8

Hydrogen Sulphide
(a) Hydrogen = 5·88
    Sulphur = 94·12
               ————
               100·00
(b) Hydrogen = 1
    Sulphur = 16

Hydrogen Chloride
(a) Hydrogen = 2·74
    Chlorine = 97·26
               ————
               100·00
(b) Hydrogen = 1
    Chlorine = 35·5

II.

Copper Oxide
(a) Oxygen = 20·15
    Copper = 79·85
             ————
             100·00
(b) Oxygen = 8
    Copper = 31·7

Lead Oxide
(a) Oxygen = 7·18
    Lead   = 92·82
             ————
             100·00
(b) Oxygen = 8
    Lead   = 103·5

Thallium Oxide
(a) Oxygen   = 3·77
    Thallium = 96·23
               ————
               100·00
(b) Oxygen   = 8
    Thallium = 204

III.

Copper Sulphide
(a) Sulphur = 33·53
    Copper = 66·47
              ————
              100·00
(b) Sulphur = 16·
    Copper = 31·7

Lead Sulphide
(a) Sulphur = 13·39
    Lead   = 86·61
              ————
              100·00
(b) Sulphur = 16
    Lead   = 103·5

Thallium Sulphide
(a) Sulphur = 7·27
    Thallium = 92·73
               ————
               100·00
    Sulphur = 16
    Thallium = 204

IV.

Copper Chloride
(a) Chlorine = 52·86
    Copper   = 47·14
               ————
               100·00
(b) Chlorine = 35·5
    Copper   = 31·7

Lead Chloride
(a) Chlorine = 25·54
    Lead   = 74·46
               ————
               100·00
(b) Chlorine = 35·5
    Lead   = 103·5

Thallium Chloride
(a) Chlorine = 14·82
    Thallium = 85·18
               ————
               100·00
    Chlorine = 35·5
    Thallium = 204

These numbers represent parts by weight of each element which combine with 8 parts by weight of oxygen, 16 of sulphur, or 35·5 of chlorine; in other words *these numbers represent the parts by weight of each element which combine with one combining weight of oxygen, or sulphur, or chlorine.*

The conception of *combining weight* may be extended to all the elements. The combining weight of an element which forms a compound with hydrogen must be regarded by us at present as a number expressing the mass of the element which combines with unit mass of hydrogen. The combining weight of an element which does not form a compound with hydrogen we shall for the present regard as the mass of that element which combines with one combining weight of oxygen, or of sulphur, or of chlorine; i.e. with that mass of oxygen, sulphur, or chlorine, which combines with unit mass of hydrogen, i.e. with 8 parts by weight of oxygen, 16 of sulphur, or 35·5 of chlorine.

75   The laws of multiple, and reciprocal, proportions may now be put into one statement.

The elements combine in the ratios of their combining weights, or in ratios which bear a simple relation to these.

To illustrate this mode of expressing the laws of multiple and reciprocal proportions, let us tabulate (1) the combining weights of several elements, (2) the compositions of several compounds of these elements stated as so many combining weights of each element.

Combining weights of some elements.

A. *Determined from composition of compounds with hydrogen.*	B. *Determined from composition of compounds with elements in column A.*
Nitrogen = 4·6	Chromium = 17·46
Oxygen = 8	Tin = 29·5
Sulphur = 16	Copper = 31·7
Chlorine = 35·5	Antimony = 40
Bromine = 80	Mercury = 100
Iodine = 127	

Composition of some compounds; stated as number of combining weights of each element.

Oxides of nitrogen. c. ws. of oxygen : c. ws. of nitrogen = 1 : 1 in compound *a*, 1 : 3 in compound *b*, 2 : 3 in compound *c*, 4 : 3 in compound *d*.

Oxides of chromium.   c. ws. of oxygen : c. ws. of chromium $= 2 : 3$ in compound a, $8 : 9$ in b, $1 : 1$ in c, $4 : 3$ in d, $2 : 1$ in e.

Chlorides of antimony.   c. ws. of chlorine: c. ws. of antimony $= 1 : 1$ in compound a, $5 ; 3$ in b.

Bromides of tin.   c. ws. of bromine : c. ws. of tin $= 1 ; 2$ in compound a, $1 ; 1$ in b.

Iodides of mercury.   c. ws. of iodine : c. ws. of mercury $= 1 : 2$ in compound a, $1 : 1$ in b.

Sulphides. of copper.   c. ws. of sulphur : c. ws. of copper $= 1 : 2$ in compound a, $1 : 1$ in b.

The composition of all compounds may be stated in this way. Let us use a *symbol* to *represent one combining weight of an element*. Let N represent one combining weight of nitrogen; N_2, two c. ws. of nitrogen; N_3, three c. ws. of nitrogen; generally N_x, x c. ws. of nitrogen: let O represent one c. w. of oxygen; O_x, x c. ws. of oxygen: Cr, one c. w. of chromium: Sb, one c. w. of antimony: Sn, one c. w. of tin: Hg, one c. w. of mercury: Cu, one c. w. of copper: Cl, one c. w. of chlorine: Br, one c. w. of bromine: I, one c. w. of iodine: and S, one c. w. of sulphur. Then the compositions of the above compounds may be represented thus ;—

Oxides of nitrogen.   ON, ON_3, O_2N_3, O_4N_3.

Oxides of chromium. O_2Cr_3, O_3Cr_3, OCr, O_4Cr_3, O_2Cr.

Chlorides of antimony.   ClSb, Cl_5Sb_3.

Bromides of tin. $BrSn_2$, BrSn.

Iodides of mercury. IHg_2, IHg.

Sulphides of copper. SCu_2, SCu.

CHAPTER VI.

SYMBOLS AND FORMULAE.

76   IT is customary to express the composition of compounds in a kind of shorthand by a method the principle of which is the same as that we are at present illustrating.

A symbol is given to each element; this symbol is formed either of the first letter, or of the first and some other letter, of the name of the element. When the names of several elements begin with the same letter that element which has been longest known and best studied generally gets a symbol formed of the first letter only; but there is no universally applicable rule. Some of the symbols are derived from the names by which the elements were known to the ancients or in the middle ages. The symbols of two elements, potassium (K), and sodium (Na), are derived from the names *kalium* and *natrium* by which these elements are known to German chemists. The symbol W is given to the element tungsten, it is derived from the name (*Wolfram*) of the mineral from which tungsten was first obtained.

It is of the utmost importance to remember that each of these symbols represents a definite mass of the element; it represents either one, two, three, four, five, or six, *combining weights*, as we are at present using the term combining weight, of the element. The following table gives the names and symbols of the elements.

Elements.

Name.	Symbol.	Mass of element expressed by symbol[1].	Name.	Symbol.	Mass of element expressed by symbol[1].
Aluminium	Al	27	Molybdenum	Mo	96
Antimony	Sb	120	Nickel	Ni	58·6
Arsenic	As	75	Niobium	Nb	94
Barium	Ba	137	Nitrogen	N	14
Beryllium	Be	9	Osmium	Os	193
Bismuth	Bi	208	Oxygen	O	16
Boron	B	11	Palladium .	Pd	106
Bromine	Br	80	Phosphorus	P	31
Cadmium	Cd	112	Platinum	Pt	194
Caesium	Cs	133	Potassium	K	39
Calcium	Ca	40	Rhodium	Rh	104
Carbon	C	12	Rubidium	Rb	85·4
Cerium	Ce	140	Ruthenium	Ru	104·6
Chlorine	Cl	35·5	Scandium	Sc	44
Chromium	Cr	52·2	Selenion	Se	79
Cobalt	Co	59	Silicon	Si	28
Copper	Cu	63·2	Silver	Ag	108
Didymium	Di	144	Sodium	Na	23
Erbium	Er	166	Strontium	Sr	87
Fluorine	F	19	Sulphur	S	32
Gallium	Ga	69·9	Tantalum	Ta	182
Germanium	Ge	72·2	Tellurium	Te	125
Gold	Au	197	Terbium	Tr	148
Hydrogen	H	1	Thallium	Tl	204
Indium	In	113·4	Thorium	Th	232
Iodine	I	127	Tin	Sn	118
Iridium	Ir	192·6	Titanium	Ti	48
Iron	Fe	56	Tungsten	W	184
Lanthanum	La	139	Uranium	U	240
Lead	Pb	207	Vanadium	V	51·2
Lithium	Li	7	Yttrium	Y	89
Magnesium	Mg	24	Ytterbium	Yb	173
Manganese	Mn	55	Zinc	Zn	65
Mercury	Hg	200	Zirconium	Zr	90

[1] The values in this table are given in round numbers; they are only approximately correct.

77   That collocation of symbols which expresses the composition of a compound is called the *formula* of that compound. The formulae BaO, B_2O_3, Cr_2Cl_6, HI, tell, that barium and oxygen combine to form barium oxide in the ratio $137 : 16$ by weight, that boron and oxygen combine in the ratio $22 : 48 (= 11 \times 2 : 16 \times 3)$, that chromium and chlorine combine in the ratio $104 \cdot 4 : 213 (= 52 \cdot 2 \times 2 : 35 \cdot 5 \times 6)$, and that hydrogen and iodine combine in the ratio $1 : 127$.

Or, the facts concerning composition which these formulae express may be thus stated; 153 parts by weight of barium oxide are formed by the combination of 137 parts by weight of barium with 16 parts by weight of oxygen; 70 parts by weight of boron oxide are formed by the combination of 22 parts by weight of boron with 48 parts by weight of oxygen; $317 \cdot 4$ parts of chromium chloride are produced by the combination of $104 \cdot 4$ parts of chromium with 213 parts of chlorine; 128 parts of hydrogen iodide are formed by the union of 1 part of hydrogen with 127 parts of iodine.

78   The numbers in the third column of the preceding table are sometimes called the combining weights of the elements. We have already given a meaning to the term combining weight (*s.* par. 74). If that meaning is adopted, the mass of an element expressed by its symbol is seldom the same as the value obtained for the combining weight of that element; but when it is not the same, it is a simple multiple of the combining weight.

We are not yet in a position to go fully into this matter of combining weights. We have already used the expression combining weight to mean, that mass of an element which combines with unit mass of hydrogen, or, in the cases of elements which do not combine with hydrogen, that mass which combines with 8 parts by weight of oxygen, or 16 of sulphur, or $35 \cdot 5$ of chlorine. But when we come to apply this definition we meet with many difficulties. Thus, nitrogen and phosphorus each form one compound with hydrogen ; nitrogen forms 5 compounds, and phosphorus 2 compounds, with oxygen. From the composition of each of these compounds a value may be deduced for the combining weight of nitrogen, or for that of phosphorus. Similarly iron forms 3 compounds with oxygen, and 2 with chlorine ; from the composition of these, values are found for the combining weight of iron. The values are these.

Combining weights of nitrogen, phosphorus, and iron.

Deduced from composition

	of hydrides.	*of oxides.*	*of chlorides.*
Nitrogen	4·6	2·8, 3·5, 4·6, 7, 14	
Phosphorus	10·3	6·2, 10·3	6·2, 10·3
Iron	—	18·6, 21, 28	18·6, 28

This list might be largely extended; in very few cases should we find but one value for the combining weight (as defined) of an element.

To adopt. several combining weights for each element would introduce endless confusion into our system of representing the composition of compounds. It is absolutely necessary to adopt one value and one value only, not merely for convenience but also for cogent reasons which will be given later. Sometimes the highest value found by the method already stated is adopted, e.g. for nitrogen ($N = 14$; comp. above results with the table in par. 76) ; sometimes a simple multiple of this highest value is adopted, e.g. for iron ($Fe = 56$: *s.* table in par. 76). If we define combining weight as has been already done, then the definition generally leads to several values for the combining weight of each element. If we call the numbers in the table in par. 76 combining weights, then we cannot accurately define the term combining weight.

The best compromise, at any rate for us at present, is to **79** say, that *the actually used combining weight of an element is a number which expresses either the largest mass of the element which combines with 1 part by weight of hydrogen, or 8 parts of oxygen, or 16 of sulphur, or 35·5 of chlorine, or it expresses a simple multiple of this mass.*

The following table presents; in column I., the largest mass of each element which is known to combine with either 1 part by weight of hydrogen, or 8 of oxygen, or 16 of sulphur, or 35·5 of chlorine; and in column II., the actually used values for what are generally called the combining weights of the elements.

	I.	II.		I.	II.
Aluminium	9	27	Bismuth	104	208
Antimony	40	120	Boron	3·6	11
Arsenic	25	75	Bromine	80	80
Barium	68·5	137	Cadmium	56	112
Beryllium	4·5	9	Caesium	133	133

	I.	II.		I.	II.
Calcium	20	40	Oxygen	16	16
Carbon	12	12	Palladium	106	106
Cerium	46·6	140	Phosphorus	10·3	31
Chlorine	35·5	35·5	Platinum	97	194
Chromium	26·1	52·2	Potassium	39	39
Cobalt	29·5	59	Rhodium	52	104
Copper	63·2	63·2	Rubidium	85·4	85·4
Didymium	48	144	Ruthenium	52·3	104·6
Erbium	58·6	166	Scandium	14·6	44
Fluorine	19	19	Selenion	39·5	79
Gallium	23·3	69·9	Silicon	7	28
Germanium	36·1	72·2	Silver	108	108
Gold	197	197	Sodium	23	23
Hydrogen	1	1	Strontium	43·5	87
Indium	37·8	113·4	Sulphur	16	32
Iodine	127	127	Tantalum	45·5	182
Iridium	96·3	192·6	Tellurium	62·5	125
Iron	28	56	Terbium	49·3	148
Lanthanum	46·6	139	Thallium	204	204
Lead	103·5	207	Thorium	58	232
Lithium	7	7	Tin	59	118
Magnesium	12	24	Titanium	24	48
Manganese	27·5	55	Tungsten	46	184
Mercury	200	200	Uranium	60	240
Molybdenum	48	96	Vanadium	51·2	51·2
Nickel	29·3	58·6	Yttrium	29·6	89
Niobium	47	94	Ytterbium	57·6	173
Nitrogen	14	14	Zinc	32·5	65
Osmium	96·5	193	Zirconium	45	90

80   As we advance in our study of chemical events we shall learn that there is no purely chemical, and general, method, by using which a decision may be arrived at regarding the best value to be given to the combining weight of an element. Each case must be discussed by itself; the result is at best a compromise. But we shall also find that the application of certain physical conceptions to chemical phenomena leads to a generally applicable method, based on one definite principle, whereby values may be obtained for what we at present call the combining weights of the elements.

81   The symbol of an element, then, expresses a definite mass of that element. The formula of a compound expresses the masses of the elements, stated as a certain number of combining weights of each element, which combine to form a specified mass of the compound. A number placed beneath (or sometimes above) the symbol of an element in the formula of a compound tells that the symbol is to be multiplied by this number. A number placed at the beginning of the formula of a compound multiplies the whole of the formula, or if a full stop occurs in the formula the number multiplies all as far as that stop; sometimes the formula is put in brackets and the multiplier is placed outside the bracket. The following formulae will illustrate these points.

$Fe = 56$, $O = 16$, $S = 32$.

FeO means $56 + 16 = 72$ parts by weight of a compound called *ferrous oxide*; this formula also tells that one c. w. of iron combines with one c. w. of oxygen to form ferrous oxide.

Fe_2O_3 means $(56 \times 2) + (16 \times 3) = 160$ parts by weight of a compound called *ferric oxide*; also that 2 c. ws. of iron combine with 3 c. ws. of oxygen to form ferric oxide.

$FeSO_4$ means $56 + 32 + (16 \times 4) = 152$ parts by weight of a compound called *ferrous sulphate*; also that one c. w. of iron, one c. w. of sulphur, and four c. ws. of oxygen, combine to form ferrous sulphate.

$Fe_2.3SO_4$ or $Fe_2(SO_4)_3$ means $(56 \times 2) + 3(32 + 64) = 400$ parts by weight of a compound called *ferric sulphate*; also that two c. ws. of iron, three c. ws. of sulphur, and twelve c. ws. of oxygen, combine to form ferric sulphate.

$3Fe_2.3SO_4$ or $3Fe_2(SO_4)_3$ means $3\{(56 \times 2) + 3(32 + 64)\} = 1200$ parts by weight of ferric sulphate.

Chemical changes are also expressible in formulae, so far **82** at least as the composition of the elements or compounds before and after such changes is concerned. Thus, we have learned that

(1) Sulphur and iron combine when heated in the ratio $1 : 1.75$, to form iron sulphide;

(2) Hydrogen and oxygen combine in the ratio $1 : 8$ to form water.

These chemical reactions may be shortly expressed thus;—

(1) $S + Fe = FeS$.   (2) $2H + O = H_2O$.

$Fe = 56$, $S = 32$, $O = 16$. The ratio $32 : 56 = 1 : 1.75$; the ratio $2 : 16 = 1 : 8$.

The sign + signifies *reacts chemically with*; the sign = signifies *with production of.*

The total mass of matter on one side of the sign = is equal to the total mass of matter on the other side.

Let us consider one or two rather more complex reactions.

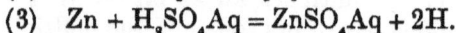

(1) $Na + H_2O + Aq = NaOHAq + H.$

(2) $3Fe + 4H_2O = Fe_3O_4 + 8H.$ ·

(3) $Zn + H_2SO_4Aq = ZnSO_4Aq + 2H.$

$Na = 23, O = 16, Fe = 56, Zn = 65, S = 32.$

(1) When sodium and water interact, 23 parts by weight of sodium and 18 parts by weight of water disappear, and there are produced 40 parts of sodium hydroxide, which remains dissolved in the water that has not been changed, and 1 part by weight of hydrogen.

(2) When iron and water interact, 168 parts of iron and 72 of water are changed to 232 parts of iron oxide and 8 parts of hydrogen.

(3) When zinc and a solution in water of sulphuric acid interact, 65 parts by weight of zinc and 98 of the acid are changed into 161 parts of zinc sulphate, which remains in solution, and 2 parts of hydrogen.

These *chemical equations*, as they are called, also represent the compositions of the compounds before and after the change, expressed as so many combining weights of each elementary constituent of each compound; when elements take part in the reactions, the equations also express the number of combining weights of this or that element which interacts with a certain mass of a compound, or with a certain number of combining weights of another element, and the number of combining weights of this or that element which is produced by the interaction. Thus (1) states, more shortly than can be done in words, the fact that one c. w. of sodium interacts with 18 parts of water to produce 40 parts of sodium hydroxide and 1 c. w. of hydrogen; and (3) states that one c. w. of zinc interacts with 98 parts by weight of sulphuric acid dissolved in water (which 98 parts are composed of 2 c. ws. of hydrogen, 1 c. w. of sulphur, and 4 c. ws. of oxygen) to produce 161 parts of zinc sulphate (composed of 1 c. w. of zinc, 1 c. w. of sulphur, and 4 c. ws. of oxygen) which remain in solution, and 2 c. ws. of hydrogen.

The symbol Aq is used here, and generally in this book, to mean a large (indefinite) quantity of water; when placed after the formula of a compound or element it means that that body is in solution in a large quantity of water.

· Chemical formulae express other facts regarding chemical **83** changes ; these we shall learn as we advance. It is advisable to note here that these formulae and equations do not say anything regarding the conditions under which the chemical interactions occur. Thus $Fe + S = FeS$ only tells us that a certain mass of iron combines with a certain mass of sulphur to produce the sum of these masses of· iron sulphide. So $Na + H_2O + Aq = NaOHAq + H$ tells that certain masses of sodium and water interact to produce certain masses of sodium hydroxide [which is dissolved in the *excess* of water (*s. ante* par. 56)] and hydrogen, and that the sum of the masses of sodium and water is equal to the sum of the masses of sodium hydroxide and hydrogen. The equations in no way indicate the facts that iron and sulphur only combine when heated, but that sodium and water interact at ordinary temperatures. The equation $Zn + H_2SO_4Aq = ZnSO_4Aq + 2H$ expresses certain definite quantitative facts (*s. ante*) ; but it does not indicate or even suggest that the compositions of the products of the interaction of zinc and sulphuric acid vary with variations in the temperature at which the interaction occurs, and that the interaction proceeds according to the representation given by the equation only at the ordinary temperature.

Chemical equations evidently give very incomplete representations of chemical changes. But nevertheless chemical formulae are of the greatest value, inasmuch as they enable us to exhibit, in a simple and intelligible way, the composition of compounds, and those changes of composition, the study of which forms one part of chemical science.

We have learned that the symbol of an element represents **84** a definite mass, and also one combining weight, of that element.

The formula of a compound also represents a definite mass of the compound, and tells the composition of that definite mass, both in parts by weight, and also in combining weights, of each of the elements by the combination of which the compound has been formed. The following are the formulae of some well-known compounds ;—

Water	$H_2O.$	Formic acid	$H_2CO_2.$
Hydrogen peroxide	$H_2O_2.$	Hydrogen sulphide	$H_2S.$
Oxalic acid	$H_2C_2O_4.$	Sulphur chloride	$S_2Cl_2.$

Hydrogen chloride HCl.   Antimony chloride $SbCl_3$.
Ferric chloride Fe_2Cl_6.
              Benzene H_6C_6.
              Acetylene H_2C_2.
              Methane H_4C.

These formulae suggest a question, the answer to which is of the utmost importance, but a question to which a satis-factory answer cannot yet be given.

Why should we choose to represent hydrogen peroxide as composed of 2 combining weights of hydrogen with 2 c. ws. of oxygen? Water is represented as produced by the union of 1 c. w. of oxygen with 2 c. ws. of hydrogen; why should not the composition of peroxide of hydrogen be represented by the formula HO? The ratio H : O is the same as the ratio H_2 : O_2. Again, why should the formula of ferric chloride be Fe_2Cl_6 rather than $FeCl_3$? Chloride of antimony, $SbCl_3$, is represented as formed by the union of 1 c. w. of antimony with 3 c. ws. of chlorine; why should we choose a formula for ferric chloride which represents the composition of that mass of this compound which is formed by the union of 2 c. ws. of iron with 6 c. ws. of chlorine?

Similar questions are suggested by the other formulae. In some cases the formula appears to be the simplest that could be given to the compound, e.g. H_2O, H_2S, HCl, $SbCl_3$; in other cases a needless and foolish complication seems to be introduced. Why not HCO_2 in place of $H_2C_2O_4$; HC in place of H_6C_6; SCl in place of S_2Cl_2; HO in place of H_2O_2; $FeCl_3$ in place of Fe_2Cl_6? Or if the more complex formulae are to be used, why should such formulae not be always used? Why not H_4O_2 or H_6O_3 in place of H_2O; H_4S_2 or H_6S_3 or $H_{12}S_6$ in place of H_2S; Sb_2Cl_6 in place of $SbCl_3$; &c.?

85   There must be some reason for these apparent incon-sistencies. There are several reasons; but we are not yet in a position fully to understand and appreciate these reasons. We may however gain some notion of the kind of reasoning employed in determining which of several possible formulae best represents the composition and reactions of a compound. The gist of the matter, as we shall hereafter find, is in the conception expressed by the words *composition and reactions*. So long as we look only at the composition of compounds we cannot find answers to our questions. If we disregard the composition and look only at the reactions of compounds we cannot find answers to our questions.

The symbol of an element represents a certain mass of **86** that element usually called its combining weight. Elements combine in the ratios of their combining weights, or in ratios bearing a simple relation to these.

The formula of a compound represents the composition of a certain mass of that compound; *this mass we propose to call the reacting weight of the compound.* Compounds interact in the ratios of their reacting weights, or in ratios bearing a simple relation to these.

The reacting weight of water is 18 ($H_2 = 2 + O = 16$).

The combining weight of sodium is 23 (this, as a matter of fact, is the mass of sodium which combines with 8 parts by weight of oxygen). Let us examine the interaction of water and sodium.

When sodium is thrown into water a reaction immediately occurs; the sodium rapidly disappears and hydrogen gas is produced. When the reaction is finished, let the solution be evaporated; water passes away as steam, and a white solid (caustic soda) remains. The composition of this solid is represented by the formula NaOH ($Na = 23$, $O = 16$), that is to say, this compound is produced by the combination of one combining weight of sodium, one c. w. of hydrogen, and one c. w. of oxygen. We know that water is a compound of hydrogen and oxygen, and that sodium is an element. Hence the oxygen and hydrogen which form part of the caustic soda must have come from the water. But besides caustic soda, hydrogen was produced; this must also have come from the water. Hence when sodium and water interact, a portion of the hydrogen which was combined with oxygen is evolved as hydrogen gas, and another portion enters into combination with the sodium and the oxygen to produce caustic soda. When this experiment is made quantitative, it is found that 23 parts by weight of sodium interact with 18 parts by weight of water, and there are produced 40 parts by weight of caustic soda and 1 part by weight of hydrogen. The 40 parts of caustic soda are composed of 23 parts of sodium, 16 parts of oxygen, and 1 part of hydrogen.

The conclusion from these experiments is, that, as regards the interaction of water with sodium, 18 is the *reacting weight* of water, and that the decomposition of one reacting weight of water results in the production of 2 combining weights of hydrogen and 1 c. w. of oxygen. But if $H = 1$ and $O = 16$, the formula H_2O (18) summarises the results of this experiment.

A quantitative study of the reactions of water, carried out in the way thus briefly indicated, leads to the conclusion that the mass of water which interacts with other compounds and with elements can always be represented as 18, or as a whole multiple of 18.

The composition of the hydrocarbon benzene is most simply represented as one c. w. of carbon combined with one c. w. of hydrogen; therefore the smallest value that can be given to the reacting weight of benzene is 13 (CH; $C = 12$, $H = 1$).

Is this the best value to adopt for the *reacting weight* of benzene?

Benzene and chlorine react to form a series of compounds, each composed of carbon, hydrogen, and chlorine; the formation of each of these is accompanied by the formation of hydrogen chloride (HCl). The first of these compounds is composed of 35·5 parts by weight of chlorine, 72 of carbon, and 5 of hydrogen; therefore (as $C = 12$, and $Cl = 35·5$) the simplest formula to be given to this compound is C_6H_5Cl. The composition of the next compound cannot be represented by a simpler formula than $C_6H_4Cl_2$. The other compounds have compositions which cannot be expressed by formulae simpler than $C_6H_3Cl_3$, $C_6H_2Cl_4$, C_6HCl_5, and C_6Cl_6, respectively. Now, as $C = 12$, and $H = 1$, and as carbon and hydrogen combine to form benzene in the ratio $12 : 1$, the simplest formula which we can use to express the composition of the *reacting weight* of benzene is $C_6H_6 = 78$. When we extend our quantitative study of the reactions of benzene we find that the mass of this compound which interacts with other compounds and with elements is either 78 or a whole multiple of 78.

These examples give some notion of the methods used for determining the value to be given to the reacting weight of a compound. There is no generally applicable chemical method. Each compound must be considered apart from other compounds. The object of the inquiry is to find the relative weight of the smallest mass of the compound which interacts with other compounds, or with elements, in chemical changes. The composition of this mass is then expressed in the formula of the compound.

It will be noticed that in this inquiry the combining weights of the elements are assumed to be known. But we know that great difficulties have to be overcome before the combining weights of the elements can be determined; indeed it was stated that the only satisfactory principle on which a method for finding these combining weights has been based is physical rather than chemical. We shall see later on that the same physical principle gives us a means for determining the reacting weights of compounds.

In addition to the three laws of chemical combination now **87** considered—the law of fixity of composition, the law of multiple proportions, and the law of reciprocal proportions— there is another generalised statement regarding the volumes of gaseous elements or compounds which interact and the volumes of the gaseous products of these interactions.

The law of volumes, or *the law of Gay Lussac,* states that *the volume of a gaseous compound produced by the interaction of gaseous elements or compounds bears a simple relation to the volumes of the gases from which it is produced, and the volumes of the interacting gaseous elements or compounds bear a simple relation to each other.*

All volumes are measured under the same conditions of temperature and pressure.

Thus :—

Vols. of reacting gaseous elements or compounds.		Vols. of gaseous products.
1 vol. hydrogen and 1 vol. chlorine	produce $H + Cl = HCl.$	2 vols. hydrogen chloride.
2 vols. hydrogen and 1 vol. oxygen	produce $H_2 + O = H_2O.$	2 vols. water-gas.
3 vols. hydrogen and 1 vol. nitrogen	produce $3H + N = H_3N.$	2 vols. ammonia.
2 vols. carbon oxide and 2 vols. chlorine	produce $CO + Cl_2 = COCl_2.$	2 vols. carbonyl chloride.
2 vols. hydrogen iodide and 1 vol. chlorine	produce $HI + Cl = HCl + I.$	2 vols. hydrogen chloride and 1 vol. iodine-gas.
2 vols. ethane and 2 vols. chlorine	produce $C_2H_6 + Cl_2 = C_2H_5Cl + HCl.$	2 vols. chlorethane and 2 vols. hydrogen chloride.

5—2

2 vols. alcohol-gas and 2 vols. hydrogen iodide produce 2 vols. iodo-ethane and 2 vols. water-gas.

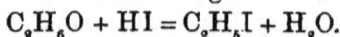

$$C_2H_6O + HI = C_2H_5I + H_2O.$$

88   Hydrogen is taken as the standard gas to which the others are referred. Any specified volume, say 1 litre, is adopted as the standard volume, and this is called *one volume.*

If the weight of this one volume of hydrogen is taken as unity, then it is found that the weight of 1 volume of

chlorine	is	35·5 :	that is, 1 vol.	of chlorine weighs	35·5	
oxygen	„	16 :	„ 1 „	oxygen „	16	times
nitrogen	„	14 :	„ 1 „	nitrogen „	14	
iodine-gas	„	127 :	„ 1 „	iodine-gas „	127	

more than 1 vol. of hydrogen.

But the combining weights of chlorine, oxygen, nitrogen, and iodine, are 35·5, 16, 14, and 127, respectively. Hence the numbers which represent the combining weights of these elements also represent the specific gravities of these elements in the gaseous state referred to hydrogen as unity.

This statement is applicable to many of the gaseous elements.

The composition of the reacting weights of hydrogen chloride, water, ammonia, carbonyl chloride, hydrogen iodide, ethane, chlorethane, alcohol, and iodo-ethane, are represented by the formulae HCl, H_2O, NH_3, $COCl_2$, HI, C_2H_6, C_2H_5Cl, C_2H_6O, C_2H_5I, respectively. But these formulae also represent the composition of 2 volumes of each compound in the gaseous state; i.e. they represent the composition of that volume of each gaseous compound which is equal to twice the volume occupied by 1 part by weight of hydrogen.

This statement is applicable to all gaseous compounds.

The formula of a gaseous compound represents the composition of the reacting weight of that compound, and this is that weight which occupies twice the volume occupied by 1 part by weight of hydrogen.

These statements assume that all volumes are measured under the same conditions of temperature and pressure.

89   Let us now glance back at what we have learned regarding chemical composition.

We have learned that chemical changes involve changes of composition and changes of properties; that these changes occur when elements interact with elements or compounds, or compounds with compounds; that the composition of every compound is definite and unchangeable; that elements combine, or interact, in the ratios of their combining weights, and compounds in the ratios of their reacting weights, or in ratios bearing a simple relation to these; and that the volumes of gaseous elements and compounds which combine, or interact, are simply related to each other and to the gaseous products of the reactions. We have also gained some notion of the meanings of the terms *combining weight*, and *reacting weight*, as applied to elements and compounds respectively; and we have seen how difficult it is to determine the values of these quantities by purely chemical considerations.

The study of the composition of compounds has necessitated some study of the properties of elements and compounds. The properties we have found it incumbent on us to examine have not been those exhibited by elements or compounds considered apart from each other, but rather those exhibited in the mutual interactions of elements and compounds. To arrive at any just generalisations regarding the connexions between changes of composition and changes of properties— the study of which connexions is the business of chemistry— we have found it necessary to study the relations between classes or groups of chemical events. The study of isolated occurrences, or the study of isolated elements or compounds, cannot lead to far-reaching conclusions concerning chemical change. We have repeatedly found that chemical changes are accompanied by physical changes: we have tried to keep our attention fixed on the chemical parts of the phenomena; but we should always remember that nothing in nature is "defined into absolute independent singleness."

CHAPTER VII.

CHEMICAL STUDY OF WATER AND AIR.

90   THAT we may become better acquainted with the kind of phenomena which chemistry studies, and that we may apply the principles gained in our study of chemical changes, so far as that study has gone, let us examine some of the phenomena presented to the chemist by the two kinds of matter, *air* and *water*.

91   **Water.**   To which of the classes, *Element* or *Not-Element*, does water belong?

We have already had an answer to this question.   In par. 27 we separated a specified mass of water into two kinds of matter different from, and each weighing less than, itself.

This separation, or *analysis*, was effected (1) by passing an electric current through water ; (2) by the interaction between water and sodium.   In the first process, the gases into which water was decomposed, hydrogen and oxygen, were collected separately and examined.   In the second process, a portion of one of the gases, hydrogen, was obtained, but the other gas, and the rest of the hydrogen, combined with the sodium to form caustic soda.   These proofs that water is a *not-element* were supplemented by the *synthesis* of water (1) by passing electric sparks through a mixture of 1 part by weight of hydrogen with 8 of oxygen ; and (2) by the interaction of hydrogen with hot copper oxide, whereby water and copper were produced.

92   Having proved water to belong to the class *Not-Element*, the next question to be answered is ; is water a *compound* or a *mixture*?   The experiments whereby water is proved not to be an element afford an answer to this question.   Water is so completely unlike either of its constituents, hydrogen and oxygen, that we must consider it a compound, not a mixture, of these. Of course it might be urged that a compound may be formed by the union of these two gases, but that water may be a mixture of this compound with some other substance or substances.

In describing the experiments whereby the composition of water has been demonstrated it was assumed (the assumption was noted at the time) that the hydrogen and oxygen used for the synthesis of water were perfectly pure, and that every precaution was taken in the experiments.

The statement that water is a compound of hydrogen and oxygen, and that these gases combine to form water in the ratio 1 : 8 by weight, implies, that the whole of the hydrogen and the whole of the oxygen disappear, that water is the only substance produced, and that the mass of the water thus produced is exactly equal to the sum of the masses of the hydrogen and oxygen. Details of the experimental methods by which each of these statements is proved to be correct were not given. Nor need these details be given here. But it will be well briefly to recapitulate the stages in Davy's proof of the fact that when pure water is decomposed by an electric current, hydrogen and oxygen are the only kinds of matter produced.

Priestley had proved that when a mixture of air and **93** hydrogen was exploded in a closed vessel, water was found in the vessel after the explosion. Knowing that air contained oxygen, Cavendish thought it probable that the water noticed by Priestley was a product of the union of hydrogen with oxygen in the air. Cavendish proved that this was really the case; he exploded mixtures of hydrogen and oxygen in various proportions; the loudest explosion was obtained when two volumes of hydrogen were used to one volume of oxygen, and in this case no trace of either gas remained in the vessel after the explosion. Cavendish found that the water produced by exploding air with hydrogen always contained a little acid; the production of this acid he traced to a constituent of the air other than oxygen; when he used pure oxygen in place of air, the water produced contained no acid.

Davy decomposed water which had been purified by distillation, in glass vessels, by passing an electric current through it; in every case a little acid was produced at the positive electrode, and a little alkali at the negative electrode. He re-distilled the water and again electrolysed it, but the result was the same. He noticed that the glass vessels in which the water was decomposed were slightly corroded, so he placed the re-distilled water in agate vessels and passed the electric current through it. But the result was as before; traces of acid and alkali were produced. He used electrodes made of different materials; the results were the same. He distilled the water again; there was rather less alkali, but as much acid as before. After another distillation the alkali had further diminished.

Davy concluded that the source of the alkali was some substance in the water itself. He placed the water in gold vessels, and electrolysed it; a very little alkali appeared at the negative electrode; after the current had passed for some minutes the production of alkali nearly ceased; but the acid was still produced, and the quantity of it slowly increased as the process of electrolysis continued. By evaporating some of the water used to dryness in a silver dish, Davy obtained a small quantity of a white solid substance, which, after being heated, was distinctly alkaline. A small quantity of the same water was now electrolysed in a gold vessel; after a few minutes, when the production of alkali had nearly ceased, a little of the white solid obtained by evaporating the water was placed in the water being electrolysed; the alkali was at once produced in some quantity at the negative electrode. Davy then distilled some of the water he had been using in a silver retort, and electrolysed a portion of the distillate; no alkali appeared; he placed a little bit of glass in the water, alkali began to be formed. He had thus traced the production of alkali to the action of the water and the current on the glass or agate vessels used to contain the water; and he had conclusively proved that water is not changed into alkali by the action of the electric current.

But it still remained to account for the production of acid at the positive electrode. The acid he found to be nitric acid. He knew that this acid is a compound of three elements; hydrogen, oxygen, and nitrogen. Hydrogen and oxygen Davy could confidently affirm to be constituents of water; he knew that nitrogen is present in air. On these facts Davy framed an hypothesis.

As the decomposition of the water proceeds in every experiment in contact with air, and as hydrogen and oxygen are produced in this decomposition, the conditions for the production of nitric acid are realised; the hydrogen and oxygen combine with the nitrogen of the air to produce nitric acid, and this dissolves in the water. If this hypothesis is correct, removal of nitrogen from contact with the decomposing water should be attended with cessation of the production of nitric acid; re-introduction of nitrogen should be accompanied by re-appearance of nitric acid.

Davy placed a gold vessel containing pure water on a plate of glass, and covered it with a strong glass jar connected with an air-pump; he exhausted the air from the jar, admitted hydrogen, again exhausted, and again filled the jar with hydrogen; he continued this treatment until he could feel sure that the whole of the air had been withdrawn from the jar. He then filled the jar with hydrogen, and passed the electric current; not a trace of acid was produced; hydrogen and oxygen, and these gases only, appeared at the electrodes. He admitted air into the jar; the acid began to form at the positive electrode. But he had already proved that the production of acid was not connected with the presence of any substance in the water, nor with the nature of the vessels containing the water, nor with the material of the electrodes; hence the production of acid always accompanied the presence of nitrogen. The latter was the cause of the former. "It seems evident then," says Davy, "that water, chemically pure, is decomposed by electricity into gaseous matter alone, into oxygen and hydrogen."

This remarkable research is a type of all scientific inquiry. **94** Facts were noticed and verified, conclusions were drawn and tested by experiments; hypotheses were framed on the basis of the experimentally determined facts, and were used to explain these facts by suggesting fresh lines of inquiry. The result which Davy obtained was not a barren fact; it at once prompted him to further discoveries. The electric current had slowly decomposed the glass vessels; probably it would also decompose other substances more or less resembling glass in composition. Water was electrolysed in cups of gypsum; lime appeared at one electrode and sulphuric acid at the other. Other substances were employed; he generally obtained an alkaline body at the negative, and an acid at the positive, electrode. This led Davy to regard many compounds as built up of two parts, one positively, the other negatively, electrified.

This conception prompted him to make more experiments; these furnished him with new hypotheses; these in turn led to further inquiry ; and this reaction of experiment on theory and theory on experiment proceeded until he had framed a general conception of the composition of salts, and of the relations between salts, acids, and alkalis, which had a most important influence on the development of chemistry. The facts noticed during the electrolysis of water also led Davy to investigate the action of the current on various substances which were included in the class of elements ; many of these he succeeded in decomposing ; he obtained new elements, and to a large extent changed the whole course of the chemical study of matter.

95 The *synthesis* of water by passing hydrogen over hot copper oxide has been already mentioned. The synthesis was carried out in the most accurate manner by Dumas. Fig. 14 represents the apparatus employed. A is a flask in which hydrogen is produced by the interaction of zinc and dilute sulphuric acid (the cylinder to the left of A contains mercury ; it, and the tube dipping into it, serve as a means for allowing the hydrogen to pass away without taking the apparatus to pieces) : the seven large U tubes contain materials by passing through which the hydrogen is purified and dried: the small U tube B contains phosphorus pentoxide, a substance which greedily absorbs moisture ; this tube is weighed before and after each experiment, should it increase in weight after an experiment, the results of that experiment are rejected, as the increase in the weight of B tells that the hydrogen was not perfectly dry when it passed into C : C is a bulb of hard glass--con-

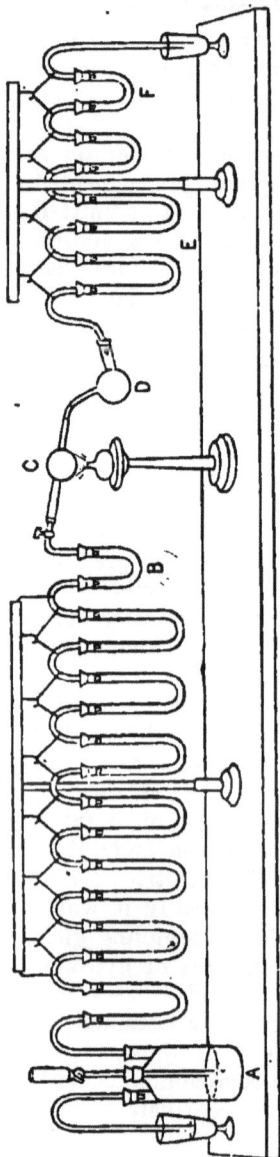

Fig. 14.

taining a weighed quantity of perfectly pure and dry copper
oxide; the neck of this bulb is drawn out and passes into the
next bulb D: D is a dry bulb of glass destined to contain
the water produced in the reaction; it is weighed before and
after each experiment: the U tubes E contain materials to
absorb any traces of water which may not be retained in D:
the small tube F also contains drying materials; it is weighed
before and after each experiment; should it shew an increase
in weight after an experiment is finished the results of that
experiment are rejected, because a doubt arises as to whether
the whole of the water produced has been retained in the
apparatus and weighed.

Let the weight of the copper oxide before an experiment
$= x$; the weight of the copper remaining in C after the experiment $= y$; the weight of water produced (that is the increase
in weight of D and E) $= z$: then $x - y$ gives the weight of
oxygen which has combined with hydrogen to produce the
weight z of water; let this weight of oxygen $= a$: then $z - a$
gives the weight of hydrogen which has combined with a of
oxygen to produce z of water.

Dumas' result was that 1 part by weight of hydrogen
combines with 7·9804 parts by weight of oxygen, to produce
8·9804 parts by weight of water.

The volumetric synthesis of water has already been briefly **96**
described: when this synthesis is conducted with all precautions
the result is that one volume of hydrogen combines with half
a volume of oxygen to produce water; as careful determinations
have shewn that oxygen is 15·96 times heavier than hydrogen,
the result of the volumetric synthesis, altogether confirms that
of the gravimetric synthesis, of water.

When two volumes of hydrogen are caused to combine
with one volume of oxygen in a vessel the temperature of
which is above the boiling point of water, that is when the
conditions are arranged so that the water produced is main-
tained in the state of gas, the result is that two volumes of
hydrogen combine with one volume of oxygen to produce two
volumes of water-gas.

Water then is a compound, not a mixture, since it has
been shewn to be of constant composition, and to conform to
the laws of chemical combination.

The results of experiments on the composition of water   **97**
are summed up in the formula H_2O, and in the equation
$H_2 + O = H_2O$.

Assuming that the *combining weight* of oxygen is 16 (in round numbers), and that the symbol O represents 16 parts by weight of oxygen, then this formula, and this reaction, tell us, that the masses of hydrogen and oxygen which combine to produce water are in the ratio $1 : 8$ (in round numbers); that the *reacting weight* of water is 18, that this reacting weight is composed of two combining weights of hydrogen and one combining weight of oxygen, and that this quantity of water can be decomposed and either one or two combining weights of hydrogen removed, but that if oxygen is removed the whole of the oxygen must be removed; and that two volumes of water-gas are formed by the union of, or can be resolved into, two volumes of hydrogen-gas and one volume of oxygen-gas (comp. pars. 86 and 87).

We may here inquire, what would be the composition of other compounds of hydrogen and oxygen if such compounds existed? The law of multiple proportions tells us that the composition of such compounds would be expressed by the general formula H_xO_y, where H represents a combining weight of hydrogen, O represents a combining weight of oxygen, and x and y are whole numbers.

Two compounds of hydrogen and oxygen are known; one is water, H_2O; the other is hydrogen peroxide H_2O_2.

But neither the formula H_2O, nor the equation $H_2 + O = H_2O$, says anything as to the conditions under which water is produced by the union of hydrogen and oxygen; they do not tell, or even suggest, the properties of water; nor do they indicate the physical changes which accompany the chemical change from a mixture of hydrogen and oxygen to the compound water.

We have already learned something of the conditions under which the reactions expressed by the equations

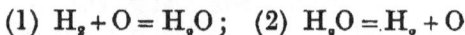

$$(1) \ H_2 + O = H_2O ; \quad (2) \ H_2O = H_2 + O$$

proceed; we know a few of the properties of water; and we are not wholly ignorant of the fact that the production of water is an event which has a physical as well as a chemical aspect. But we ought now to look a little more closely at some of these points.

Conditions under which the equations (1) $H_2 + O = H_2O$ *and* (2) $H_2O = H_2 + O$ *are realised.*

(1) When two volumes of hydrogen are mixed with one

volume of oxygen, and an electric spark is passed through the mixture, or a lighted taper is applied.

(2)   When an electric current is passed through 18 parts by weight (say 18 grams) of water until the whole of the water has disappeared.   Or the water is raised to a very high temperature (say $1500°$—$2000°$) under conditions such that the oxygen and hydrogen are removed from contact with the yet undecomposed water as quickly as they are produced. Or the water in the form of steam is passed over hot finely divided iron, or hot magnesium (or certain other metals); the hydrogen is collected; the oxygen combines with the iron or magnesium to form an oxide of either metal; by decomposing this oxide by suitable means the oxygen may be obtained.

Physical changes which accompany the chemical change from the mixture $H_2 + O$ *to the compound* H_2O.

(1)   A large quantity of heat is produced.   When 2 grams of hydrogen combine with 16 grams of oxygen to form 18 grams of water, the heat produced is sufficient to raise the temperature of (in round numbers) 68,000 grams of water from $0°$ to $1° C.$; in other words 68,000 *gram-units of heat* are produced.

This statement tells that the chemical change $H_2 + O = H_2O$, i.e. the change of certain masses of two definite kinds of matter into a mass of another kind of matter equal to the sum of the masses of the two kinds of matter, is accompanied by the degradation of a large quantity of energy (*s.* Chap. XIV.). The system H_2O possesses much less energy, it can do much less work, than the system $H_2 + O$; the difference between the energies of the two systems is approximately represented by 68,000 gram-units of heat.

(2)   A contraction of volume occurs.   The mixture of 2 grams of hydrogen with 16 grams of oxygen occupies about 44,000 c.c. at $0°$ and 760 mm.; the 18 grams of water produced occupy about 18 c.c.   If the temperature is kept a little above $100°$, the 18 grams of water-gas produced occupy about 30,000 c.c.

(3)   The mixture of hydrogen and oxygen is gaseous; the water formed is a liquid below $100°$ at 760 mm.

(4)   The change is accompanied by the production of a flash of light.

These are some of the more important conditions under which the changes represented by the chemical equations we are considering occur, and some of the more important physical changes which accompany the chemical changes.

100   *Properties of water.* Here we must distinguish the physical from the chemical properties (*v. ante;* pars. 2 to 13, and 35 to 41). The physical properties are those which characterise the substance water considered as a definite kind of matter apart from other kinds of matter. The chemical properties are those which are exhibited by water when it interacts with other kinds of matter. If this definition is accepted, then, strictly speaking, water can have no chemical properties; the properties we shall have to examine are properties exhibited by a changing system composed of water *plus* something else; the properties called chemical only come into play when water interacts with other bodies, hence they are the properties neither of the water *per se*, nor of the other body or bodies *per se*, but of the system of which water forms a part. The simplest chemical occurrence is at least two-sided.

One of the chemical properties of hydrogen is that 1 part by weight of this gas combines with 8 parts by weight of oxygen to produce 9 parts by weight of water. But this chemical occurrence may be stated in terms of oxygen, hydrogen, or water. A chemical property of oxygen is that 8 parts by weight of oxygen combine with 1 part of hydrogen to produce 9 parts of water. A chemical property of water is that 9 parts by weight of it are produced by the union of 1 part of hydrogen with 8 parts of oxygen.

101   We can here only enumerate a few of the more important physical properties of water. At temperatures below 0° C. water is a solid, from 0° to 100° it is a liquid, and above 100° it is a gas. The change from solid water to liquid water is accompanied by a slight decrease of volume; the change from liquid water to water-gas is accompanied by an increase of volume; each change is accompanied by the disappearance of a considerable quantity of heat. The reverse change from water-gas to liquid water is accompanied by a large condensation of volume, and the change from liquid water to ice is accompanied by a slight increase of volume; during both changes much heat is produced. The temperature at which each change occurs depends upon the pressure of the atmosphere on the surface of the water; the temperature at which the change from liquid to gaseous water occurs freely varies very considerably with variations of pressure.

To trace the relations of volume, pressure, and temperature, for a given mass of water, is a typical physical inquiry. Under ordinary conditions of pressure, solid water melts at 0^0, and liquid water boils at 100^0.

Water dissolves very many and very different substances. **102** The solution is sometimes unattended with any chemical change; e.g. common salt or sugar dissolves in water, on evaporating off the water the salt or sugar remains (*v. ante*, par. 10). In some cases solution in water is accompanied by chemical change; e.g. sodium dissolves in water, but on evaporating off the water, not sodium, but a different substance, caustic soda, is obtained (*v. ante*, par. 86). The phenomena presented during solution of bodies in water are extremely complex; they cannot be classed wholly as physical or wholly as chemical. We are not yet in a position to examine these phenomena with any prospect of approximately understanding them. Let us rather turn to some of the more important phenomena presented by the interactions of water with other kinds of matter, that is, to the chemical properties of water.

Water combines with many compounds, and with one or two **103** elements.

When the gaseous element chlorine is passed into ice-cold water, after a time the whole becomes semi-solid; by filtering off the liquid water at a temperature under 0^0, crystals having the composition $Cl . 5H_2O$ ($Cl = 35 \cdot 5$) are obtained. Heated to a little above 0^0 chlorine hydrate is wholly decomposed into chlorine and water.

The compound cobalt chloride is a blue solid; its composition is expressed by the formula $CoCl_2$ ($Co = 59$, $Cl = 35 \cdot 5$). When this salt is dissolved in water, a reddish liquid is produced from which, after partial evaporation, red crystals separate having the composition $CoCl_2 . 6H_2O$. When these red crystals are heated they separate into water, which passes off as steam, and blue cobalt chloride, which remains. When a little water is added to the blue solid, the red compound is reproduced. The equations

(1) $CoCl_2 + 6H_2O = CoCl_2 . 6H_2O$, occurring at the ordinary temperature;

(2) $CoCl_2 . 6H_2O = CoCl_2 + 6H_2O$, occurring at about 100^0; represent the two chemical changes.

When water is poured on to the compound calcium oxide, or lime, the water disappears, the lime swells up, much heat is produced, and the new compound calcium hydroxide, or slaked lime, is formed. The equation $CaO + H_2O = CaO . H_2O$ represents the change of composition $(Ca = 40)$. When the solid compound calcium hydroxide is strongly heated—say to 400° or 500°—it is decomposed into calcium oxide and water ; thus $CaO . H_2O = CaO + H_2O$; the water passes away and the calcium oxide remains as a solid.

Copper sulphate—$CuSO_4$ $(Cu = 63·2,\ S = 32)$—is a white solid ; when it is exposed to moist air it begins to turn blue ; this change is more quickly effected by pouring a little water on to the solid. The blue substance thus produced is a com-pound of copper sulphate and water, called hydrated copper sulphate ; its composition is $CuSO_4 . 5H_2O$. When this com-pound is heated to about 220° water passes off as steam, and $CuSO_4$ remains as a white solid.

When solid potassium oxide, K_2O $(K = 39)$, is placed in water it instantly dissolves ; when the solution is boiled to dryness a white solid remains having the composition $K_2O . H_2O$, called potassium hydroxide, or hydrated potassium oxide, or (more commonly) caustic potash. This solid is not chemically changed by heat ; at a high temperature it melts, and at a very high temperature it is volatilised ; but the melted sub-stance, or that which is volatilised, has the same composition, $K_2O . H_2O$, as the solid.

If an aqueous solution of caustic potash is added to an aqueous solution of copper sulphate, a greenish blue solid compound is formed ; when this is collected, washed, and dried at about 40° to 50°, it has the composition $CuO . H_2O$ $(Cu = 63·2)$. When this greenish blue solid is heated to 100° or 150° water passes off as steam and black copper oxide, CuO, remains. If a little water is now added to this copper oxide no chemical change occurs ; the water and the copper oxide remain mixed.

104   If we glance back over the statements made regarding the compounds of water with chlorine, cobalt chloride, &c. we see that the compounds may be divided into three classes, as follows :—

(i) *Compounds formed by bringing together water and the other constituent of the compound, and also decomposed by heat into water and the other constituent ;—*

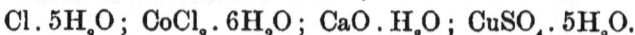

$$Cl . 5H_2O ;\ CoCl_2 . 6H_2O ;\ CaO . H_2O ;\ CuSO_4 . 5H_2O.$$

(ii) *Compound formed by bringing together water and the other constituent, but not decomposed by heat ;—*

$$K_2O . H_2O.$$

(iii) *Compound decomposed by heat into water and another compound, but not formed by bringing together water and that other compound;*

$$CuO.H_2O.$$

We also notice that the compounds under (i) are decomposed by heat at different temperatures, varying from a little above $0°$ in the case of $Cl.5H_2O$, to $400°$ or $500°$ in the case of $CaO.H_2O$.

Inasmuch as each substance we have been considering is either formed by the combination of water with another substance, or is resolved into water and another substance, we are justified in calling them all compounds of water with other elements or compounds.

But in many of its interactions with elements and com- **105** pounds water is decomposed, and new bodies are formed which cannot be regarded as compounds of water.

We have learned that when sodium and water interact the products are hydrogen and sodium hydroxide; the equation $Na + H_2O + Aq = NaOHAq + H$ expresses the composition of the system before and after this interaction.

A similar change occurs when potassium is thrown into water; $K + H_2O + Aq = KOHAq + H$. In each of these reactions much heat is produced; in the case of potassium the reaction proceeds very rapidly, and the temperature of the hydrogen produced is raised so much that this gas takes fire.

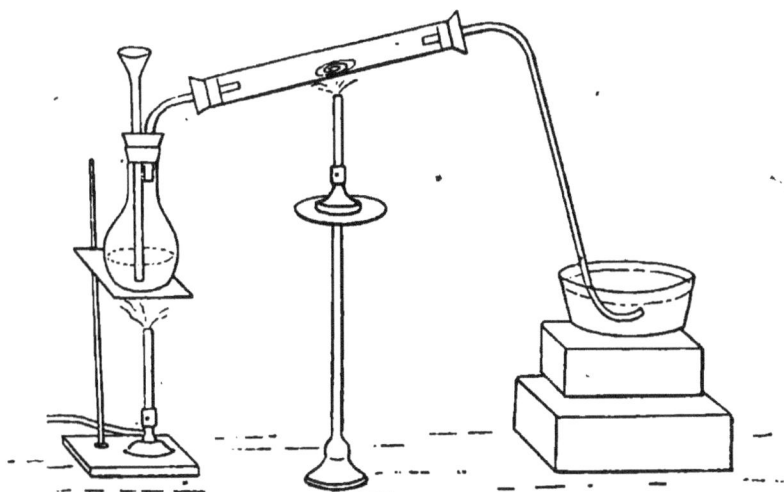

Fig. 15.

When water-gas is passed over hot magnesium, or hot finely divided iron, in an apparatus as represented by fig. 15, hydrogen is obtained, and oxide of magnesium or iron is formed and remains in the tube in which the magnesium or iron was heated. Quantitative experiments have proved that for every 18 parts by weight of water decomposed 2 parts by weight of hydrogen are obtained, and an oxide of magnesium or iron is formed by the union of the 16 parts by weight of oxygen, formerly combined with the 2 parts of hydrogen, with 24 parts of magnesium or 42 parts of iron. The combining weights of magnesium and iron are 24 and 56, respectively; the reacting weights of the oxides formed in the process just described are 40 and 232, respectively, and the compositions of these oxides are represented by the formulae MgO and Fe_3O_4. Knowing that the combining weight of oxygen is 16, and the reacting weight of water is 18 (H_2O), we can summarise the changes of composition which occur in the reactions between steam and heated magnesium or iron in these equations ($Mg = 24$, $Fe = 56$, $O = 16$);—

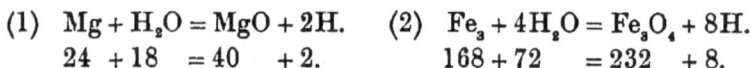

(1) $Mg + H_2O = MgO + 2H.$ (2) $Fe_3 + 4H_2O = Fe_3O_4 + 8H.$

   24 + 18   = 40   + 2.   168 + 72   = 232   + 8.

When the gaseous element chlorine is passed into boiling water and the mixture of steam and chlorine thus obtained is passed through a porcelain tube, loosely packed with pieces of porcelain, and heated to bright redness, oxygen and hydrogen chloride are produced. The apparatus represented in fig. 16 may be used. The exit end of the porcelain tube is connected with a vessel containing caustic potash solution (A). The gases coming from the tube bubble through this solution. Hydrogen chloride is absorbed by caustic potash, but oxygen is not. The gas which is not absorbed by the caustic potash is collected and proved to be oxygen. Quantitative experiments shew that the compositions of the interacting substances and of the products of the interaction are expressed by the equation ($Cl = 35.5$, $O = 16$);—

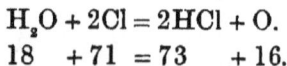

$$H_2O + 2Cl = 2HCl + O.$$
$$18   + 71 = 73   + 16.$$

For every 18 parts by weight of water decomposed, 71 parts of chlorine are used, and 73 parts of hydrogen chloride and 16 parts of oxygen are produced. As all the substances taking part in this reaction are gases under the conditions

of the experiment, and as we know (i) that the symbol of an
elementary gas (with a few exceptions) represents the mass of

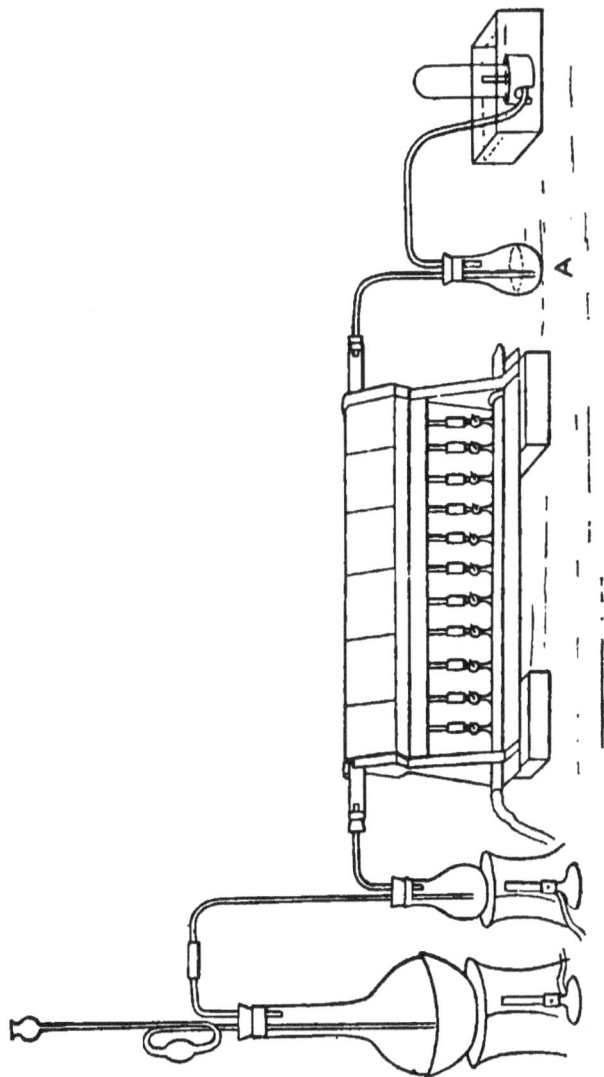

Fig. 16.

it which occupies 1 volume (i.e. the volume occupied by unit
weight of hydrogen), and (ii) that the formula of a compound

gas represents the mass of it which occupies 2 volumes (*v. ante*, par. 88), the foregoing equation tells that 2 volumes of water-gas react with 2 volumes of chlorine gas to produce 4 volumes of hydrogen chloride gas and 1 vol. of oxygen gas.

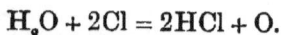

$$H_2O + 2Cl = 2HCl + O.$$
$$\text{vols. } 2 + 2 \text{ give } 4 + 1.$$

The *volumes* on each side of the sign $=$ are not the same; the *masses* on each side of the sign $=$ are, and in chemical equations always are, the same.

106 Our study of the properties of water has served to illustrate the nature of chemical change; to emphasise the distinctions between mixtures, elements, and compounds; to shew the importance of the laws of chemical combination; to familiarise us with the use of chemical formulae and equations; and to illustrate the meanings of the terms analysis and synthesis. This study has kept before us the notion of each element and compound interacting chemically with other elements and compounds in certain definite masses which are all simple multiples of one and the same mass. It seems as if a quantity of water, for instance, were composed of a vast number of little particles of water the masses of all of which are the same, and as if chemical interaction occurred between 1, 2, 3,......n of these little particles and a definite number of little particles of the element or compound with which the water interacts. The chemical conception of every element or compound having its own reacting weight leads to some such physical conception as this of small definite particles. Finally, the slight examination we have given to the chemical properties of water has shewn very clearly how closely interwoven chemical changes are with physical changes, and how impossible it is to arrive at any trustworthy conclusions regarding either otherwise than by quantitative experiments and accurate reasoning.

107 **Air.** When magnesium is burnt in air magnesia is produced; but magnesia is a compound of magnesium and oxygen, therefore, the chemical change which occurs during the burning of magnesium in air consists in the combination of magnesium and oxygen. Therefore, in all probability, oxygen is a constituent of air. When mercury is heated in a measured quantity of air, mercury oxide is produced, and some of the air disappears; when the oxide is collected and strongly heated, oxygen and mercury are formed, and the quantity of oxygen is equal to the quantity of air which disappeared.

Therefore the heated mercury combined with a part of the air in which it was heated, and this part was oxygen (*v. ante*, par. 18). From these experimental results we may conclude that if magnesium or mercury is burnt in air, the air which remains after burning will almost certainly differ in properties and composition from the air which was present before burning began. And if this is so we may further conclude that when any element which is known to combine with oxygen is burnt in an enclosed volume of air, the whole or a part of the oxygen in the air will combine with the element, and the air which remains will most probably differ from the original air.

Phosphorus is an element which is easily burnt, and which **108** very readily combines with oxygen.

Let an apparatus be arranged as shewn in fig. 17. A is a glass jar; the space from the cork to within about 3 or 4 inches of the open end is divided into 5 equal parts. The jar is placed, open end downwards, in such a quantity of water that the level of the water stands at the point where the graduation of the jar begins. B is an iron cup supported on an iron pillar with a broad foot. Let a piece of dry phosphorus be placed on B; let the jar be put over the phosphorus and iron stand; let the end of the brass chain be highly heated, and then let the chain be brought quickly into the jar, as shewn in the figure, so that the heated part of the chain touches the phosphorus. The phosphorus begins to burn, white clouds of phosphorus oxide fill the jar, and the water slowly rises in the jar. When the burning is finished and the clouds have disappeared—the phosphorus oxide produced dissolves in the water—let water be poured into the outer vessel until the level of the

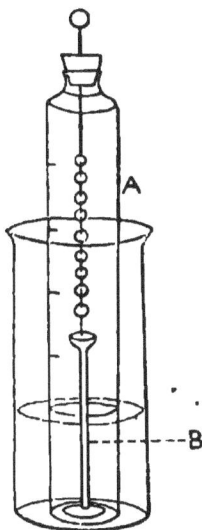

Fig. 17.

water inside and outside A is the same. It is seen that $\frac{1}{5}$ of the air has disappeared. Withdraw the cork, and plunge a lighted taper into A; the flame is instantly extinguished. Therefore the air in A after the burning of phosphorus is not the same as the air before the phosphorus was burnt.

By a little careful manipulation, portions of the air which remain after the phosphorus is burnt may be transferred from A to glass tubes or bottles, and the properties of this air may be examined. It is found to be a colourless, odourless, gas, a very little lighter, bulk for bulk, than ordinary air; it does not support combustion, nor is it combustible; it reacts chemically with but few elements and compounds. Every attempt to separate a specified mass of this gas into unlike parts has failed. But very many compounds are known of each of which this gas is a constituent. The gas is an element; it is called *nitrogen*.

We know that oxygen is also an element. Hence we have obtained from air two elements nitrogen and oxygen.

109   Is air a compound or a mixture of these gases? If it is a compound, the properties of air must differ considerably from the properties of either oxygen or nitrogen; and these elements must be united in air in a ratio expressed by the formula $N_x O_y$ where x is 1, 2, 3, 4...n times the combining weight of nitrogen (14), and y is 1, 2, 3, 4...n times the combining weight of oxygen (16). If air is a mixture of nitrogen and oxygen, it must be possible to recognise both of these elements in air by making use of the properties which each possesses when unmixed with other kinds of matter.

Whichever hypothesis is adopted as a guide in experimental inquiry, we must begin by determining the properties of air, the properties of oxygen, and the properties of nitrogen.

110   We already know some of the properties of oxygen and nitrogen. Both are colourless, odourless, gases; nitrogen is 14 times, and oxygen is 16 times, heavier than hydrogen. Combustible bodies burn rapidly and brilliantly in oxygen, but they cease to burn in nitrogen. The ratios of diffusion of both are nearly equal; but oxygen passes through a thin sheet of india-rubber about $2\frac{1}{2}$ times more rapidly than nitrogen. Oxygen is slightly soluble, nitrogen is less soluble, in water; 1 vol. of water at 16° dissolves ·0295 vols. of oxygen, and ·0145 vols. of nitrogen. The combining weight of oxygen is 16, and the combining weight of nitrogen is 14.

111   The prominent physical properties of air are known to all. Accurate analyses of air have shewn that 100 parts by weight of dry air freed from carbon dioxide (*v. infra*, par. 113) are composed of 23 parts of oxygen and 77 parts of nitrogen, by weight. The simplest formula which will fairly accurately represent this composition, assuming air to be a compound, is $N_{51}O_{13}$; this compound, if it existed, would be composed of 22·5 parts of oxygen by weight, and 77·5 parts of nitrogen, per 100 parts. Five definite compounds of nitrogen and oxygen are known;

their compositions are represented by the formulae N_2O, NO, N_2O_3, NO_2, N_2O_5. It is improbable that a sixth compound of these elements should exist having as complex a composition as $N_{51}O_{13}$. But this is the simplest formula which can be given to air if air is a compound of nitrogen and oxygen. Hence the argument based on analogy of composition leads to the conclusion that air is probably a mixture and not a compound.

Assuming air to be a mixture of nitrogen and oxygen, we next inquire, what volume of air ought to be dissolved by 1 vol. of water, say at 16°? The solution of a mixture of gases by a liquid between which and the gases there is no chemical interaction follows the same course as if each gas were dissolved separately in the liquid. The solution of a gaseous compound, on the other hand, in a liquid which does not interact chemically with the compound follows a course of its own; the vol. dissolved is independent of the vols. of the gaseous constituents of the compound dissolved under the same conditions.

1 vol. of water at 16° dissolves ·0295 vols. of oxygen, and ·0145 vols. of nitrogen; now 1 vol. of dry air freed from carbon dioxide (v. infra, par. 113) is composed of ·2096 vols. of oxygen and ·7904 vols. of nitrogen; therefore, if air is a mixture, 1 vol. of water will dissolve $(·0295 \times ·2096) + (·0145 \times ·7904)$ = ·01765 vols. of air, at 16°.

Experiment proves that 1 vol. of water dissolves ·0177 vols. of air at 16°.

1 vol. of water at 16° dissolves ·7535 vols. of nitrous oxide (N_2O); but if this gas were a mixture of nitrogen and oxygen in the ratio in which these gases unite to form 1 vol. of nitrous oxide, water would dissolve ·02925 vols. of the nitrous oxide.

These calculations and experiments shew that air is dissolved by water exactly as if the air were a mixture of oxygen and nitrogen and not a compound of these elements. In other words: one of the properties of oxygen is to dissolve in water to a certain definite extent, and one of the properties of nitrogen is to dissolve in water to a certain definite extent; but both oxygen and nitrogen retain this property when they are present in air; therefore air is a mixture, and not a compound, of oxygen and nitrogen.

We may carry the inquiry further on the same lines. If air is a mixture of oxygen and nitrogen, and if oxygen passes through a thin sheet of india-rubber about 2½ times quicke than nitrogen it ought to be possible to effect a partial separation of air into its constituent gases by passing it through a sheet of india-rubber. Experiment proves that when all, or almost all, the air is pumped out of an india-rubber bag, and the bag is closed and left in the atmosphere, air passes into the bag through the walls, and that the composition of the air found in the bag is approximately 40 p. ct. of oxygen and 60 p. ct. of nitrogen, by volume. But the composition of ordinary air is approximately 21 p. ct. oxygen and 79 p. ct. nitrogen, by volume. Therefore the air has been partially separated into its constituents by passing it through a sheet of india-rubber; therefore air is a mixture, not a compound, of oxygen and nitrogen.

112   The argument may be extended to chemical events. If air is a mixture, it ought to interact chemically with other substances both as oxygen interacts and also as nitrogen interacts. If air is a compound, its interactions with other substances ought to be different from those of either oxygen or nitrogen. We have learned that nitrogen is a very inert substance; it does not support combustion, it is not combustible, it combines directly with only a few elements, and it does not react chemically with many compounds. On the assumption that air is a mixture, we should, therefore, expect its chemical properties to resemble those of oxygen, but to be less strongly marked because of the presence of the inert nitrogen.

For instance, we should expect substances which burn rapidly and brilliantly in oxygen to burn in air but to burn more slowly and less brilliantly. If we can find an element which combines directly with nitrogen when heated in that gas, we should expect that element to form a compound with nitrogen when strongly heated for some time in air. .

We need not go into details regarding individual experiments, but suffice it to say that these expectations are realised; that the chemical behaviour of air is exactly what the hypothesis of its being a mixture asserts ought to be its behaviour. Air then is a mixture, not a compound, of oxygen and nitrogen.

113   But besides these gases, air contains small quantities of the compound gases, carbon dioxide, ammonia, and water-vapour.

The composition of air varies within narrow limits. Thus air has not that fixity of composition which as we have seen characterises chemical compounds.

The experiment described in par. 108 shewed that the composition of air is, roughly, 4 vols. of nitrogen to 1 vol. of oxygen. In order accurately to determine the volume-composition of air, a quantity of air is passed into a graduated glass tube fitted with two platinum wires passing through the glass near the closed end; the tube is filled with mercury, and is then inverted in a trough containing mercury. The air to be analysed is freed from carbon dioxide and ammonia, and is then passed into the tube, and the volume is measured by reading off the level of the mercury. A quantity of hydrogen equal in volume to nearly $\frac{3}{5}$ of the volume of air is passed into the tube, and the level of the mercury is again read off; the tube is pressed down on a pad of india-rubber· and securely clamped; an electric spark is then sent from one platinum wire to the other; ·the effect of this is that the whole of the oxygen in the air combines with a portion of the hydrogen to produce water which condenses. After a little time the tube is slowly raised from the india-rubber pad; mercury rushes in; the level of the mercury is read off. As we know that 2 vols. of hydrogen combine with 1 vol. of oxygen, we conclude that $\frac{1}{3}$ of the diminution of volume which occurs when the spark is passed represents the volume of oxygen in the volume of air employed. The volume occupied by the small quantity of water produced is so small that it may be neglected. Many precautions are necessary in carrying out such an analysis as this; corrections must be made for temperature and pressure; the volumes of wet air and wet gas after the explosion must be reduced to the corresponding volumes of dry gases, &c.

The carbon dioxide in air may be determined, (1) by slowly passing a large measured volume of air, freed from ammonia and water-vapour, through a series of weighed U tubes filled with caustic potash, and determining the increase in the weight of these tubes; or (2) by adding, to a measured volume of air, a known quantity of barium oxide dissolved in water, and determining the quantity of this oxide which remains when the carbon dioxide in the air has all been absorbed by a portion of the barium oxide. The chemical reactions on which these methods are based may be represented in equations thus;—

(1) $x\mathrm{KOH}$ (moist) $+ \mathrm{CO_2} = \mathrm{K_2CO_3} + \mathrm{H_2O} + (x-2)\,\mathrm{KOH}$.

(2) $x\mathrm{BaOAq} + \mathrm{CO_2} = \mathrm{BaCO_3} + (x-1)\,\mathrm{BaOAq}$.

The potassium carbonate $(\mathrm{K_2CO_3})$ and water $(\mathrm{H_2O})$ formed in (1) remain in the weighed U tubes along with the potash (KOH) which has not been changed by the carbon dioxide (CO_2): the barium carbonate ($BaCO_3$) formed in (2) is a solid, it settles down in the liquid, and the unchanged barium oxide (BaO) remains in solution and is determined by a method which need not be described here.

114   The quantities of oxygen and nitrogen in average country air freed from water-vapour, ammonia, and carbon dioxide, are :—

Percentage

	by volume.	by weight.
Oxygen =	20·96	23·0
Nitrogen	79·04	77·0
	100·00	100·0

The quantity of carbon dioxide averages about ·03 volumes per 100 vols. of air. The quantity of ammonia varies very much; it may perhaps be taken as about 1 part in 10,000,000 parts of air, by weight. The quantity of aqueous'vapour also varies with variations in the season, the district, &c. &c.

The fact that the quantities of oxygen and nitrogen in country air vary, although within very narrow limits, has been definitely established. The oxygen sometimes amounts to 20·999 vols. per 100 e.g. in air from the seashore or from inland moors; in towns the oxygen sometimes falls to 20·82 vols.; in inhabited rooms and crowded halls it may be as little as 20·28; in mines it averages about 20·26. A decrease in the volume of oxygen is usually accompanied by an increase in that of carbon dioxide; in crowded rooms the volume of this gas may be as large as ·3 to ·5 vols. per 100. Air which contains as much as ·1 vol. carbon dioxide per 100 is unpleasant, and harmful to health. The air of towns contains many gases, liquids, and solids, produced by the changes which go on among the living beings, and also by the manufactures conducted in the towns.

115   Our examination of air has afforded an application of the statements made in Chap. III. regarding the differences between mixtures and compounds; it has shewn us how we may determine to which of these classes a given substance belongs; it has also made us acquainted with some of the prominent characters of air; and it has a little familiarised us with the methods pursued in chemical inquiries.

CHAPTER VIII.

HAVING now gained a fairly clear notion of the kind of **116** material phenomena which form the subject matter of chemistry, and of the methods by which the chemical aspects of these phenomena are investigated ; and having arrived at certain fundamental generalisations from facts established by quantitative experiments and quantitative reasoning, we are in a position to proceed with the main subject of our inquiry, which is to establish the relations which exist between changes of composition and changes of properties of the definite kinds of matter we call compounds, and the relations which exist between the properties of the elementary constituents of compounds and those of the compounds themselves. This inquiry branches out in two directions ; it requires us to study (1) the properties of compounds, and the properties of elements as exhibited in their compounds ; and (2) the composition of compounds. To do this we must classify ; we must group together those compounds which have similar properties, and those which have similar compositions.

We shall begin our attempt to learn how elements and **117** compounds are classified, and to become acquainted with the more important results of this classification, by considering the two elements hydrogen and oxygen and some of the compounds of these elements.

Occurrence. Oxygen, as we know, forms about $\frac{1}{5}$ of the **118** atmosphere. Hydrogen is sometimes found in small quantities in volcanic gases. Numerous compounds of each element occur in nature ; of these water (H_2O) is the most abundant. Oxides of aluminium, iron, calcium, magnesium, silicon, and many other elements, are found widely distributed and in

large quantities. Ammonia—a compound of hydrogen and
nitrogen—and compounds of ammonia, exist in the air and in
the soil; and most of the substances which form the parts of
plants and softer tissues of animals are compounds of hydrogen,
with carbon, oxygen, and nitrogen.

119   *Physical properties.*

	Hydrogen.	Oxygen.
Sp. gr.	1	16
,, *air* =1	·0693	1·105 for gaseous oxygen. ·979 for liquid oxygen (water = 1).
Sp. hts. (constant pressure; Sp. ht. of equal mass of water =1)	3·405	·218
Vols. dissolved by 1 *vol. water at* 16⁰	·0193	·0295
Colour, appearance, &c.	Colourless, tasteless, odourless gas. Liquefied (and ? solidified) at very low temp. and great pressure; approximately −140⁰ and 600 atmospheres. Rate of diffusion about 4 times that of oxygen. Very bad conductor of sound.	Colourless, tasteless, odourless, gas. Liquefied at about − 140⁰ under pressure of 320 atmos.

120   *Chemical properties.* Hydrogen and oxygen readily combine
to form water. If a stream of hydrogen is allowed to flow
into oxygen, or into air, and a light is brought to the jet the
hydrogen takes fire and burns in the oxygen, and water is
produced. If a stream of oxygen is allowed to flow into
hydrogen from a narrow tube, and a light is brought to the
jet the oxygen takes fire and burns in the hydrogen, and water is produced. In each case the chemical change is the
same; $2H + O = H_2O$.
     Fig. 18 shews a simple arrangement for exhibiting these
reactions. A and B are stoppered glass jars; each is fitted

with a cork through which passes a tube narrowed at the end which is to go into the jar; A is filled with oxygen, B with hydrogen; each stands in a little water whereby the

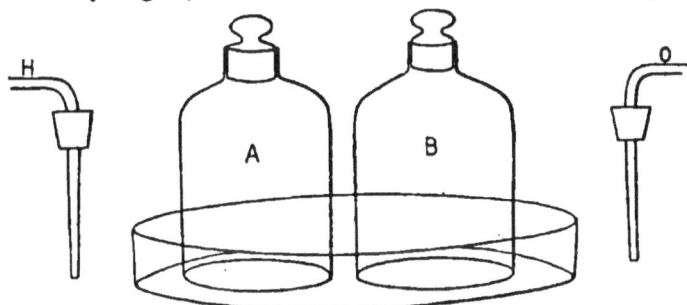

Fig. 18.

gas inside the jar is isolated from the air outside; the tube passing through the cork which fits jar A is connected with a gasholder containing hydrogen, the other tube is connected with a gasholder containing oxygen. Hydrogen is caused to pass slowly through one tube and oxygen slowly through the other; after a minute or so (when the air is all driven out of these tubes) the hydrogen jet is lighted, the stopper of A is withdrawn and the cork with its tube is quickly inserted; the hydrogen burns brilliantly; the stopper of B is withdrawn and a light is brought near the opening of B, the hydrogen in B burns; the cork is now very quickly pressed into its place, and the jet of oxygen is seen to burn in the atmosphere of hydrogen.

A little consideration shews that the chemical reaction $2H + O = H_2O$ must occur, for the most part, at or near the surface of that gas which is flowing into the other, which other is, comparatively, at rest. If the inflowing gas is hydrogen, then, as the flame is visible along the surface of the inflowing gas, we say that the hydrogen burns in the oxygen; that the hydrogen is burnt and the oxygen supports the combustion. If the inflowing gas is oxygen, the flame being as before visible along the surface of the inflowing gas, we say that the oxygen is burnt, and the hydrogen supports the combustion.

Oxygen combines directly with many elements; compounds **121** of oxygen with every other element, except bromine and fluorine, have been prepared, either by direct combination, or as the results of several chemical changes.

I. The elements *sodium, potassium, lithium, thallium, phosphorus,* and some others, combine with oxygen more or less rapidly at ordinary temperatures. II. *Antimony, arsenic, carbon, lead, sulphur,* and many other elements, combine with oxygen at temperatures above the ordinary. III. Oxides of *calcium, bismuth, chromium, copper,* &c. &c. are usually prepared by (i) obtaining compounds of these metals with oxygen and hydrogen, and (ii) heating these hydroxides, and so decomposing them into oxides and water. IV. Oxides of *lead, manganese, bismuth,* and some other metals—composed of much oxygen relatively to the mass of lead &c.—are obtained by bringing these metals, or oxides of them composed of the metal united with relatively small masses of oxygen, in contact with two or more compounds which interact to produce oxygen. V. Oxides of *nitrogen, sulphur, tellurium,* &c. are obtained by decomposing, by heat or otherwise, compounds of these elements with oxygen and some other element or elements.

The following equations present examples of each of the foregoing methods of preparing oxides :—

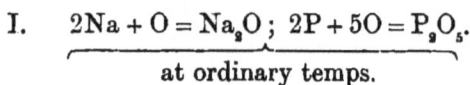

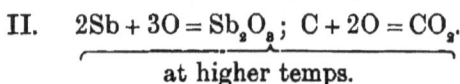

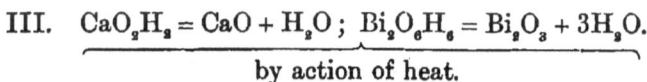

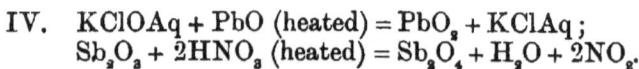

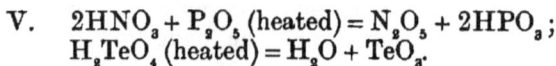

I. $2Na + O = Na_2O$; $2P + 5O = P_2O_5$.

at ordinary temps.

II. $2Sb + 3O = Sb_2O_3$; $C + 2O = CO_2$.

at higher temps.

III. $CaO_2H_2 = CaO + H_2O$; $Bi_2O_6H_6 = Bi_2O_3 + 3H_2O$.

by action of heat.

IV. $KClOAq + PbO$ (heated) $= PbO_2 + KClAq$;
$Sb_2O_3 + 2HNO_3$ (heated) $= Sb_2O_4 + H_2O + 2NO_2$.

V. $2HNO_3 + P_2O_5$ (heated) $= N_2O_5 + 2HPO_3$;
H_2TeO_4 (heated) $= H_2O + TeO_3$.

Many elements form more than one compound with oxygen. Thus, five oxides of nitrogen are known, viz. N_2O, NO, N_2O_3, NO_2, N_2O_5; four oxides of lead have been prepared, viz. PbO, Pb_3O_4, Pb_2O_3, PbO_2.

122   Hydrogen combines directly with a few elements; the combination usually occurs at moderately high temperatures : thus, $2H + S$ (molten) $= H_2S$; $H + Br$ (heated) $= HBr$; $2C + 2H$ (by passing electric sparks) $= C_2H_2$; &c. $H + Cl = HCl$

Compounds of hydrogen with other elements are sometimes formed in chemical interactions between several elements or compounds; thus when phosphorus is heated with an aqueous solution of caustic potash, phosphorus hydride (PH_3) is one of the products of the reaction; when a solution of arsenic oxide in water is brought into contact with dilute sulphuric acid and zinc, arsenic hydride (AsH_3), zinc sulphate, hydrogen, and water are formed.

Very many compounds of oxygen and hydrogen, each with two or more other elements have been prepared.

Compounds of oxygen and one other element are called **123** *oxides;* compounds of hydrogen with one other element are called *hydrides.*

The physical properties of oxides and hydrides vary much; some are solids, others are liquids, others are gases, at ordinary temperatures and pressures. The chemical properties of these compounds also vary much, but a great many oxides may be placed in one or other of two classes.

As representatives of these classes let us take the oxides of **124** sulphur and magnesium whose compositions are expressed by the formulae SO_3 and MgO, respectively ($S = 32$, $Mg = 24$, $O = 16$).

Sulphur trioxide (SO_3) is a white, crystalline, solid; it dissolves very easily in water forming a colourless solution. This solution has the following (among other) properties:— (1) a sour taste; (2) it turns blue litmus solution bright red; (3) it dissolves many metals—e.g. zinc, iron, aluminium, magnesium, cadmium, barium, &c. &c.—with production of two or more substances, one of which is hydrogen, and another is a compound of sulphur, oxygen, and the metal used; (4) it interacts with oxides of many metals—e.g. oxide of zinc, iron, aluminium, magnesium, cadmium, barium, &c. &c.—to produce two or more substances one of which is the same compound of sulphur, oxygen, and the metal of the oxide used, which was produced in reaction (3), and another of which is water. Further, if the solution of sulphur trioxide in water is evaporated considerably a thick oily liquid is obtained; if this liquid is cooled below 0° crystals separate having the composition H_2SO_4. This compound is called *sulphuric acid.* If sulphuric acid is dissolved in water the solution exhibits the same properties as a solution in water of sulphur trioxide. It seems therefore fair to conclude that an aqueous solution of this oxide contains the compound H_2SO_4.

Assuming that this is so, the chemical changes enumerated above may be represented in equations as follows :—

$$SO_3 + H_2O + Aq = H_2SO_4Aq.$$

(3) $Zn + H_2SO_4Aq = ZnSO_4Aq + 2H$;
     $Fe + H_2SO_4Aq = FeSO_4Aq + 2H$;
     $2Al + 3H_2SO_4Aq = Al_23SO_4Aq + 6H$;
     $Mg + H_2SO_4Aq = MgSO_4Aq + 2H$;
     $Cd + H_2SO_4Aq = CdSO_4Aq + 2H$;
     $Ba + H_2SO_4Aq = BaSO_4 + 2H + Aq.$

(4) $ZnO + H_2SO_4Aq = ZnSO_4Aq + H_2O$;
     $FeO + H_2SO_4Aq = FeSO_4Aq + H_2O$;
     $Al_2O_3 + 3H_2SO_4Aq = Al_23SO_4Aq + 3H_2O$;
     $MgO + H_2SO_4Aq = MgSO_4Aq + H_2O$;
     $CdO + H_2SO_4Aq = CdSO_4Aq + H_2O$;
     $BaO + H_2SO_4Aq = BaSO_4 + H_2O + Aq.$

The compounds of zinc, iron, &c. produced in these reactions are called *salts.* If the composition of each of these salts is compared with that of the sulphuric acid, H_2SO_4, by the interaction of an aqueous solution of which with a metal or the oxide of a metal the salt is produced, it is seen that the salt is composed of a metal together with all the sulphur and oxygen which were combined, before the chemical change began, with hydrogen, forming the compound H_2SO_4. The compound H_2SO_4 is called an *acid.* The solution of this acid in water has a sour taste ; turns blue litmus red ; reacts with zinc, iron, aluminium, and many other metals, to produce salts, and hydrogen; and reacts with oxides of zinc, iron, aluminium, magnesium, and other metals, to produce salts and water.

The oxide SO_3 is an *acidic,* or *acid-forming, oxide* ; that is to say, it reacts with water to produce a compound which is characterised by the properties enumerated in the preceding sentence.

Many oxides resemble sulphur trioxide in that they react with water to produce compounds of oxygen, hydrogen, and the other element of the oxide, which compounds have a sour taste, turn blue litmus red, and interact with metals and metallic oxides to produce salts. These oxides are placed in one class and are called *acid-forming,* or *acidic, oxides.* Thus, the oxides whose compositions are represented by the formulae P_2O_5, CrO_3, N_2O_5, SeO_2, respectively, are acidic oxides. The interactions of these oxides with water may be thus represented in equations :—

Acidic oxide.   *Acid.*

$$P_2O_5 + 3H_2O = 2H_3PO_4$$
$$CrO_3 + H_2O = H_2CrO_4$$
$$N_2O_5 + H_2O = 2HNO_3$$
$$SeO_2 + H_2O = H_2SeO_3.$$

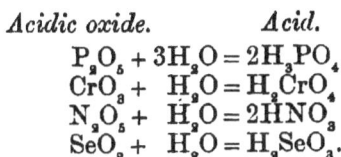

Let us now turn to the other oxide—MgO. *Magnesium* **125** *oxide* is a white solid; it dissolves in a large quantity of water; this solution has not a sour taste; it turns red litmus blue. Magnesium oxide interacts with sulphuric acid, and with other acids, to produce salts and water (*s. par.* 124).

Many oxides resemble magnesium oxide in that they interact with acids to form salts; some of these oxides further resemble magnesium oxide in being more or less soluble in water and thus forming solutions which turn red litmus blue.

Those oxides which interact with acids to produce salts are placed in one class and are called *basic oxides*, or sometimes *salt-forming oxides*. Those basic oxides which easily dissolve in water producing liquids which turn red litmus blue are usually placed in a sub-class to which the name *alkaline*, or *alkali-forming, oxides* is given. Thus the oxides Al_2O_3, ZnO, CdO, FeO, BaO are basic oxides (*s.* reactions represented in equations in par. 124).

The *oxides* of *boron, chlorine, iodine, nitrogen, phosphorus,* **126** *selenion, sulphur,* and several other elements, are *acidic oxides.*

The *oxides* of *aluminium, barium, beryllium, cadmium, copper, iron, lithium, magnesium, mercury, nickel, palladium, silver, sodium,* and many other elements, are *basic oxides.*

Some of the *oxides* of *chromium, molybdenum, tin, tungsten, uranium, vanadium,* and a few other elements, are *basic,* while *other oxides of the same elements* are *acidic.*

Can we classify the hydrides by a method similar to that **127** by which we have roughly arranged the oxides in classes?

The only element which forms many compounds with hydrogen is carbon. Some of the hydrides, other than those of carbon, interact with water to produce acids; among these are HBr, HCl, HF, HI, H_2S. One or two hydrides interact with water to produce compounds which again react with acids to form salts; the best marked hydride of this class is ammonia, NH_3. Several hydrides are either unchanged by water, or dissolve in it without producing either an acid or a salt-forming compound; e.g. H_3Sb, H_3As, H_2Cu_2, H_3P, H_2Te.

Hydrides cannot therefore be wholly classified by arranging them as acidic (or acid-forming), and basic (or salt-forming),

M. E. C.   7

hydrides; but nevertheless several hydrides belong to one or other of these classes.

128 But there may of course be other properties of the hydrides on which a classification might be based.

We might, for instance, classify the hydrides by looking to the composition of their reacting weights, and arranging them in classes according as the reacting weight is composed of one, two, or more, combining weights of hydrogen, and one, two, or more, combining weights of the other element. Thus the formula M_xH_y would represent the composition of all the hydrides, where H represents one combining weight of hydrogen, M one combining weight of the element other than hydrogen, and x and y vary.

The following table represents a classification of the greater number of the hydrides, excepting those of carbon, framed on this basis.

Hydrides; M_xH_y.

MH.   *MII_2*.

BrH, ClH, FH, IH. OH_2, SeH_2, SH_2, TeH_2.

MII_3.   *MII_4*.   *M_2II_2*.

SbH_3, AsH_3, NH_3, PH_3. SiH_4. Cu_2H_2, O_2H_2.

This classification is evidently based altogether on composition. Before deciding whether it is, or is not, a good classification we should study the properties of the hydrides, with the view of finding whether those placed in each class, MH, MH_2, &c., are marked by some common property which cuts them off from those in each of the other classes.

129 A slight examination of the hydrides in the foregoing table shews that many are gaseous at ordinary temperatures and pressures; some of these are easily decomposed when mixed with air or oxygen and heated. A classification might be based on the study of these decompositions. In most cases the products of the decomposition are oxide of hydrogen (water; H_2O), and oxide, or oxides, of the other element. The following table summarises this method of classifying gaseous hydrides.

Gaseous Hydrides; M_xH_y

Easily decomposed, by mixing with oxygen and heating, into M_xO_y and H_2O.

SeH_2, SH_2, TeH_2;
SbH_3, AsH_3, PH_3;
SiH_4; Cu_2H_2, O_2H_2.

Not easily decomposed by mixing with oxygen and heating.

BrH, ClH, FH, IH; NH_3

This classification is based on the chemical properties of the bodies classified, Before deciding whether it is, or is not, a good classification we should examine the composition of the hydrides, with the view of determining whether those placed in one class have similar compositions.

A comparison of this with the preceding table shews a certain connexion between composition and readiness or unreadiness to interact with oxygen.

The classification which we made of oxides into acidic and **130** basic oxides was based, for the most part, on chemical properties.

Let us now look a little to the compositions of the basic and the acidic oxides. As all are oxides, that is compounds of oxygen each with one other element, it is evident that one of the circumstances which conditions the basic or acidic character of an oxide is the nature of the element with which the oxygen is combined.

What then are the chemical properties of those elements which combine with oxygen to produce basic oxides? And what are the chemical properties of the elements which combine with oxygen to produce acidic oxides?

In par. 9 the decomposition of water by the electric current was described. The products of the electrolysis of water are hydrogen and oxygen; the hydrogen is always produced in contact with the terminal of the wire in connexion with the zinc plate of the battery, and the oxygen is always produced in contact with the terminal of the wire in connexion with the carbon, or copper, or platinum, plate of the battery. These terminals are called the *negative electrode* (wire coming from zinc plate), and *positive electrode* (wire coming from copper plate), respectively. As 'electricities of opposite sign attract each other,' *hydrogen* is called an *electro-positive element*, and *oxygen* an *electro-negative element*.

If hydrogen chloride (HCl) is electrolysed, hydrogen separates at the negative, and chlorine at the positive, electrode: *chlorine* is therefore said to be an *electro-negative element*. If hydrogen sulphide (H_2S) is electrolysed, sulphur separates at the positive electrode; *sulphur* is therefore said to be an *electro-negative element*. Sulphur, chlorine, and oxygen are electro-negative elements: if a compound of sulphur and chlorine is electrolysed sulphur separates at the positive electrode; sulphur is therefore negative relatively to chlorine, and chlorine is positive relatively to sulphur, although, as we

7—2

have seen, it is negative relatively to hydrogen. Electrolysis of a compound of sulphur and oxygen results in the separation of sulphur at the negative electrode ; sulphur is therefore positive relatively to oxygen, although it is negative relatively to chlorine. The terms electro-positive and electro-negative are therefore purely relative terms ; an element A may be positive towards an element B, but it may be negative towards another element C.

131 The following list indicates the arrangement of the commoner elements in electrical order ; each element is positive to all that precede it and negative to all that come after it : the order is only approximately correct. *Negative;* O, S, N, F, Cl, Br, I, P, As, B, C, Sb, Si, H, Pt, Hg, Ag, Cu, Bi, Sn, Pb, Co, Ni, Fe, Zn, Mn, Al, Mg, Ca, Ba, Sr, Na, K, Rb, Cs, ; *Positive.*

132 Taking the elements on the negative side of hydrogen as a class it is found that they vary much in physical properties; some are gases, some are solids, one (bromine) is a liquid ; their colours vary much ; some are lustrous, some are not ; those which are solids are more or less brittle, none malleable, ductile, or tenacious ; they are bad conductors of heat and electricity ; their emission-spectra are generally very complex. Some of these elements interact with steam to produce oxygen and a hydride of the element used (*s.* par. 105 where the reaction of steam with the negative element chlorine is described). The elements relatively positive to hydrogen more nearly resemble each other in many physical properties ; most of them are white, or grey, lustrous solids ; very many are malleable; several are tenacious and ductile ; they are generally hard, and many of them are heavy ; they are good conductors of heat and electricity ; their emission-spectra, as a rule, are less complex than those of the negative elements. Many of these elements interact with water or steam to produce hydrogen and an oxide of the element used (*s.* par. 105 where the reaction of steam with iron, and of water with sodium, is described).

The members of the first class are usually called *electro-negative* elements, and those of the second class *electro-positive*. Hydrogen occupies a position between these classes, but it is more closely related to the positive than to the negative elements.

The physical characters of the positive elements are summed up in the word *metal*, the physical characters of the negative elements are expressed by the term *non-metal*.

With these physical characters are associated certain distinct chemical characters; e.g. nature of the products of the reactions of each class of elements with water. Of these chemical characters we shall learn more as we proceed, meanwhile we have to look especially to the chemical properties of the oxides of each class. Looking at the matter broadly we may assert that the *oxides of the non-metallic elements* are *acidic*, and the *oxides of the metallic elements* are *basic*. Further we may assert that most of the non-metals combine with hydrogen to form hydrides, but that very few hydrides of the metals are known.

We have now associated certain properties of several elements with one definite chemical characteristic of the oxides of these elements.

Non-metallic elements are electro-negative as regards metallic elements; their oxides are generally acidic; most of them form hydrides. Metallic elements are electro-positive as regards non-metallic elements; their oxides are generally basic; few of them form hydrides.

CHAPTER IX.

133   THE sketch of the chemical properties of oxygen and hydrogen in Ch. VIII. has shewn that, in order to learn anything of these properties we have been obliged to examine the properties of many of the compounds of these elements, and to do this we have found it necessary to study the properties of very many other elements. Our attempt to classify oxygen and hydrogen has taught us, more fully than before, that the chemical properties of this or that definite kind of matter are the properties exhibited in the interactions of the specified kind of matter with other kinds, both elements and compounds.

We have also learned that a classification of elements carries with it a classification of compounds, and *vice versa;* and we have found that the method of chemical classification is based on the study both of the composition and the properties of the kinds of matter to be classified.

134   The study of the chemical properties of hydrogen and oxygen led to a consideration of the meaning of the terms *acidic* and *basic oxides,* and this brought with it a consideration of the properties of *acids* and *salts.* These classes of compounds are of great importance in the classification of elements and compounds. We must now consider them somewhat more fully.

Sulphur trioxide (SO_3) is a typical *acidic oxide* ; *potassium oxide* (K_2O) is a typical *basic oxide.* Let each be dissolved in water and each solution be evaporated; the first solution, until it becomes thick ; 'the second, until all the water is removed and a solid remains. Then let the thick oily liquid obtained by evaporating the first solution be cooled to $-10°$ or so; crystals separate ; let these be collected with suitable precautions and analysed ; their composition is expressed by the formula

H_2SO_4 (S = 32, O = 16). Let the white solid obtained by evaporating the solution in water of potassium oxide be analysed; its composition is expressed by the formula KOH (K = 39, O = 16). Each of these solids is then dissolved in water; the solution of H_2SO_4 has a very sour taste, corrodes animal and vegetable membranes, and turns blue litmus bright red; the solution of KOH has a burning, but not a sour, taste, corrodes the skin, and turns red litmus deep blue.

The compound H_2SO_4 is an *acid*, the compound KOH is an *alkali*.

A measured portion of the aqueous solution of H_2SO_4 is placed in a basin, a drop or two of litmus solution is added, and the solution of KOH is run in from a graduated vessel, drop by drop, with constant stirring, until the red colour of the litmus changes to a purplish-blue tint. The liquid is then evaporated to dryness, and the white solid which remains is analysed; its composition is expressed by the formula K_2SO_4. This solid is dissolved in water; the solution has no pronounced taste, it is without action on vegetable or animal membranes, and it does not affect the colour of either blue or red litmus. A comparison of the compositions of the three compounds

(1) KOH   (2) H_2SO_4 (3) K_2SO_4

shews that (3) differs from (2) in that hydrogen does not enter into its composition, but in place of two combining weights of hydrogen—combined with one combining weight of sulphur and four combining weights of oxygen—there are two combining weights of potassium.

Compound (1) is an *alkali*, (2) is an *acid*, (3) is a *salt*. The salt is produced by the interaction of the alkali and the acid. The interaction in question is represented by the equation

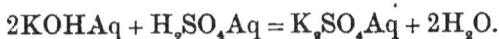

$$2KOHAq + H_2SO_4Aq = K_2SO_4Aq + 2H_2O.$$

Oxide of nitrogen N_2O_5 is *acidic*; *oxide of sodium* Na_2O is **135** *basic*; N_2O_5 dissolves in water, and it may be proved that the solution contains the compound HNO_3; Na_2O dissolves in water, and on evaporation the compound NaOH is obtained. If a solution of NaOH is added to a solution of HNO_3 until the liquid just ceases to affect the colour of litmus, and this solution is evaporated nearly to dryness, crystals separate having the composition $NaNO_3$ (Na = 23, N = 14, O = 16). An aqueous solution of this compound does not affect the colour of litmus and is without any of the marked characteristics of an acid or an alkali. The interaction by which this compound has been produced is represented in an equation thus ;—

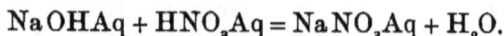

$$NaOHAq + HNO_3Aq = NaNO_3Aq + H_2O.$$

NaOH is an *alkali*, HNO_3 an *acid*, and $NaNO_3$ a *salt*.

136 When an alkali reacts with an acid to form a salt, the *acid is said to be neutralised by the alkali*, and *the alkali to be neutralised by the acid*. The salt does not exhibit the properties of either acid or alkali.

The following equations represent interactions of alkalis and acids with production of salts :—

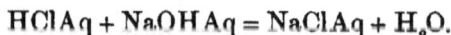

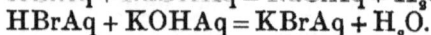

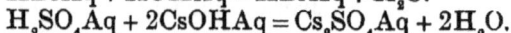

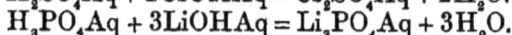

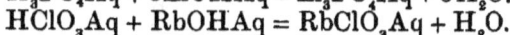

$$HClAq + NaOHAq = NaClAq + H_2O.$$
$$HBrAq + KOHAq = KBrAq + H_2O.$$
$$H_2SO_4Aq + 2CsOHAq = Cs_2SO_4Aq + 2H_2O.$$
$$H_3PO_4Aq + 3LiOHAq = Li_3PO_4Aq + 3H_2O.$$
$$HClO_3Aq + RbOHAq = RbClO_3Aq + H_2O.$$

In each case the composition of the salt is, a metallic (electro-positive) element combined with the elements of the acid except hydrogen. This relation between the compositions of the acid and the salt is usually expressed by saying that *the hydrogen of the acid is replaced (or displaced) by the metal of the alkali.*

137 But we have already learned that all basic oxides are not alkali-forming. Let us consider the interaction between a basic, but non-alkaline, oxide and an acid. The oxide of iron whose composition is expressed by the formula Fe_2O_3 $(Fe = 56)$ is a basic oxide. This oxide is insoluble in water, but it dissolves in solutions of various acids, and on evaporating these solutions salts are obtained : thus,

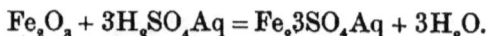

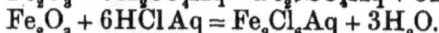

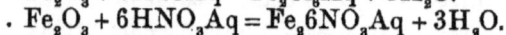

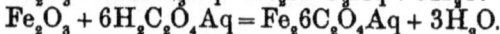

$$Fe_2O_3 + 3H_2SO_4Aq = Fe_23SO_4Aq + 3H_2O.$$
$$Fe_2O_3 + 6HClAq = Fe_2Cl_6Aq + 3H_2O.$$
$$Fe_2O_3 + 6HNO_3Aq = Fe_26NO_3Aq + 3H_2O.$$
$$Fe_2O_3 + 6H_2C_2O_4Aq = Fe_26C_2O_4Aq + 3H_2O.$$

In each case the relation between the composition of the salt and the acid may be expressed by saying that the metal of the basic oxide (iron) has replaced or displaced the hydrogen of the acid. An aqueous solution of the salt produced in each interaction does not affect the colour of litmus ; it has not a sour or burning taste, and it does not corrode animal or vegetable membranes.

Salts then are produced by the interactions between basic **138**
oxides and aqueous solutions of acids. Salts are also generally
produced when metals interact with aqueous solutions of acids.
We have had examples of these reactions before (par. 124). The
following may be added ;—

Metal	Solution of acid	Salt	Other products of reaction
Iron	Sulphuric	Iron sulphate	Hydrogen

$$Fe + H_2SO_4Aq = FeSO_4Aq + 2H$$

Metal	Solution of acid	Salt	Other products of reaction
Iron	Nitric	Iron nitrate	Water and ni-

$$2Fe + 8HNO_3Aq = Fe_2 6NO_3Aq + 4H_2O + 2NO \quad \text{tric oxide}$$

Metal	Solution of acid	Salt	Other products
Iron	Hydrochloric	Iron chloride	Hydrogen

$$2Fe + 6HClAq = Fe_2Cl_6Aq + 6H$$

Metal	Solution of acid	Salt	Other products
Sodium	Nitrous	Sodium nitrite	Hydrogen

$$Na + HNO_2Aq = NaNO_2Aq + H$$

Metal	Solution of acid	Salt	Other products
Sodium	Oxalic	Sodium oxalate	Hydrogen

$$2Na + H_2C_2O_4Aq = Na_2C_2O_4Aq + 2H$$

Metal	Solution of acid	Salt	Other products
Barium	Perchloric	Barium perchlorate	Hydrogen

$$Ba + 2HClO_4Aq = Ba2ClO_4Aq + 2H$$

Metal	Solution of acid	Salt	Other products
Copper	Nitric	Copper nitrate	Water and ni-.

$$3Cu + 8HNO_3Aq = 3Cu\,2NO_3Aq + 4H_2O + 2NO \quad \text{tric oxide}$$

Metal	Solution of acid	Salt	Other products
Zinc	Nitric	Zinc nitrate	Water, and

$$Zn + xHNO_3Aq = Zn2NO_3Aq + \&c. \quad \text{ammonia or}$$

nitric oxide and sometimes other compounds of nitrogen,
according to temperature and concentration of acid used.

The relation between the composition of the salt and the
acid which interacted with a metal to produce the salt may in
each case be expressed by saying that the hydrogen in one or
more reacting weights of the acid has been replaced by the
metal ; in many reactions the displaced hydrogen is obtained,
but in some cases further chemical change occurs, and water
and compounds formed of the elements formerly combined in
the acid are produced.

Sometimes the whole of the hydrogen combined with other
elements to form a reacting weight of an acid is not displaced
by metal when a salt is produced ; thus

(1) $KOHAq + H_2SO_4Aq = KHSO_4Aq + H_2O.$

(2) $2KOHAq + H_2SO_4Aq = K_2SO_4Aq + 2H_2O.$

(1) $LiOHAq + H_3PO_4Aq = LiH_2PO_4Aq + H_2O.$

(2) $2LiOHAq + H_3PO_4Aq = Li_2HPO_4Aq + 2H_2O.$

(3) $3LiOHAq + H_3PO_4Aq = Li_3PO_4Aq + 3H_2O.$

Nevertheless each of the compounds in the third column ($KHSO_4$ &c.) is called a salt.

139 So far as composition goes we may at present regard a salt as a compound of a metal with the elements of an acid except the whole or a part of the hydrogen of that acid; as regards mode of formation we may say, that a salt is one of the products of the interaction between an acid (in solution) and a metal, a basic oxide, or an alkali.

140 The meaning of the term *salt* then evidently includes the meanings of the terms *acid, metal, basic oxide*, and *alkali*; and any one of these terms can be understood only by considering the meanings of them all.

We already know the general characteristics of those elements which are called metals; we also know that most basic oxides are oxides of metals; and that alkalis are produced by the interaction of certain basic oxides with water.

The meaning given above to the term salt implies that an acid is a compound of hydrogen. But all compounds of hydrogen are not acids. The characteristic of those compounds of hydrogen which are acids is that when they interact with metals, basic oxides, or alkalis, they exchange the whole or a part of their hydrogen for metal, and thereby form a salt, or salts. Alkalis also are compounds of hydrogen with oxygen and a metal. The composition of a reacting weight of the alkalis is represented by the formula MOH; where $M = Li$, Na, K, Rb, or Cs. These elements, Li, &c. are the most positive of all the elements.

If the composition of acids is compared with the properties of the element or elements which combine with hydrogen to produce these acids, it is found that the non-hydrogenous constituents of acids, as a rule, are distinctly negative elements. This is shewn by comparing the compositions of the following well known acids with the arrangement of elements in electrical order given in par. 131.

HNO_3, H_2SO_4, HCN, $H_2C_2O_4$, $HClO_3$, H_3PO_4, HI, HCl, HF, H_2SiF_6, H_3BO_4, H_2SiO_3.

But the relations between the compositions and properties of acids, alkalis, and salts cannot be properly elucidated until a later stage.

The term *base* is frequently used to include basic oxides, alkaline hydroxides, and alkalis; as thus used it indicates a compound which interacts with an acid to produce a salt and water.

Looking back for a moment we see that in attempting **141**
to classify oxides we have been led to classify very many
elements.   We have divided these elements into electro-
positive and electro-negative.   With the characters expressed
by each of these terms we have connected several other physical,
and some chemical, characteristics.

The electro-positive elements, as a class, are metallic; their
oxides are basic; the oxides of the more electro-positive elements
interact with water to form compounds of the composition MOH
which are alkalis.   Many of these elements decompose water,
or steam, with production of hydrogen and a metallic oxide.

The electro-negative elements, as a class, are non-metallic;
their oxides are generally acidic, that is they interact with
water to produce acids. ˉ Some of these elements decompose
steam with production of oxygen and a non-metallic hydride.*

* For a fuller treatment of the subject of acids and salts s. Chap. XI.

CHAPTER X.

CHEMICAL NOMENCLATURE.

142   BEFORE proceeding further with the examination of the properties of classes of elements and compounds with the view of tracing connexions between changes of composition and changes of properties we must acquaint ourselves with the system of nomenclature used in chemistry.

Many names of elements, and such names of classes of compounds as oxides, hydrides, &c. have been incidentally employed.

143   A name is given to each element. Sometimes the name expresses a characteristic chemical or physical property of the element; e.g. *oxygen*=acid-producer, *hydrogen*=water-producer, *bromine*, because of its powerful and obnoxious smell ($\beta\rho\omega\mu os$), *iodine*, because of the violet colour of its vapour ($\iota\omega\delta\eta s$), *chromium*, because of its many-coloured compounds ($\chi\rho\omega\mu a$). Sometimes the name is that which was used by the ancients or is a modification of this name; e.g. *arsenic* ($\dot{a}\rho\sigma\epsilon\nu\iota\kappa o\nu$), *copper* (*cuprum*). Sometimes the name is derived from the name used by the alchemists, many of which were derived from the names of the planets; e.g. *mercury*. The names of many recently discovered elements are derived from the names of the minerals from which they were first obtained, or from the names of the districts, or in some cases countries, in which these minerals were found; thus *strontium* (from the mineral *strontianite* found near the village of *Strontian* in Argyleshire), *beryllium* (from the mineral *beryll*), *ytterbrium, yttrium, erbium* (from *Ytterby* the district in Sweden where the minerals were found from which the three elements were obtained), *gallium, germanium* (the former was discovered by a French, the latter by a German, chemist).

Some names are purely fanciful; e.g. *tellurium, selenion, uranium, vanadium*, (from *tellus* = the earth, $\sigma\epsilon\lambda\eta\nu\eta$ = the moon, the planet *Uranus*, and the Scandinavian deity *Vanadis*, respectively). The names of the more recently discovered metals all end in *um*.

The name given to a compound expresses the qualitative **144** composition of that compound; if more than one compound of the same elements is known, names are given indicative of the relative quantities of the elements which unite to form reacting weights of the compounds.

The name of every compound of two elements ends in *ide.* Thus all compounds of oxygen with one other element are called *oxides*. The variety of oxide is indicated by prefixing the name of the element united with oxygen; thus we have *iron oxides, zinc oxides, sulphur oxides,* &c.

Similarly we have *sulphides*, i.e. compounds of sulphur with one other element; *chlorides*, i. e. compounds of chlorine with one other element; *bromides, fluorides, hydrides,* &c. &c. We say hydrogen oxide, hydrogen chloride, hydrogen sulphide; not oxygen hydride, chlorine hydride, sulphur hydride: oxygen, chlorine, and sulphur, are all more negative, or more nonmetallic, elements than hydrogen. The name of the more negative of the two elements of a binary compound is changed into a qualifying term ending in *ide*. Thus it is better to say oxide of chlorine—or chlorine oxide—than chloride of oxygen (or oxygen chloride).

When two oxides, chlorides, &c. of one element are known, the general custom is to form an adjective from the name of the element other than oxygen, &c. and to modify the termination of this adjective so as to express that one compound is composed of more, or less, oxygen, relatively to a fixed mass of the other element, than the other compound is. A comparison of the following names with the composition of each compound as expressed by its formula will illustrate this method of naming pairs of oxides, &c. of the same element.

Iron oxides	Sulphur oxides
Ferrous oxide FeO	Sulphurous oxide SO_2
Ferric ,, Fe_2O_3.	Sulphuric ,, SO_3.
Chromium chlorides	Tin bromides
Chromous chloride $CrCl_2$	Stannous bromide $SnBr_2$
Chromic ,, $CrCl_3$.	Stannic ,, $SnBr_4$.

Copper sulphides
Cuprous sulphide Cu_2S
Cupric   ,, CuS.

The termination -ous always indicates less of the non-metallic or negative element than the termination -ic, relatively to the same mass of the metallic or positive element.

When more than two oxides, &c. of the same element are known, two methods of naming are adopted. Four oxides of bismuth are known. Their compositions, and the names given to each, are as follows :—

Hypobismuthous oxide; or Bismuth dioxide . . Bi_2O_2
Bismuthous oxide; or Bismuth trioxide . . . Bi_2O_3
Hypobismuthic oxide; or Bismuth tetroxide . . Bi_2O_4
Bismuthic oxide; or Bismuth pentoxide . . . Bi_2O_5.

Five oxides of nitrogen are known :—

Nitrous oxide; or Nitrogen monoxide . . N_2O
Nitric oxide . . . . NO
Nitrogen trioxide . . . N_2O_3
Nitrogen tetroxide . . N_2O_4 NO_2
Nitrogen pentoxide . . N_2O_5.

One oxide gets the termination -ic; another, with relatively less oxygen, the termination -ous; the prefix hypo- is used to express relatively less oxygen than that of the -ic or -ous oxide. The prefix per- is sometimes employed to designate that oxide of a series which has relatively most oxygen.

Or the number of combining weights of oxygen in a reacting weight of each oxide is expressed by the prefixes mono, di, tri, &c. Unfortunately neither system is very strictly carried out. We shall have further examples of each system as we proceed.

The composition of one oxide of a series is sometimes expressed by a name formed from the names of other two oxides of the series; thus FeO is ferrous, Fe_2O_3 is ferric, and Fe_3O_4 is ferroso-ferric, oxide.

The name sesquioxide is frequently used; it implies that a positive element and oxygen are united in the ratio of $1 : 1\frac{1}{2}$ combining weights; thus Fe_2O_3 is often called iron sesquioxide, Cr_2O_3 chromium sesquioxide, &c.

Acidic oxides are sometimes named so as to indicate the acid obtained by interaction of each with water; thus SO_3 is called sulphuric anhydride, because it interacts with water to produce sulphuric acid. On this system of naming, the term anhydride means an acidic oxide. Lastly an oxide, chloride, &c. is sometimes distinguished from another oxide, chloride, &c. of the same element by a term indicating some prominent physical character, usually colour; thus one sometimes speaks of *the brown oxide of chromium*, the *black*, or *the magnetic, oxide of iron*.

The various systems of naming binary compounds are summarised in the following examples.

Sulphur oxides.

SO_2 Sulphurous oxide; Sulphur dioxide; Sulphurous anhydride.

SO_3 Sulphuric oxide; Sulphur trioxide; Sulphur peroxide; Sulphuric anhydride.

Chromium oxides.

CrO Chromous oxide.

Cr_3O_4 Chromo-chromic oxide.

Cr_2O_3 Chromic oxide; Chromium sesquioxide; Green oxide of chromium.

CrO_2 Chromium dioxide; Brown oxide of chromium.

CrO_3 Chromium trioxide; Chromium peroxide; Chromic anhydride; Red oxide of chromium.

The nomenclature of many compounds of three or more **145** elements is based on the relations which exist between acids and salts. To each acid is given a name indicative, as far as possible, of its composition. Prefixes and terminations are used as in the naming of binary compounds. Thus all acids obtained by combining sulphur with hydrogen and oxygen are called *sulphur acids*; those formed by the combination of chlorine with hydrogen and oxygen are called *chlorine acids*; those produced by uniting nitrogen with hydrogen and oxygen are called *nitrogen acids*; and so on.

The following examples shew how one acid is distinguished from others of the same series.

Sulphur acids.		Chlorine acids.	
H_2SO_2	Hyposulphurous acid.	$HClO$	Hypochlorous acid.
H_2SO_3	Sulphurous ,,	$HClO_2$	Chlorous ,,
H_2SO_4	Sulphuric ,,	$HClO_3$	Chloric ,,
		$HClO_4$	Perchloric ,,

Nitrogen acids.		Phosphorus acids.	
HNO	Hyponitrous acid.	H_3PO_2	Hypophosphorous acid.
HNO_2	Nitrous ,,	H_3PO_3	Phosphorous ,,
HNO_3	Nitric ,,	H_3PO_4	
		HPO_3 }	Phosphoric · ,,
		$H_4P_2O_7$	

The three acids H_3PO_4, HPO_3, $H_4P_2O_7$ are all called phosphoric acid because they are all obtainable from the same oxide or anhydride, P_2O_5. The composition of the acid formed by the interaction of this oxide with water varies according to the relative masses of the interacting compounds, and the temperature ; thus

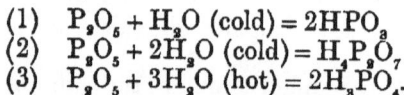

(1) $P_2O_5 + H_2O$ (cold) $= 2HPO_3$
(2) $P_2O_5 + 2H_2O$ (cold) $= H_4P_2O_7$
(3) $P_2O_5 + 3H_2O$ (hot) $= 2H_3PO_4$.

That acid of the three from which the greater number of well known and stable salts are obtained, viz. H_3PO_4, is called *ortho-phosphoric acid* ($\acute{o}\rho\theta os$ = right or true) ; the acid HPO_3 may be obtained from H_3PO_4 by removing water ($H_3PO_4 - H_2O = HPO_3$), it is generally called *metaphosphoric* (*meta* implies change of composition) ; the third acid $H_4P_2O_7$ may be produced by heating H_3PO_4 ($2H_3PO_4$ heated $= H_4P_2O_7 + H_2O$), it is called *pyrophosphoric acid.*

146   The names of the salts obtained from a given acid by causing it to interact with metals, basic oxides, or alkalis, are derived from the name of the acid ; each salt is distinguished from others by the name of the metal or metals which form part of its composition. Thus the salts obtained from sulphur*ous* acid are called sulph*ites*, those from sulphur*ic* acid are called sulph*ates*, and so on.

Hypochlorous acid; $HClO$.	Chlorous acid; $HClO_2$.
$KClO$ Potassium hypochlorite.	$AgClO_2$ Silver chlorite.
$NaClO$ Sodium ,,	$Pb2ClO_2$ Lead ,,

Perchloric acid; $HClO_4$.

$Ba(ClO_4)_2$ ·Barium perchlorate.
$RbClO_4$ Rubidium perchlorate.

Sulphuric acid; H_2SO_4.

Iron sulphates.	*Mercury sulphates.*
$FeSO_4$ Ferrous sulphate. .	Hg_2SO_4 Mercurous sulphate.
Fe_23SO_4 Ferric ,,	$HgSO_4$ Mercuric ,,

Nitric acid; HNO_3.

Tin nitrates.	*Iron nitrates.*
$Sn2NO_3$ Stannous nitrate.	$Fe2NO_3$ Ferrous nitrate.
$Sn4NO_3$ Stannic ,,	$Fe3NO_3$ Ferric ,,

These examples shew the use of the adjectival form of the name of the metal, and the meaning of the terminations -*ous* and -*ic*, in naming salts. A salt whose name ends in -*ous* is composed of less of the non-metallic elements, relatively to a fixed mass of the metal, than a salt of the same acid and the same metal whose name ends in -*ic*.

Ternary compounds (compounds of three elements) which **147** are not salts, as we are using this term, are generally named on the same principle as that which guides the nomenclature of binary compounds. Thus BiOCl is called *bismuth oxychloride*; BiSCl, *bismuth sulphochloride*; HgBrI, *mercury bromoiodide*, or *iodobromide*.

The nomenclature of carbon compounds cannot be discussed **148** here; suffice it to say that a name is usually given to each class of these compounds and that the individual members of this or that class are distinguished according to their composition. Thus, as we have a large class of *acids*, so we have a class of carbon compounds shewing certain common properties and certain well marked analogies in composition called *alcohols*; to another group of carbon compounds the name *aldehydes* is given; and so on.

CHAPTER XI.

CHEMICAL CLASSIFICATION.

149 ' WE have now gained some fairly clear notions of the methods adopted in chemistry for classifying elements and compounds. Similar elements are put into the same class. Similar elements are those which interact with other elements and with compounds under similar conditions to produce similar compounds.

Compounds again are said to be chemically similar or analogous when their compositions and their properties are similar.

The questions to be answered with regard to any element before its position in a scheme of chemical classification can be determined are such as these ;—Does the element combine with oxygen? Under what conditions are its oxides formed? What is the composition of each of these oxides? Are its oxides basic or acidic? Does it combine with hydrogen? What is the composition of its hydride or hydrides? Does it form any acids? Under what conditions are these acids produced, and what are their compositions and properties? Does it interact with acids to form salts? What are the products of its interaction with water? Or is it unchanged when brought into contact with water or steam? Does it form chlorides, bromides, oxychlorides, &c.? Under what conditions are these compounds produced? What are their compositions, and how do they interact with other bodies?

As chemistry is the study of the connexion between changes of composition and changes of properties, the subject of classification must be all important in chemistry. We must therefore proceed to examine a few classes of elements that we may learn how answers are gained to such questions as these just proposed, and what kind of answers they are which are gained.

The three elements *chlorine, bromine, iodine,* are placed in **150** the same class.

	Chlorine.	Bromine.	Iodine.´
Combining Weights.	35·5	80	127
Specific gravities as gases; air=1.	2·45	5·5	8·7
Specific gravities as gases; hydrogen = 1.	35·5	80	127
Specific gravities as solids or liquids; water = 1.	—	3·18 (liquid)	4·94 (solid)
Melting points (approxi- *Boiling* mate). *points*		−22⁰	115⁰
	−35⁰	63⁰	above 200⁰
Appearance and prominent physical properties.	Yellow-green gas, liquified at 0⁰ under pressure of 6 atmos. to a yellow, very, refractive, liquid. Very penetrating odour. Poisonous.	Dark-reddish-black liquid; solidifying at about −25⁰ to reddish-brown crystals. Odour very strong; rapidly corrodes animal and vegetable tissues. Poisonous.	Grey, lustrous solid; vapour is deep violet colour. Odour less marked than that of the two other elements. Poisonous.

Occurrence. None of these elements is found uncombined **151** with others. The commoner compounds are those with sodium, potassium, calcium, and magnesium. Chloride, bromide, and iodide of potassium, sodium, calcium, or magnesium occur in sea water, in mineral waters, and in some rocks. Chlorides of some or all the metals named are widely distributed ; bromides occur in smaller quantities; and iodides, with iodates (salts of the acid HIO_3), are found only in very small quantities.

Preparation. Sodium (or potassium) chloride, bromide, or **152** iodide, is mixed with manganese dioxide and sulphuric acid, and the mixture is heated ; manganese sulphate, sodium sulphate, water, and chlorine, bromine, or iodine, are formed. Putting X as $= Cl$, Br, or I, the change may be thus expressed ;—

$$2NaX + 2H_2SO_4 + MnO_2 = MnSO_4 + Na_2SO_4 + 2H_2O + 2X.$$

If NaX is heated with sulphuric acid only, hydrogen chloride, bromide, or iodide is formed; if one of these compounds is heated with manganese dioxide, chlorine, bromine, or iodine is formed along with manganese chloride, bromide, or iodide : hence the change represented in the above equation probably takes place in two parts occurring simultaneously. Thus

$\left\{\begin{array}{l} (1) \quad 2\mathrm{NaX} + \mathrm{H_2SO_4} = \mathrm{Na_2SO_4} + 2\mathrm{HX}. \\ (2) \quad 2\mathrm{HX} + \mathrm{MnO_2} + \mathrm{H_2SO_4} = \mathrm{MnSO_4} + 2\mathrm{H_2O} + 2\mathrm{X}. \end{array}\right\}$

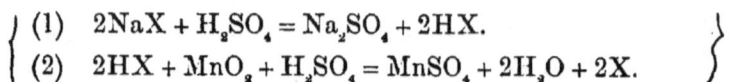

153   *Chemical properties.* **Compounds with hydrogen.** The elements chlorine, bromine, and iodine, combine directly with hydrogen, to form the compounds HX, where $\mathrm{X} = \mathrm{Cl}$, Br, or I. Hydrogen chloride, HCl, is formed by exposing a mixture of equal volumes of hydrogen and chlorine to diffused daylight; hydrogen bromide is formed by passing a mixture of hydrogen and bromine vapour through a hot tube; hydrogen iodide is formed, but only in small quantities, by passing a mixture of hydrogen and iodine vapour through a tube containing powdered glass or finely divided platinum, heated to 300° or 400°. These compounds, HX, are also produced by decomposing NaX or KX by an acid; e.g.

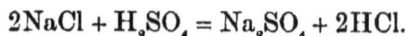

$$2\mathrm{NaCl} + \mathrm{H_2SO_4} = \mathrm{Na_2SO_4} + 2\mathrm{HCl}.$$

They are also formed, along with phosphorous acid ($\mathrm{H_3PO_3}$), when water interacts with chloride, bromide, or iodide, of phosphorus; thus $\mathrm{PX_3} + 3\mathrm{H_2O} = 3\mathrm{HX} + \mathrm{PO_3H_3}$.

The compounds HX are gaseous at ordinary temperatures and pressures; they are very soluble in water and the solution in each case is markedly acid (*s.* chap. IX.). Hydrogen iodide is decomposed by heat into hydrogen and iodine; the change begins at about 180°; hydrogen bromide is also decomposed by heat but the change does not begin until a temperature higher than 180° is reached; hydrogen chloride is not decomposed by heat.

154   **Compounds with oxygen.** None of the elements, chlorine, bromine, iodine, combines directly with oxygen. No oxide of bromine has been isolated; one oxide of iodine, and two (or perhaps three) oxides of chlorine are known. The compositions of these oxides are as follows;—$\mathrm{Cl_2O}$, $\mathrm{ClO_2}$, ($? \mathrm{Cl_2O_3}$), $\mathrm{I_2O_5}$. The oxides $\mathrm{Cl_2O}$ and $\mathrm{I_2O_5}$ are prepared by carrying out chemical changes in which oxygen is produced in presence of chlorine and iodine, respectively. To prepare $\mathrm{Cl_2O}$, mer-

curic oxide (HgO) is heated in a stream of dry chlorine. When mercuric oxide is heated it is decomposed into mercury and oxygen ; therefore by passing chlorine over heated mercuric oxide we carry out a reaction in which oxygen is produced in presence of chlorine. When a concentrated solution of nitric acid (HNO_3) is heated, oxygen, water, and oxides of nitrogen are produced. If iodine is heated with concentrated nitric acid, a part of the oxygen coming from the nitric acid combines with the iodine to form iodine pentoxide (I_2O_5). The two reactions are represented in equations thus ;—

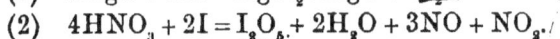

(1) $2HgO + 4Cl = HgCl_2 . HgO + Cl_2O.$

(2) $4HNO_3 + 2I = I_2O_5 + 2H_2O + 3NO + NO_2.$

In the interaction of water with chlorine, oxygen is evolved; if iodine is suspended in water and chlorine is passed into the water hydrated iodine pentoxide is produced ; thus

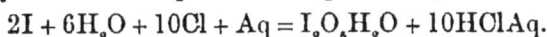

$$2I + 6H_2O + 10Cl + Aq = I_2O_5H_2O + 10HClAq.$$

When the compound $I_2O_5 . H_2O$ thus produced is heated, water is removed and I_2O_5 remains.

The other oxide of chlorine, ClO_2, is produced by decomposing chloric acid, $HClO_3$, by heat. Chloric acid is very easily decomposed ; when the potassium salt of the acid interacts with sulphuric acid, chloric acid and potassium sulphate are formed, but the heat produced in the reaction suffices to decompose the chloric acid into chlorine dioxide (ClO_2) and other products. These reactions may be represented thus in equations ;—

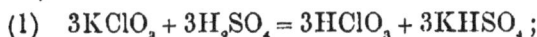

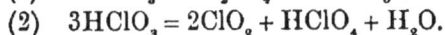

(1) $3KClO_3 + 3H_2SO_4 = 3HClO_3 + 3KHSO_4 ;$

(2) $3HClO_3 = 2ClO_2 + HClO_4 + H_2O.$

It is still undecided whether the body described as chlorine trioxide, Cl_2O_3, is a definite compound or a mixture of chlorine dioxide and chlorine.

The oxides of chlorine and iodine are acidic. Each interacts with water to form an acid or acids ; thus

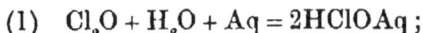

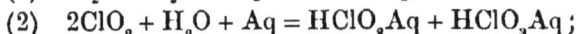

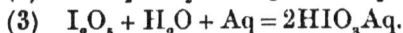

(1) $Cl_2O + H_2O + Aq = 2HClOAq ;$

(2) $2ClO_2 + H_2O + Aq = HClO_4Aq + HClO_3Aq ;$

(3) $I_2O_5 + H_2O + Aq = 2HIO_3Aq.$

The three acids hypochlorous (HClO), chlorous ($HClO_2$), chloric ($HClO_3$), are known only in aqueous solutions ; when these solutions are concentrated to remove water the acids are decomposed into oxides of chlorine and water. Iodic acid

(HIO_3) is known as a solid body; when heated it decomposes into iodine pentoxide (I_2O_5) and water.

155   **Compounds with non-metallic elements other than hydrogen or oxygen.** Chlorine, bromine, and iodine, combine directly with many non-metallic elements. The compositions of these compounds can generally be expressed by formulae in which X represents one combining weight of any one of the three elements. The following table shews the compositions of some of the best known of these compounds.

with arsenic; AsX_3 where $X = Cl$ or Br only.
  „ antimony; SbX_3: also $SbCl_5$.
  „ boron; BX_3 where $X = Cl$ or Br only.
  „ phosphorus; PX_3: also PX_5 where $X = Cl$ or Br only: also P_2I_4.
  „ selenion; Se_2X_2: also SeX_4: also SeX_2 where $X = Cl$ or Br only.
  „ silicon; SiX_4 where $X = Cl$ or I only.
  „ sulphur; S_2X_2.
  „ tellurium; TeX_2: also TeX_4.

A few other compounds of chlorine, bromine, and iodine with non-metallic elements are formed indirectly; e.g. CX_4, C_2X_4, C_2X_6, $SiBr_4$, Si_2Cl_6, &c.

156   **Compounds with metallic elements.** The elements we are considering combine with most metals to form compounds with similar compositions. The compositions of some of these are represented in the following formulae

$NaCl$, KBr, KI; CdX_2, CaX_2, BaX_2, ZnX_2; BiX_3; Cr_2X_6; Cu_2X_2 and CuX_2; Fe_2X_4 and Fe_2X_6; Hg_2X_2 and HgX_2: PtX_2 and PtX_4; SnX_2 and SnX_4.

These compounds are usually produced by heating the metal in contact with chlorine, bromine, or iodine; in some cases however it is necessary to use indirect methods of preparation.

The binary compounds of the three elements we are considering are usually called *haloid compounds* (i.e. compounds resembling common salt, NaCl); the elements themselves are often called *halogens*. Many of the haloid compounds of the non-metals other than hydrogen and oxygen, and some of those of the metals, can be gasified without decomposition. Most of the non-metallic haloid compounds interact with water to produce solutions of HX (where X = Cl, Br, or I), and generally either an oxide or an oxygen-containing acid of the non-metal formerly combined with halogen. The following equations represent some of these changes:—

$$2AsCl_3 + 3H_2O + Aq = As_2O_3Aq + 6HClAq.$$
$$BCl_3 + 3H_2O + Aq = H_3BO_3Aq + 3HClAq.$$
$$PBr_3 + 3H_2O + Aq = H_3PO_3Aq + 3HBrAq.$$
$$2S_2Br_2 + 3H_2O + Aq = H_2SO_3Aq + 3S + 4HBrAq.$$
$$2Se_2I_2 + 3H_2O + Aq = H_2SeO_3Aq + 3Se + 4HIAq.$$

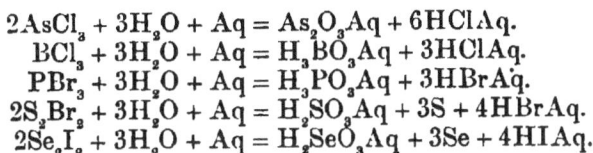

In some cases the products of the interaction of a non-metallic haloid compound and water are HX and a compound of the non-metal with oxygen and halogen ; thus

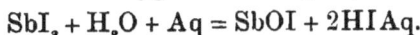

$$SbI_3 + H_2O + Aq = SbOI + 2HIAq.$$

Most of the haloid compounds of the metallic elements are chemically unchanged when brought into contact with water ; several dissolve in water. In some cases however chemical change occurs ; the usual products are haloid compounds of. hydrogen (HX) and an oxychloride, oxybromide, or oxyiodide, of the metal :—thus,

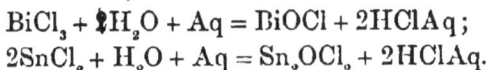

$$BiCl_3 + \tfrac{3}{2}H_2O + Aq = BiOCl + 2HClAq ;$$
$$2SnCl_2 + H_2O + Aq = Sn_2OCl_2 + 2HClAq.$$

Interactions with water. The three elements dissolve in **157** water, chlorine very freely, bromine less freely, and iodine only in small quantities. By cooling aqueous solutions of chlorine or bromine,crystals separate having the composition $Cl.5H_2O$ and $Br.5H_2O$ respectively : no *hydrate* of iodine—i.e. compound of iodine with water—has been obtained. Aqueous solutions of the three elements contain small quantities of hydrochloric, hydrobromic, and hydriodic acids, respectively ; i.e. the water and chlorine &c. interact as shewn by the equation

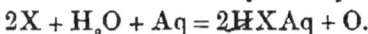

$$2X + H_2O + Aq = 2HXAq + O.$$

This reaction proceeds more rapidly when $X = Cl$ than when $X = Br$. When $X = I$ but very little reaction occurs. These reactions are hastened by sunlight. If some easily oxidised substance is dissolved in water and chlorine is passed into the liquid the substance is usually oxidised ; thus a solution of sulphur dioxide reacts with chlorine to produce sulphur tri-oxide, a solution of phosphorous oxide reacts with chlorine to produce phosphoric oxide :—or, in equations

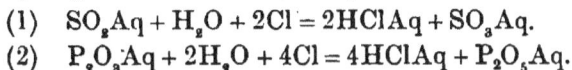

(1) $SO_2Aq + H_2O + 2Cl = 2HClAq + SO_3Aq.$

(2) $P_2O_3Aq + 2H_2O + 4Cl = 4HClAq + P_2O_5Aq.$

The bleaching action of chlorine depends upon its inter-acting with water to produce oxygen. Dry chlorine does not bleach a piece of madder-dyed cloth ; but if water is present the cloth is bleached. The colourless bodies produced are the results of the interaction of oxygen with the colouring matter of the cloth; this oxygen is produced from the water by inter-action with chlorine as already described. An aqueous solu-tion of bromine bleaches more slowly than a solution of chlorine, and a solution of iodine bleaches very slowly indeed : the bleaching action is more or less rapid according as the element decomposes water rapidly or slowly (*v. supra*).

158   **Interactions with solutions of alkalis.** Chlorine, bromine, and iodine interact with cold aqueous solutions of caustic potash, soda, &c. to produce potassium or sodium (&c.) chloride, bromide, or iodide, and also potassium (&c.) hypochlorite, hypobromite, or (probably) hypoiodite. Thus in equations ($X = Cl$, Br, or I)

$$6KOHAq + 6X = 3KXAq + 3KXOAq + 3H_2O.$$

The interaction which occurs between one of the halogens and a hot solution of caustic potash, soda, &c. is expressed thus :—

$$6KOHAq + 6X = 5KXAq + KXO_3Aq + 3H_2O.$$

The products are potassium (&c.) chloride (bromide or iodide), potassium (&c.) chlorate (bromate or iodate), and water.

Solutions in water of potassium, sodium, (&c.) hypochlorite or hypobromite are changed by heat into potassium (&c.) chloride or bromide, and chlorate or bromate, thus

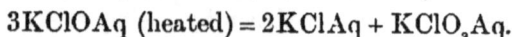

$$3KClOAq \text{ (heated)} = 2KClAq + KClO_3Aq.$$

If an easily oxidised substance is present it is oxidised and only potassium chloride or bromide is produced. From these facts it follows that if an easily oxidised substance is dissolved in an aqueous solution of caustic potash, the solution is heated and chlorine is passed in, oxidation ought to occur. Experi-ment shews that this conclusion is correct; experiment further shews that an element or compound which is not soluble in aqueous caustic potash may often be oxidised by suspending it in hot potash solution and passing in chlorine. Examples of such interactions are these :—

(1) $SeO_2 + 2KOHAq + 2Cl = SeO_3Aq + 2KClAq + H_2O$,

(2) $Bi_2O_3 + 4KOHAq + 4Cl = Bi_2O_5 + 4KClAq + 2H_2O$,

(3) $MnSO_4Aq + 2KOHAq + 2Cl$

$$= MnO_2 + K_2SO_4Aq + 2HClAq,$$

(or this reaction may be thus expressed,

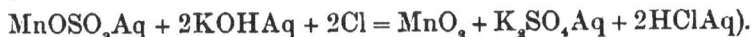

$$MnOSO_3Aq + 2KOHAq + 2Cl = MnO_2 + K_2SO_4Aq + 2HClAq).$$

**Interactions between one of the halogens and binary com- 159
pounds of the others.**

Chlorine reacts with most bromides to form a chloride and bromine ; bromine generally reacts with iodides to form iodine and a bromide. These changes occur most readily when the aqueous solutions are employed ; thus

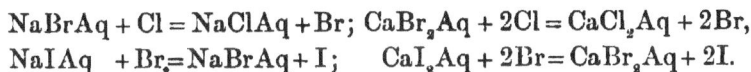

$$NaBrAq + Cl = NaClAq + Br; \quad CaBr_2Aq + 2Cl = CaCl_2Aq + 2Br,$$
$$NaIAq + Br = NaBrAq + I; \quad CaI_2Aq + 2Br = CaBr_2Aq + 2I.$$

Hence it follows that an aqueous solution of an iodide will be decomposed by chlorine ; e.g.

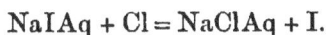

$$NaIAq + Cl = NaClAq + I.$$

The chemical changes described in the foregoing paragraphs shew that the three elements chlorine, bromine, and iodine, are chemically similar. They are produced from similar compounds under similar conditions. They combine with the same elements to form compounds similar in composition and in properties. The reactions described also shew that chlorine bromine and iodine are markedly negative or non-metallic elements; their oxides are acidic; they decompose water to produce oxygen and, in each case, a hydride ; they form numerous compounds with oxygen and another element ; none of them interacts with acids to produce salts. The facts we have learned concerning the three elements also shew a gradation of properties from chlorine to iodine, and exhibit a connexion between this gradation and the combining weights of the three elements. As the combining weight increases the elements become heavier, darker in colour, and more solid ; the oxides and oxygen compounds generally become more stable, and the hydrides become less stable, as regards the action of heat ; the rate at which water is decomposed decreases. The binary compounds of the element with largest combining weight are generally decomposed by the other elements of the group.

The elements *lithium, sodium, potassium, rubidium,* and 160
caesium, form a group or family. Let us briefly consider their properties.

	Lithium.	Sodium.	Potassium.	Rubidium.	Caesium.
Combining weights.	7	23	39	85·2	133
Specific gravities (water =1)	·59	·98	·87	1·52	1·88
Melting points.	180⁰	95⁰·5	58⁰—62⁰	38⁰	26⁰—27⁰
Appearance, &c.	Silver-white: very soft; not volatile at red heat. Electro-positive to all other elements except Na, K, Rb, and Cs.	Silver-white: soft as wax: can be distilled at red heat. Positive to all other elements except K, Rb, and Cs.	White: brittle at 0⁰; pasty at 15⁰; can be distilled at 700⁰—800⁰. Positive to all other elements except Rb and Cs.	White: soft as wax at – 10⁰. Positive to all other elements except Cs.	Silver-white; soft at ordinary temp. Positive to all other elements.

161 *Occurrence.* None of these elements is found in nature uncombined with others. Nitrates, chlorides, silicates, and some other compounds of sodium and potassium, occur in large quantities in rocks and mineral waters. Silicates and phosphates &c. of lithium and rubidium are very widely distributed but occur only in very small quantities; caesium compounds are found in very minute quantities in several rocks and mineral waters.

162 *Preparation.* Sodium, potassium, and rubidium, are prepared by heating a mixture of their carbonates (M_2CO_3; M = Na, K, or Rb) and carbon to a high temperature. The chemical changes may be thus represented:

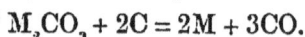

$$M_2CO_3 + 2C = 2M + 3CO.$$

Lithium is prepared by passing an electric current through fused lithium chloride (LiCl) mixed with ammonium chloride; and caesium by electrolysing fused caesium-barium cyanide [$CsCN . Ba(CN)_2$].

163 *Chemical properties.* These five elements are very easily oxidised; when exposed to air at ordinary temperatures the surface of the element at once becomes covered with a film of oxide. They decompose cold water rapidly with formation of hydrogen and a compound of oxygen, hydrogen, and the element; thus

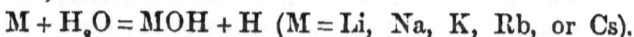

$$M + H_2O = MOH + H \quad (M = Li, Na, K, Rb, or Cs).$$

The compound MOH—called a *hydroxide*—dissolves in

the *excess* of water; the chemical change is therefore better represented by the equation

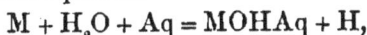

$$M + H_2O + Aq = MOHAq + H,$$

where Aq means a large, indeterminate, quantity of water.

That we may learn the exact meaning of this equation, let a weighed piece of sodium—say 1 gram—be thrown into a large quantity of water, weighing say x grams; the sodium moves about on the surface of the water with a hissing sound, and hydrogen is rapidly evolved; after a little time the sodium has entirely disappeared. The mass of hydrogen produced weighs ·044 gram; the mass of liquid remaining weighs $1 + x - $·044 grams; this liquid is evaporated so that the water which boils off may be collected and weighed, $x - $·783 grams of water are obtained, and 1·74 grams of a white solid remain. This white solid is analysed; its composition is expressed by the formula NaOH (Na = 23, O = 16). Hence 1 gram of sodium has interacted with ·783 gram of water to produce ·044 gram of hydrogen and 1·74 grams of sodium hydroxide (or caustic soda); the whole of the hydrogen produced has been evolved as gas, and the 1·74 grams of sodium hydroxide have dissolved in the *excess* of water, that is, in the water which did not chemically interact with the sodium. But as the combining weight of sodium is 23, and the reacting weight of water is 18 (O = 16), the experimental results are expressed by the equation

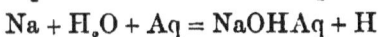

$$Na + H_2O + Aq = NaOHAq + H$$
$$(23 : 18 = 1 : \cdot783, \text{ and } 40 : 1 = 1\cdot74 : \cdot044).$$

The elements we are considering do not combine with hydrogen either directly or, indirectly. They combine with oxygen, with oxygen and hydrogen, with the halogens, and with many other, chiefly non-metallic, elements. The compositions of the more important compounds are represented by the formulae: M_2O, MOH, M_2S, MSH, MX (X = F, Cl, Br, I), M_2SO_4 and MHSO$_4$, M_2CO_3 and MHCO$_3$, MNO$_3$ (M = Li, Na, K, Rb, or Cs).

Compounds with oxygen. Oxides M$_2$O. The five elements **164** combine directly with oxygen at ordinary temperatures; but caesium oxide has not yet been obtained approximately pure. The oxides are white solids, which are unchanged by the action of heat; they dissolve very rapidly in water, and these solutions turn red litmus deep blue. The oxides are decidedly basic; they interact with aqueous solutions of acids to produce salts and water: thus—

Oxide.   Acid.   Salt.

$$M_2O + 2HClAq = 2MClAq + H_2O,$$
$$M_2O + H_2SO_4Aq = M_2SO_4Aq + H_2O,$$
$$M_2O + 2HNO_3Aq = 2MNO_3Aq + H_2O,$$
$$M_2O + H_2C_2O_4Aq = M_2C_2O_4Aq + H_2O,$$
$$M_2O + H_3PO_4Aq = M_2HPO_4Aq + H_2O,$$
$$M_2O + 2HClO_3Aq = 2MClO_3Aq + H_2O.$$

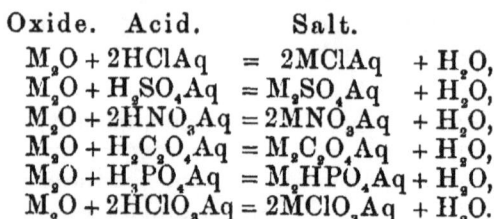

These oxides interact with water to form hydroxides which dissolve in the excess of water : thus—

$$M_2O + H_2O + Aq = 2MOHAq.$$

165   Compounds with oxygen and hydrogen. Hydroxides MOH. Prepared by evaporating aqueous solutions of the oxides to dryness and heating the residual solids; also by the interactions between solutions of the carbonates of sodium &c. and lime, thus—

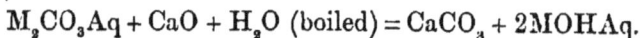

$$M_2CO_3Aq + CaO + H_2O \text{ (boiled)} = CaCO_3 + 2MOHAq.$$

The hydroxides are white solids which melt at high temperatures without undergoing any chemical change. They cannot be decomposed into oxides and water by the action of heat alone. When molten each hydroxide interacts with its own metal to produce hydrogen and metallic oxide; thus—

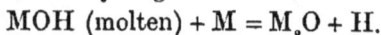

$$MOH \text{ (molten)} + M = M_2O + H.$$

The hydroxides (MOH) dissolve rapidly in water, forming *alkaline* solutions. These solutions interact with acids to produce salts and water, thus

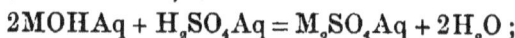

$$2MOHAq + H_2SO_4Aq = M_2SO_4Aq + 2H_2O;$$

they readily combine with carbon dioxide (gas) to produce carbonates and water, thus

$$2MOHAq + CO_2 = M_2CO_3Aq + H_2O;$$

they interact with solutions of salts of iron, manganese, chromium, zinc, mercury, and many other metals, to produce salts of sodium, potassium, &c. and compounds of iron, manganese, &c. with hydrogen and oxygen, thus

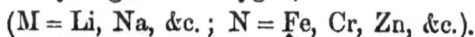

(M = Li, Na, &c. ; N = Fe, Cr, Zn, &c.).

Hydroxide;   Salt of   Hydroxide   Salt of
   MOH.   N.   of N.   M.

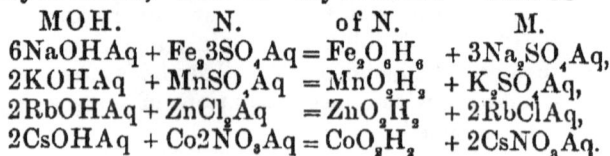

$$6NaOHAq + Fe_23SO_4Aq = Fe_2O_6H_6 + 3Na_2SO_4Aq,$$
$$2KOHAq + MnSO_4Aq = MnO_2H_2 + K_2SO_4Aq,$$
$$2RbOHAq + ZnCl_2Aq = ZnO_2H_2 + 2RbClAq,$$
$$2CsOHAq + Co2NO_3Aq = CoO_2H_2 + 2CsNO_3Aq.$$

In some cases an oxide, not a hydroxide, of the metal N is produced; thus with mercury salts and potash we have

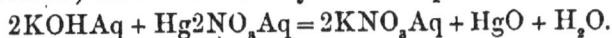

$$2KOHAq + Hg2NO_3Aq = 2KNO_3Aq + HgO + H_2O.$$

Aqueous solutions of the hydroxides MOH have a soap-like, but corrosive, action on the skin, and a burning, but not sour, taste; they interact with fats to form soaps and glycerine. Solutions having these properties are said to be *alkaline*. The oxides M_2O are said to be *alkali-forming* because they interact with water to produce the *alkalis* MOH.

The elements lithium, sodium, potassium, rubidium, and caesium, are called the *alkali-metals*.

The word alkali is of Arabic origin; it was originally applied to the ashes of sea-plants, and was afterwards extended to include all substances which more or less resembled these ashes in being very soluble in water, and feeling somewhat soapy to the touch, and reacting with acids to produce substances which exhibited neither the properties of the alkalis nor of the acids.

Compounds with the halogens. MX. These compounds are **166** produced (1) by the direct union of the elements; thus sodium heated in chlorine forms sodium chloride (NaCl): (2) by reactions between aqueous solutions of HX ($X = Cl$, Br, I) and the oxides M_2O or hydroxides MOH; thus sodium hydroxide dissolves in aqueous hydrochloric acid to produce sodium chloride and water ($NaOHAq + HClAq = NaClAq + H_2O$). The haloid compounds MX ($M = Na$ &c., $X = Cl$ &c.) are white solids, soluble in water, unchanged by heat; they combine with many other haloid compounds to form *double compounds*, thus $HgBr_2.KBr$; $ZnCl_2.2NaCl$; $CuCl_2.2KCl$; $CdBr_2.KBr$; $CdI_2.2KI$, &c.

Interactions with water. The alkali metals interact with **167** water at ordinary temperatures to produce hydroxides (MOH) and hydrogen (*s.* par. 163). This chemical change is accompanied by a considerable running down of energy*. The system $M + H_2O + Aq$ ($M = Na$, &c.) is able to do much more work than the system MOHAq + H. The following numbers shew the gram-units of heat produced when 7 grams of lithium, 23 grams of sodium, and 39 grams of potassium, interact with water, to produce a solution of 24 grams of lithium hydroxide, 40 grams of sodium hydroxide, and 56 grams of potassium hydroxide, respectively, and, in each case, 1 gram of hydrogen;

---

* A fuller treatment of this subject will be found in chap. xiv.

$$M + H_2O + Aq = MOHAq + H \text{ (all taken in grams)},$$
$$M = Li \; ; \; 49,084 \text{ gram-units of heat produced,}$$
$$Na \; ; \; 43,450 \qquad \text{''} \qquad \text{''} \qquad \text{''}$$
$$K \; ; \; 48,100 \qquad \text{''} \qquad \text{''} \qquad \text{''}$$

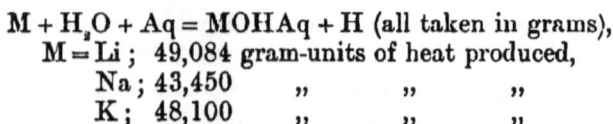

168   **Interactions with acids.** The alkali metals interact with acids to produce salts, hydrogen is generally evolved ; thus

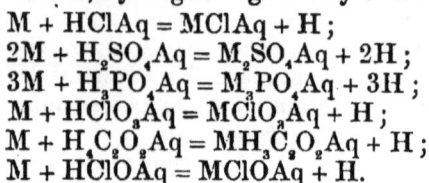

$$M + HClAq = MClAq + H \; ;$$
$$2M + H_2SO_4Aq = M_2SO_4Aq + 2H \; ;$$
$$3M + H_3PO_4Aq = M_3PO_4Aq + 3H \; ;$$
$$M + HClO_3Aq = MClO_3Aq + H \; ;$$
$$M + H_4C_2O_2Aq = MH_3C_2O_2Aq + H \; ;$$
$$M + HClOAq = MClOAq + H.$$

Very many of the salts of the alkali metals are soluble in water ; several of the lithium salts are less soluble than those of the other metals of the group. The sulphates (M_2SO_4), with the exception of lithium sulphate, combine with sulphate of aluminium to form double salts called alums, the composition of which is $M_2SO_4 . Al_2 3SO_4 . 24H_2O$.

169   The alkali metals are evidently very similar in their properties. They are all light, soft, very easily oxidised, very positive, elements ; all combine with oxygen to form oxides, which dissolve in water with production of alkaline hydroxides MOH. Their oxides and hydroxides are strongly basic ; and are unchanged by the action of heat. None of these elements combines directly or indirectly with hydrogen. They all interact with acids to form salts having similar compositions and similar properties. As the combining weights of the alkali metals increase the metals become heavier (sodium and potassium are exceptions), more positive, more easily oxidised, and more easily melted.

The halogen elements, which have been already considered, differ in the most marked way from the alkali metals ; some of the more prominent differences are presented in the following table.

Halogens. *Chlorine, Bromine, Iodine.*	Alkali metals. *Lithium, Sodium, Potassium, Rubidium, Caesium.*
One gaseous; one liquid; one solid.	All soft, light, lustrous, easily melted, solids.
Markedly electro-negative.	More electro-positive than any other elements.
Form compounds with hydrogen, HX, which dissolve easily in water producing strongly acid liquids.	Do not combine with hydrogen.

Halogens.	Alkali metals.
Oxides formed only indirectly: generally unstable; interact with water to produce acids; are markedly acidic.	Oxides formed directly at ordinary temperatures: very stable; interact with water to produce alkalis; are markedly basic.
Unite readily with many other elements, especially with positive elements.	Unite readily with negative elements.
Do not interact with acids to produce salts.	Interact with acids to produce salts.
Interact with steam to produce acids (HX) and evolve oxygen.	Interact with cold water to produce alkalis (MOH) and evolve hydrogen.

The halogens may be taken as typical electro-negative or non-metallic elements; and the alkali metals, so far as chemical properties are concerned, as typical electro-positive or metallic elements.

We shall now consider a group of elements which on the **170** whole are non-metallic and negative but in some respects shew analogies with the more positive or metallic elements. This group comprises the three elements *sulphur*, *selenion*, and *tellurium*.

	Sulphur.	Selenion.	Tellurium.
Combining weights.	32	79	125
Specific gravities (water $=1$).	about 2.	4·3 to 4·5	6·25
Melting points.	about 115°.	about 210°.	about 500°.
Boiling points.	440°	about 700°.	not accurately determined; below 1390°.
Appearance, &c.	Yellow, brittle, solid; non-conductor of electricity.	Red-grey to black lustrous solid; one form is a non-conductor of electricity, the other conducts fairly (*s.* par. 173).	Lustrous, white, metal-like, solid; very bad conductor of electricity.

Occurrence. The three elements are found uncombined **171** with others in nature; sulphur in large quantities in volcanic districts, the others only in small quantities. Very many compounds of sulphur, chiefly with metals, are found as minerals; certain compounds of selenion and tellurium with metals (copper, bismuth, &c.) also occur, but only in small quantities.

172   *Preparation.* Sulphur is obtained by purifying native sulphur, or by roasting sulphide of iron or copper out of contact with air. Selenion and tellurium are prepared by tedious and indirect methods which cannot advantageously be considered here.

173   *Chemical properties.* The three elements combine directly with hydrogen, oxygen, the halogens, and many other elements. The hydrides are feebly acidic; the oxides, as a class, are acidic; many oxyhaloid compounds are known; the elements do not interact with acids to form salts. The compositions of the more important compounds are expressed by the following formulae, where $M = S$, Se, or Te :—MH_2, MO_2, MO_3, MO_3H_2, MO_4H_2, M_2X_2, MX_4 ($X = Cl$, Br, I), MOX_2, MO_2X_2, $M_2O_yX_z$.

Sulphur and selenion exhibit differences in physical, and to some extent also in chemical, properties, according to the conditions under which they are prepared.

Ordinary native sulphur crystallises in rhombic octahedral forms. If a quantity of sulphur is melted in a crucible, and allowed to cool slowly until a crust forms on the surface, and if holes are then pierced in this crust and the still molten sulphur is poured out, it is found that the sulphur remaining in the crucible has crystallised in monoclinic prisms. These however soon change to octahedral forms. If sulphur is melted, heated to 400° or so, and then suddenly cooled, by pouring into cold water, a semi-pasty, soft, plastic, solid, having properties somewhat like caoutchouc, is formed: this soft solid is sulphur. Plastic sulphur soon becomes brittle and crystalline. If an acid is added to an aqueous solution of potassium tetra- or penta-sulphide (K_2S_4 or K_2S_5)—obtained by fusing solid potash with sulphur—sulphur is precipitated in the form of a white amorphous solid. If an acid is added to a warm aqueous solution of sodium thiosulphate ($Na_2S_2O_3$) a yellow solid is produced which is also amorphous sulphur. These five sulphurs differ in specific gravity, solubility in carbon disulphide, &c.; the more prominent differences are presented in the following table.

Sulphur.

Soluble in carbon disulphide (CS_2).

1. *Octahedra.*	Sp. grav. 2·05, melting pt. 114°·5.
2. *Rhombic prisms.*	„ 1·98 „ 120°.
3. *Amorphous white.*		

Insoluble in carbon disulphide.

1. *Amorphous yellow.*
2. *Plastic.* Sp. grav. 1·95.

The insoluble varieties of sulphur have not been obtained wholly free from the soluble. If ordinary sulphur, which is wholly soluble in carbon disulphide, is heated to 130° or 140° and quickly cooled no insoluble sulphur is produced; at ·170° a considerable quantity of the insoluble variety is formed, but not much more at 230°. According to Berthelot, the soluble varieties of sulphur are electro-negative to the insoluble varieties.

Selenion may be obtained by passing sulphur dioxide (SO_2) into an aqueous solution of selenious acid (H_2SeO_3). If the reddish black solid thus produced is melted and allowed to cool, a black, amorphous, lustrous, solid, with a fracture like glass, is obtained: if this is heated to 95°–100° the temperature suddenly rises to 200° or 230°, and a reddish-grey, metal-like, crystalline, somewhat malleable, solid is obtained. The crystalline variety of selenion is about 4·8 times, and the amorphous glassy variety about 4·3 times, heavier than water. Other varieties of selenion appear to exist. Amorphous selenion is a non-conductor of electricity: crystalline selenion conducts electricity; the conductivity increases as the temperature rises, and also increases very markedly when the selenion is exposed to sunlight. The crystalline variety is usually called metallic selenion.

Tellurium does not exhibit any changes of properties analogous to the change of one variety of sulphur, or selenion, into the other varieties. Tellurium is a very bad conductor of electricity, but the conductivity is slightly increased by raising the temperature considerably and also by exposure to sunlight.

The specific gravities of sulphur and selenion as gases vary with the temperature; the specific gravity of tellurium gas is constant. Thus;—

Specific gravities of gases; (air = 1).

Temp.	450°–500°	700°	850°	1100°	1160°	1420°	1390°–1440°
Sulphur.	6·6–6·9	2·8	2·4	2·1	2·1		
Selenion.			7·67	6·3		5·6	
Tellurium.							9·0

The proof that each of the five varieties of sulphur, and each of the two varieties of selenion, is composed of only sulphur, or only selenion, consists in the facts—(1) that a given mass of one variety is changeable into exactly the same mass of another variety; (2) that equal masses of the different varieties when oxidised produce each exactly the same mass of the same oxide. Thus when 1 gram of any variety of sulphur is wholly burnt in oxygen, 2 grams of sulphur dioxide (SO_2) are produced. The change of a specified mass of any variety of sulphur other than the octahe- dral into the same mass of the octahedral variety is accompanied by the production of heat; the change of a specified mass of amorphous selenion into the same mass of crystalline selenion is accompanied by the production of heat; therefore there is less energy in a specified mass of octahedral sulphur, or crystalline selenion, than in the same mass of any other variety of sulphur, or selenion.*

Sulphur and *selenion* are said to exhibit *allotropy;* the change from one variety of sulphur, or selenion, to another is called an *allotropic change.* *Tellurium* does *not* exhibit *allotropy.*

We shall consider allotropic changes more fully hereafter; meanwhile the student should compare what has been said regarding the existence of more than one variety of the same element with what he learned about the composition of elements in chap. II. par. 32.

174 **Compounds of sulphur, selenion, and tellurium, with hydrogen. MH₂.** These compounds are gases under ordinary conditions; they have all an extremely powerful and disagreeable smell, and are very poisonous. They are obtained (1) by passing hydrogen over the molten elements, (2) by the interaction of acids in aqueous solution on various metallic sulphides, selenides, or tellurides; thus

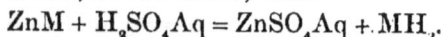

$$ZnM + H_2SO_4Aq = ZnSO_4Aq + MH_2.$$

The hydrides MH₂ dissolve in water; the solutions of the sulphur and selenion compounds redden litmus and react with alkalis as acids do; thus—

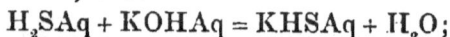

$$H_2SAq + KOHAq = KHSAq + H_2O;$$
$$KHSAq + KOHAq = K_2SAq + H_2O.$$

An aqueous solution of tellurium hydride does not exhibit acidic functions.

The aqueous solutions absorb oxygen from the air and are changed to water and the element M; in the case of H₂S,

* *s.* further Chap. XIV.

SO_2Aq is also formed, and this absorbs more ____ gen and becomes SO_3Aq. These hydrides are all decom____ by heat, thus $MH_2 = M + H_2$; the stability towards he____ reases as the combining weight of M increases.

The composition of these hydrides is determ____ by heating a measured volume of each with a weighed q____ tity of zinc, iron, or copper: zinc &c. sulphide, selenide, o____ elluride, and hydrogen, are thus produced; the volume of ____ rogen is the same as the volume of the hydride before t____ change; the increase in the weight of zinc, iron, or copper,____ epresents the mass of sulphur, selenion, or tellurium, formerly combined with hydrogen. The specific gravity of each gaseous hydride is determined. From the data thus obtained the composition of the hydrides is calculated. To take an example: 500 c.c. of hydrogen sulphide, measured at 0° and 760 mm., were heated in a closed glass tube with 3 grams of powdered zinc; when the change was completed the gaseous contents were cooled to 0° and measured, 500 c.c. of hydrogen were obtained; the mixture of zinc and zinc sulphide in the tube weighed 3·7149 grams. The specific gravity of hydrogen sulphide was found to be 17 (hydrogen = 1); as 1000 c.c. of hydrogen measured at 0° and 760 mm. weigh ·08936 gram, it follows that the 500 c.c. of hydrogen sulphide used weighed

$$\frac{17 \times \cdot 08936}{2} = \cdot 75956 \text{ gram.}$$

The experimentally determined data are then as follows :—

500 c.c. hydrogen sulphide, weighing ·75956 gram, are composed of 500 c.c. of hydrogen, weighing ·04468 gram, and ·7149 gram of sulphur.

But the combining weight of sulphur is 32. What mass of hydrogen is combined with one combining weight of sulphur? ·7149 : ·04468 = 32 : 2. Therefore the simplest formula which expresses the composition of hydrogen sulphide is H_2S. But does this formula express the composition of one *reacting weight* of hydrogen sulphide? If the volume occupied by 1 gram of hydrogen is called one volume, then 34 grams of hydrogen sulphide (i.e. the mass, taken in grams, represented by the formula H_2S) occupy 2 volumes. But we have already learned that the reacting weights of gaseous compounds are those masses of them which occupy two volumes (s. par. 88.); hence the formula H_2S expresses the composition of a reacting weight of hydrogen sulphide. Further study of the reactions of this compound confirms this conclusion; all the interactions of

hydrogen sulphide with other compounds and with elements
can be simply expressed as occurring between 34, or a whole
multiple of 34, parts by weight of hydrogen sulphide.

175   **Compounds with oxygen. MO₂ and MO₃.** When sulphur,
selenion, or tellurium, is burnt in oxygen, a dioxide (MO_2)
is produced. When a mixture of sulphur dioxide gas and
oxygen is passed over hot very finely divided platinum sulphur
trioxide (SO_3) is formed. When solid telluric acid (H_2TeO_4) is
heated tellurium trioxide (TeO_3) and water are formed. Selenion trioxide has not yet been isolated. A better yield of
tellurium dioxide is obtained by heating tellurium with concentrated nitric acid. When concentrated nitric acid is heated
a portion of it is changed to water, oxygen, and nitric oxide;
thus $2HNO_3$ (heated) $= H_2O + 3O + 2NO$; if some fairly easily
oxidised substance is present in contact with the changing
nitric acid, that substance combines with the whole or a part
of the oxygen, and is thereby oxidised. In the present instance the oxidisable body is tellurium; the tellurium is
oxidised to tellurium dioxide, and simultaneously the nitric
acid is deoxidised or *reduced* to water and nitric oxide.

Compounds and elements which do not directly combine
with oxygen may frequently be oxidised by bringing them
into contact with another substance which under the conditions
of the experiment is evolving oxygen, or into contact with
other substances which are interacting and producing oxygen.
We have already had examples of such processes of oxidation
(*s.* chap. VIII. par. 121). Another example is given by one of
the processes for forming selenic acid (H_2SeO_4). Selenion is
suspended in water and chlorine is passed into the water; the
chlorine and water interact to produce hydrochloric acid and
oxygen, and the oxygen combines with the selenion to produce
selenion dioxide, which again interacts with a portion of the
water to produce selenious acid; these changes occur simultaneously; the initial and final compositions of the whole
interacting system may be expressed by the equation

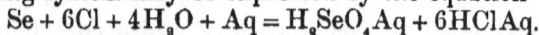

$$Se + 6Cl + 4H_2O + Aq = H_2SeO_4Aq + 6HClAq.$$

Compounds other than those of oxygen are frequently produced by arranging a chemical change so that the constituents
of the specified compound are produced in contact with each
other; thus hydrogen does not interact with an aqueous
solution of sodium sulphite (Na_2SO_3), but if this solution is
brought into contact with zinc and dilute sulphuric acid a
portion of the sodium sulphite is deoxidised, or *reduced*, to
hydrogen sulphide and water. We know that zinc and dilute

sulphuric acid interact to produce zinc sulphate ($ZnSO_4$) and hydrogen. The initial and final compositions of the whole changing system may be approximately represented thus :—

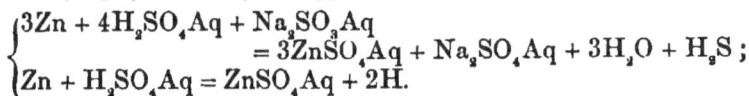

$$\begin{cases} 3Zn + 4H_2SO_4Aq + Na_2SO_3Aq \\ \qquad = 3ZnSO_4Aq + Na_2SO_4Aq + 3H_2O + H_2S \; ; \\ Zn + H_2SO_4Aq = ZnSO_4Aq + 2H. \end{cases}$$

A portion of the hydrogen evolved by the interaction of the zinc and sulphuric acid is employed in reducing the sodium sulphite, and the rest of the hydrogen is evolved as gas.

Sulphur dioxide is commonly produced by partially de-oxidising (or reducing) sulphuric acid (H_2SO_4): for this purpose a concentrated aqueous solution of the acid is heated with copper, or carbon, or sulphur ; the chemical changes which occur are these,—

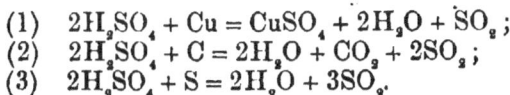

(1) $2H_2SO_4 + Cu = CuSO_4 + 2H_2O + SO_2 \; ;$
(2) $2H_2SO_4 + C = 2H_2O + CO_2 + 2SO_2 \; ;$
(3) $2H_2SO_4 + S = 2H_2O + 3SO_2.$

Sulphur dioxide (SO_2) is gaseous at ordinary temperatures and pressures, but is easily liquefied ; the other oxides of the group are solids. The oxides MO_2 and MO_3 are all acidic ; SO_2 and SO_3 dissolve in water with production of heat and formation of solutions of the acids H_2SO_3 and H_2SO_4, respectively ; SeO_2 also dissolves in water to form a solution of the acid H_2SeO_3. TeO_2 and TeO_3 are nearly insoluble in water ; they do not interact with water to produce acids ; the acids H_2TeO_3 and H_2TeO_4 however exist as definite solids, which when heated yield TeO_2 and TeO_3, (and water) respectively. The oxides MO_2 and MO_3 all interact with solutions of alkalis, or, when heated, with solid alkalis, to produce salts ; thus,—

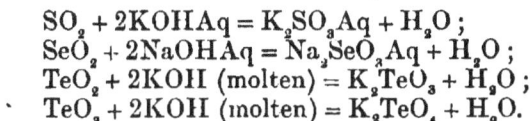

$$SO_2 + 2KOHAq = K_2SO_3Aq + H_2O \; ;$$
$$SeO_2 + 2NaOHAq = Na_2SeO_3Aq + H_2O \; ;$$
$$TeO_2 + 2KOH \text{ (molten)} = K_2TeO_3 + H_2O \; ;$$
$$TeO_3 + 2KOH \text{ (molten)} = K_2TeO_4 + H_2O.$$

Some of the oxides MO_2 and MO_3 combine with acidic oxides to form compounds, but these compounds are usually unstable, and are more or less easily decomposed by heat ; thus compounds of SO_3 with P_2O_5, As_2O_3, and B_2O_3, are known ; TeO_2 also combines with hydrogen chloride to form $TeO_2.2HCl$ and $TeO_2.3HCl$.

Compounds with the halogens. M_2X_2, MX_2, and MX_4. 176
The following compounds have been prepared by the direct union of M with halogen :—

Sulphur.

S_2Cl_2.   Liquid; can be distilled.

S_2Br_2.   Liquid; decomposes when heated.

S_2I_2.   Solid.

SCl_2.   Liquid; very unstable.

SCl_4.   Probably exists at temperatures below $-22°$; is extremely unstable.

Selenion.   Tellurium.

$\left.\begin{array}{l}Se_2Cl_2\\Se_2Br_2\end{array}\right\}$ liquids.   All definite compounds, with definite melting and boiling points; but most of them are separated by heat into selenion and halogen.

$\left.\begin{array}{l}Se_2I_2\\SeCl_4\\SeBr_4\\SeI_4\end{array}\right\}$ solids.

$\left.\begin{array}{l}TeCl_2 \text{ and } TeCl_4\\TeBr_2 \text{ and } TeBr_4\\TeI_2 \text{ and } TeI_4\end{array}\right.$ All solids with definite melting points; generally unchanged by action of heat. $TeCl_4$ has been gasified.

177   Compounds with oxygen and the halogens; oxyhaloid compounds. MOX_2 and MO_2X_2 ($X = Cl, Br, I$). Several of these compounds have been prepared : as examples we shall take $SOCl_2$ and SO_2Cl_2. The former is obtained by reacting on sodium sulphite with phosphoric chloride, the latter by a similar reaction between sulphuric acid and phosphoric chloride. The compositions of the systems before and after the changes may be approximately represented in equations as follows ;—

(1) $Na_2SO_3 + 2PCl_5 = SOCl_2 + 2POCl_3 + 2NaCl$;

(2) $H_2SO_4 + 2PCl_5 = SO_2Cl_2 + 2POCl_3 + 2HCl$.

The compounds $SOCl_2$ and SO_2Cl_2 interact with water to produce hydrochloric acid, and sulphurous or sulphuric acid ; thus—

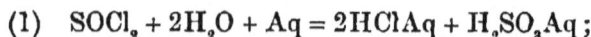

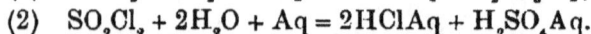

(1) $SOCl_2 + 2H_2O + Aq = 2HClAq + H_2SO_3Aq$;

(2) $SO_2Cl_2 + 2H_2O + Aq = 2HClAq + H_2SO_4Aq$.

Certain relations between the oxychloride $SOCl_2$ and the acid H_2SO_3, and between the oxychloride SO_2Cl_2 and the acid H_2SO_4, are established by the foregoing reactions : these relations are perhaps better suggested if the reactions are stated as follows ;—

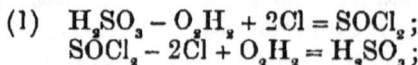

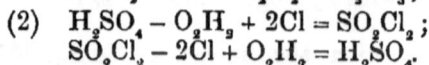

(1) $H_2SO_3 - O_2H_2 + 2Cl = SOCl_2$;
         $SOCl_2 - 2Cl + O_2H_2 = H_2SO_3$;

(2) $H_2SO_4 - O_2H_2 + 2Cl = SO_2Cl_2$;
         $SO_2Cl_2 - 2Cl + O_2H_2 = H_2SO_4$.

($SOCl_2$ is not directly obtained from H_2SO_3, because this acid only exists in aqueous solution.)

If the formulae of sulphurous and sulphuric acids are written as $SO.O_2H_2$ [or $SO(OH)_2$], and $SO_2.O_2H_2$ [or $SO_2(OH)_2$], respectively, the reactions which occur between these acids and phosphoric chloride are suggested by the formulae. (s. chap. XVII.)

Compounds with oxygen and hydrogen; Acids. Several of **178** these acids are known; the most important are those whose compositions are expressed by the general formulae H_2MO_3 and H_2MO_4. The acids H_2MO_3 where $M = S$ or Se are produced by dissolving the oxides MO_2 in water; H_2TeO_3 is formed by oxidising tellurium by nitric acid in presence of water (s. par. 175). The acid H_2SO_4 is produced by dissolving the oxide SO_3 in water; H_2SeO_4 is formed by oxidising selenion in presence of water by the interaction between water and chlorine (s. par. 175); H_2TeO_4 is formed by decomposing the barium salt of this acid by the proper quantity of sulphuric acid, and removing water by evaporation; thus

$$BaTeO_4Aq + H_2SO_4Aq = BaSO_4 + H_2TeO_4Aq.$$

Of the acids H_2MO_3 and H_2MO_4, H_2SO_3 is known only in aqueous solutions; when such a solution is evaporated water and sulphur dioxide (SO_2) are produced; the other acids have been isolated. H_2SeO_3 and H_2SO_4 are thick oily liquids at ordinary temperatures; H_2TeO_3 and H_2TeO_4 are solids. These acids are all decomposed by heat into water and the corresponding oxides; thus—

$$H_2MO_3 \text{ heated gives } H_2O + MO_2;$$
$$H_2MO_4 \quad ,, \quad ,, \quad H_2O + MO_3.$$

This change occurs at a higher temperature when $M = Te$ than when $M = Se$, and at a higher temperature when $M = Se$ than when $M = S$; in other words, the stability of the acids towards heat increases as the combining weight of M increases.

There is an oxide *corresponding to* each of these acids: this oxide, except in the cases of TeO_2 and TeO_3, interacts with water to produce the acid; it is also produced, along with water, when the acid is heated.

The nomenclature of these *oxides* and their *corresponding acids* is exhibited in the following table.

Oxide.	Corresponding acid.
MO_2. Sulphur, selenion, or tellurium, dioxide; or sulphurous, selenious, or tellurous, anhydride.	H_2MO_3. Sulphurous, selenious, or tellurous, acid.
MO_3. Sulphur, or tellurium, trioxide; or sulphuric, or telluric, anhydride	H_2MO_4. Sulphuric, selenic, or telluric, acid.

179　Compounds with electro-positive or metallic elements. The three elements combine with many metals to form sulphides, selenides, and tellurides, of similar compositions. The following are a few examples : CuM, ZnM, FeM, Bi_2M_3, &c. Most of these compounds interact with aqueous solutions of acids to give salts of the metal and a hydride of M ; thus

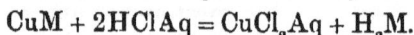

$$CuM + 2HClAq = CuCl_2Aq + H_2M.$$

180　The elements sulphur, selenion, and tellurium, are evidently similar in their properties. They all combine directly with hydrogen to form hydrides MH_2, two of which are slightly acidic. They combine with oxygen to form oxides MO_2 and MO_3, which are more or less markedly acidic. They form several haloid compounds which are not acidic; the stability of these towards heat increases as the combining weight of M increases. They form oxyhaloid compounds, each of which interacts with water to form an acid of M, and a haloid acid HX (where X = Cl, Br, or I). They do not interact with acids to form salts.

181　Some of the prominent properties of the elements of the sulphur group are compared with the more prominent properties of the halogens, and of the alkali metals, in the following table. The sulphur group of elements is evidently much more allied to the halogens than to the alkali metals.

Sulphur group. *Sulphur, Selenion, Tellurium.*	Halogens. *Chlorine, Bromine, Iodine.*	Alkali metals. *Lithium, Sodium, Potassium, Rubidium, Cæsium.*
All solids; more or less brittle; fairly high melting points.	One gaseous; one liquid; one solid.	Solids; soft, light, easily melted.
Electro - negative to many other elements.	Electro - negative to most other elements.	Electro-positive to all other elements.
Form gaseous compounds with hydrogen, MH_2; soluble in water giving feebly acid, or, in case of TeH_2, neutral solutions.	Form gaseous compounds with H, MH; very soluble in water forming markedly acid solutions.	Do not combine with hydrogen.
Form oxides by direct union with oxygen; most of these dissolve in water forming fairly stable acids.	Oxides formed indirectly; very soluble in water forming unstable acids.	Oxides formed directly, at ordinary temperatures; very soluble in water giving alkaline solutions.
Unite with many positive, or metallic, elements; compounds with negative elements,	Combine with all positive, or metallic, elements; most compounds with negative	Combine with most negative or nonmetallic, but not with metallic, elements;

Sulphur group.	Halogens.	Alkali metals.
Sulphur, Selenion, Tellurium.	Colorine, Bromine, Iodine.	Lithium, Sodium, Potassium, Rubidium, Cæsium.
as a class, not very stable as regards action of heat.	elements are unstable as regards action of heat.	compounds stable towards heat.
Do not interact with acids to produce salts.	Do not interact with acids to produce salts.	Readily interact with acids to produce salts.
Scarcely, if at all, interact with water or steam.	Interact with steam to produce acids (HX) and evolve oxygen.	Interact rapidly with cold water to produce alkalis (MOH) and evolve hydrogen.

As regards the variations of properties in each of these **182** three groups we have found that as the combining weights increase the elements become more positive. In the sulphur and halogen group the elements become heavier and their melting points increase; the oxides, oxyacids, and compounds with chlorine, become more solid, and more stable towards heat; the hydrides become less stable towards heat.

In the halogen group the interaction with water takes place more completely and at a lower temperature, the smaller is the combining weight of the element; in the alkali-metals group the reverse of this holds good. In the sulphur group, tellurium is more metal-like in appearance and general physical properties than sulphur or selenion; its oxides do not directly interact with water to form acids; its hydride is easily decomposed by heat, and its aqueous solution is not acidic; it does not exhibit allotropy: in a word, tellurium is more metallic than either sulphur or selenion.

The terms *reduction* and *oxidation*, also *reducers* or *reducing* **183** *agents* and *oxidisers* or *oxidising agents*, have been used in describing some of the chemical changes exhibited by the elements and compounds considered in preceding paragraphs. When chlorine is passed over mercuric oxide the chlorine is oxidised and the mercuric oxide is simultaneously reduced. · (*s.* par. 154.) When nitric acid is heated with tellurium the element is oxidised and the acid is simultaneously reduced. (*s.* par. 175.) When chlorine is passed into concentrated caustic potash solution potassium hypochlorite (KClO) is formed, when this is heated oxygen is evolved and potassium chloride (KCl) remains; when chlorine is passed into warm potash solution holding bismuthous oxide (Bi_2O_3) in suspension· bismuthic oxide (Bi_2O_5) and potassium chloride are formed; the bismuthous oxide is oxidised and the potassium hypochlorite (formed by the interaction of the chlorine and potash) is simultaneously reduced. (s. par. 158.) When hydrogen is produced, by the interaction of zinc and dilute sulphuric acid, in contact with sodium sulphite (Na_2SO_3) in solution, the hydrogen is oxidised to water, and simultaneously the sodium sulphite is reduced to sodium oxide (Na_2O) which reacts with the sulphuric acid present to form sodium sulphate (Na_2SO_4), and hydrogen sulphide (H_2S).

In these, and in very many other, cases, the processes of oxidation and reduction occur together as parts of a chemical change.

In the experiment with mercuric oxide and chlorine, the chlorine acted as the reducer or reducing agent, and the mercuric oxide as the oxidiser or oxidising agent. In the experiment with tellurium and nitric acid, the acid acted as the oxidiser and the tellurium as the reducer. When bismuthous oxide was oxidised by potassium hypochlorite, the latter was the oxidiser, the former the reducer: but the production of the hypochlorite was due to the interaction of chlorine with potassium hydroxide

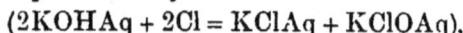

$$(2KOHAq + 2Cl = KClAq + KClOAq),$$

therefore it might be said that the chlorine was the primary oxidising agent. Similarly, if selenion is suspended in water and chlorine is passed into the liquid, selenic and hydrochloric acids are produced; thus,

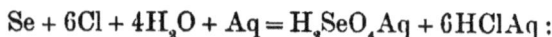

$$Se + 6Cl + 4H_2O + Aq = H_2SeO_4Aq + 6HClAq :$$

the selenion is oxidised by the oxygen which before the change began was combined with hydrogen; therefore the water is the oxidiser, and the selenion is the reducing agent; but the interaction of chlorine is required to decompose the water, therefore the primary oxidising agent is chlorine.

184　　Many elements and compounds may be classified in accordance with their actions as oxidisers or reducers; or in accordance with the conditions under which they are oxidised or reduced. Hydrogen, carbon, sodium, carbon monoxide (CO), sulphur dioxide (SO_2), nitrous acid (HNO_2Aq), stannous chloride ($SnCl_2$), aldehyde (C_2H_4O), are some of the more commonly used reducing agents. Oxygen, ozone, chlorine, nitric acid, potassium chlorate ($KClO_3$), potassium permanganate ($K_2Mn_2O_8$), are among the commonly used oxidising agents. The following equations present examples of the use of these reducers and oxidisers.

Action of reducing agents.

Original compound.	Reducer.	Oxidised product.	Reduced product.	Conditions.
Fe_2O_3	$+ 6H$	$= 3H_2O$	$+ 2Fe$	Passing hydrogen over heated ferric oxide.
ZnO	$+ CO$	$= CO_2$	$+ Zn$	Passing carbon monoxide over heated zinc oxide.
CuO	$+ C$	$= CO$	$+ Cu$	Heating carbon with copper oxide.
$BeCl_2$	$+ 2Na$	$= 2NaCl$	$+ Be$	Sodium in contact with fused beryllium chloride.
Fe_23SO_4Aq	$+ SO_2$	$= 2SO_4Aq$	$+ 2FeSO_4Aq$	Sulphur dioxide into warm solution of ferric sulphate.
H_2O_2Aq	$+ HNO_2Aq$	$= HNO_3Aq$	$+ H_2O$	Hydrogen peroxide solution in contact with solution of nitrous acid.
$HgCl_2Aq$	$+ SnCl_2Aq$	$= SnCl_4Aq$	$+ Hg$	Solutions of mercuric chloride and stannous chloride mixed.
$H_2O + 2AgNO_3Aq + C_2H_4O$		$= 2HNO_3Aq + C_2H_4O_2 + 2Ag$		Aldehyde added to warm solution of silver nitrate.

Action of oxidising agents

Original element or compound.	Oxidiser.	Reduced product.	Oxidised product.	Conditions.
H_2	$+ O$	$=$	H_2O	Electric sparks through a mixture of hydrogen and oxygen.
Hg	$+ O_3$	$= 2O$	$+ HgO$	Ozone passed into mercury.
$SbCl_3$	$+ 2Cl$	$=$	$SbCl_5$	Antimonious chloride heated in a stream of chlorine.
PbO	$+ 2KOH + 2Cl$	$= 2KCl$	$+ H_2O + PbO_2$	Lead monoxide suspended in concentrated warm potash solution and chlorine passed in.
Sn	$+ 2HNO_3$	$= N_2O_3$	$+ H_2O + SnO_2$	Tin heated with concentrated nitric acid.
$3C$	$+ 2KClO_3$	$= 2KCl$	$+ 3CO_2$	Carbon dropped into molten potassium chlorate.
$3H_2C_2O_4Aq + K_2Mn_2O_8Aq = K_2OAq + 2MnO_2 + 3H_2O + 6CO_2$				Potassium permanganate solution dropped into a warm solution of oxalic acid.

In all these instances of the action of reducing agents, the reduction of one substance is accompanied by the oxidation of another; in most of the instances of the action of oxidising agents, the oxidation of one substance is accompanied by the reduction of another.

185   In some cases,—e.g. reduction of beryllium chloride by sodium, reduction of mercuric chloride by stannous chloride, oxidation of antimonious chloride by chlorine—none of the substances taking part in the reactions is a compound of oxygen. It is customary to apply the term reduction to all chemical changes wherein the negative or non-metallic part of a compound is either wholly or partially removed, and the term oxidation to all chemical changes wherein the negative part of a compound is increased, or a negative element (or elements) is added to a more positive element (or elements).

186   Those elements which are easily oxidised might be placed in one class, and those which are oxidised with difficulty in another class; compounds which are easily and completely reduced might be classed apart from those which are only reduced at high temperatures and by indirect methods. Such a system of classification would be based, primarily at any rate, on the occurrence or non-occurrence under specified conditions of a certain chemical change; it would be based on a certain *power of doing*, rather than on the compositions, of the bodies classified. To make this classification fairly satisfactory it would be necessary to examine the compositions of the members of each class, and then to connect these compositions with the performance or non-performance of that chemical reaction which had been made the mark of each class.

187   We have had examples of classification founded on reactions rather than on composition. Oxides were divided into basic and acidic; a great many compounds were placed in one class, and called acids, because they all interacted with metals and with basic oxides to produce salts.

Can we connect the composition of those oxides which are called basic, and the composition of those which are called acidic, with the facts that the former interact with acids to produce salts, and the latter interact with water to produce acids? An answer to these questions will carry with it an answer to this; can we state in general terms the connexion between the composition of acids and the properties connoted by the term acid?

Let us begin the inquiry by learning a little more about **188**
the interactions of acids with metals, basic oxides, and alkalis,
to produce salts. Let aqueous solutions of the three acids,
hydrochloric (HCl), sulphuric (H_2SO_4), and phosphoric
(H_3PO_4), be prepared, each containing a known mass of
the acid in a specified volume; let an aqueous solution of the
alkali potassium hydroxide (KOH) be prepared containing
a known mass of the compound in a specified volume. Let
definite quantities of each acid solution be added to definite
quantities of the alkali solution, and let all the products of
each reaction be collected, examined, and analysed. The
results may be represented as follows.

Reactions between aqueous solutions of hydrochloric acid (HCl) and potassium hydroxide (KOH).
(1 gram HCl used in each case.)

Grams KOH used.	*Grams salt formed; and composition of salt.*	*Grams water formed.*	*Grams KOH remaining unchanged.*
1·54	2·04 KCl	·5	none
1·54 × 2	2·04 KCl	·5	1·54
1·54 × 3	2·04 KCl	·5	1·54 × 2
1·54 × 4	2·04 KCl	·5	1·54 × 3

Reactions between aqueous solutions of sulphuric
acid (H_2SO_4) and potassium hydroxide. (1 gram H_2SO_4
used in each case.)

Grams KOH used.	*Grams salt formed; and composition of salt.*	*Grams water formed.*	*Grams KOH remaining unchanged.*
·57	1·39 $KHSO_4$	·18	none
·57 × 2	1·77 K_2SO_4	·18 × 2	none
·57 × 3	1·77 K_2SO_4	·18 × 2	·57
·57 × 4	1·77 K_2SO_4	·18 × 2	·57 × 2

Reactions between aqueous solutions of phosphoric
acid (H_3PO_4) and potassium hydroxide. (1 gram H_3PO_4
used in each case.)

Grams KOH used.	Grams salt formed; and composition of salt.	Grams water formed.	Grams KOH remaining unchanged.
·57	1·39 KH_2PO_4	·18	none
·57 × 2	1·77 K_2HPO_4	·18 × 2	none
·57 × 3	2·13 K_3PO_4	·18 × 3	none
·57 × 4	2·13 K_3PO_4	·18 × 3	·57
·57 × 5	2·13 K_3PO_4	·18 × 3	·57 × 2

We see then (1) that hydrochloric acid and potassium hydroxide interact to produce one salt (KCl), and that the potash over and above that which interacts to produce this salt remains unchanged; (2) that two salts ($KHSO_4$ and K_2SO_4) are produced by the interaction of sulphuric acid and potash, and that the production of one or other salt depends upon the relative masses of the alkali and acid, but that if a greater mass of potash is added than is required for the production of the salt K_2SO_4 the excess of potash remains unchanged; (3) that three salts (KH_2PO_4, K_2HPO_4, and K_3PO_4) are produced by the interaction of phosphoric acid and potash, and that the production of one or other salt depends upon the relative masses of acid and alkali, but that any excess of potash beyond that which interacts to produce K_3PO_4 remains unchanged.

Similarly, if the metal potassium had been employed in place of its hydroxide we should have obtained one salt (KCl) in the case of hydrochloric acid, two salts ($KHSO_4$ and K_2SO_4) in the case of sulphuric acid, and three salts (KH_2PO_4, K_2HPO_4, and K_3PO_4) in the case of phosphoric acid.

189   If the interactions between potassium (or sodium) hydroxide and many acids are examined we find that the acids may be classified as follows :—

I.   *Monobasic acids;* {acids which interact with potash or soda to produce} only one salt.

II.   *Dibasic acids;*   „   „   „   two salts.
III. *Tribasic acids;*   „   „   „   three salts.
IV. *Tetrabasic acids;*   „   „   „   four salts.
V.   *Pentabasic acids;*   „   „   „   five salts.
VI. *Hexabasic acids;*   „   „   „   six salts.

An n-basic acid may also be defined as an acid from the reacting weight of which n combining weights of hydrogen can be displaced by sodium or potassium, when the acid interacts with sodium or potassium or the hydroxide of either of these metals.

We have thus arrived at a fairly clear notion of the meaning **190** of the term acid so far as the reactions of acids with metals, basic oxides, and alkalis, are concerned.

But what kind of compounds are acid? What are the compositions of these compounds? Have acids any common composition corresponding to their common property of forming salts under specified conditions?

The following formulae represent the compositions of several compounds which are acids—

HCl, HBr, HI, HF, H_2SO_3, H_2SO_4, HNO_2, HNO_3, $H_2C_2O_4$, $H_4C_2O_2$, $HCNO$, $H_3B_3O_4$, $HClO_3$, H_4IO_6, $HBrO_3$, HPO_3, H_3PO_4, $H_2Mn_2O_8$, HVO_3, $HSbO_3$, H_3SbO_4, H_2MoO_4, H_2WO_4, H_2SnO_3, $H_2S_2O_7$, H_2SeO_4, H_2TeO_3, $H_4Ta_4O_7$.

These acids are all compounds of hydrogen, and most of them are compounds of hydrogen with oxygen and another element.

Are all compounds of hydrogen with oxygen and another element acids?

We know that aqueous solutions of the following compounds are alkalis, i.e. exhibit properties strongly opposed to those of acids;—$LiOH$, $NaOH$, KOH, $RbOH$, $CsOH$. The following compounds are also alkaline;—CaO_2H_2, BaO_2H_2, SrO_2H_2, MgO_2H_2. Therefore all compounds of hydrogen with oxygen and another element are not acids. What then are the general characters of those elements the compounds of which with hydrogen and oxygen are acids? Most of the acids in the foregoing list are compounds of hydrogen and oxygen with non-metallic elements (S, N, C, B, Cl, I, Br, P, Se, Te); hence it would appear probable that the union of a non-metallic, or negative, element with hydrogen and oxygen would produce an acid. Further investigation gives a general confirmation to this conclusion.

By far the greater number of the compounds of the non- **191** metallic elements with oxygen and hydrogen are acids. Some of the acids in the foregoing list are compounds of hydrogen and oxygen with elements which are usually classed as · metals;—$H_2Mn_2O_8$, H_2MoO_4, H_2WO_4, $H_4Ta_4O_7$, H_2SnO_3, HVO_3, $HSbO_3$, H_3SbO_4. The elements tungsten, molybdenum, tantalum, vanadium, and antimony exhibit many of the physical, and some of the chemical, properties of metals; but they each form at least one acidic oxide; they do not form well marked and stable salts by interacting with acids ; they are electro-positive to all the distinctly non-metallic elements (O, S, N, P, Cl, Br, I, F, B, C, Si, S, Se), but they are electro-negative to most of the distinctly metallic elements. The elements manganese and tin are undoubtedly metals, both physically and chemically; they interact with acids to form salts; they form basic oxides; they do not combine with hydrogen; they react with steam at high temperatures to produce oxides and evolve hydrogen ; they are heavy, lustrous, malleable, solids which conduct heat and electricity fairly well.

The compounds (1) $H_2Mn_2O_8$ and (2) H_2SnO_3 are however acids; the ratios of the numbers of combining weights of hydrogen, metal, and oxygen in a reacting weight of each of these compounds is H : M : O in (1) $= 1 : 1 : 4$, and in (2) $= 1 : \frac{1}{2} : 1\frac{1}{2}$. The compound in a reacting weight of which there is relatively the greater quantity of oxygen ($H_2Mn_2O_8$) is decidedly an acid; the other compound in a reacting weight of which there is relatively less oxygen is an acid, but at the same time it interacts with concentrated sulphuric acid (and with some other acids) to produce a salt and water ;— thus $H_2SnO_3 + 2H_2SO_4$ (heated) $= Sn(SO_4)_2 + 3H_2O$.

Certain compounds of metals with hydrogen and a relatively large quantity of oxygen are then acids.

But in the list of acids given a little further back there appeared compounds of hydrogen with certain non-metallic elements other than oxygen ; viz. HCl, HBr, HI, HF. We have already learned something of the elements in question : we know that they are typical non-metals; that they are strongly electro-negative.

192   It appears then that acids are compounds of hydrogen, generally with markedly electro-negative or non-metallic elements of which oxygen is usually one ; but in some cases with oxygen and another element which, physically and chemically, is neither strongly metallic nor strongly negative ; and in a few cases with oxygen and another element which must be placed in the positive or metallic class. In the last mentioned cases there is usually a large quantity of oxygen combined with hydrogen and the metallic element.

193   We now see that acidic oxides are generally the oxides of the electro-negative or non-metallic elements.

The oxides corresponding to the acids HVO_3, $HSbO_3$, H_2MoO_4, H_2WO_4, $H_2Ta_4O_7$, are V_2O_5, Sb_2O_5, MoO_3, WO_3, and Ta_2O_5, respectively : these acidic oxides are the oxides of elements which are both metallic and non-metallic ; they are all composed of a large quantity of oxygen combined with a comparatively small quantity of the other element. The compositions of these oxides, as regards ratio of oxygen to other elements, may be compared with the compositions of the oxides Cl_2O, N_2O, P_2O_3, all of which are acidic, and all of which are oxides of strongly negative elements.

Oxides of the less positive metals formed by the union of much oxygen with comparatively small masses of the metals may then be acidic.

The oxides of the very positive elements (Li, Na, K, Rb, Cs, Ca, Sr, Ba, Mg, &c.) are all basic ; the oxides of the less positive, but still distinctly positive, elements (Fe, Zn, Ni, Cd, Hg, &c.) are generally basic ; the lower oxides—i.e. those with comparatively little oxygen—of the more negative of the metals (SnO, PbO, Pb_2O_3, MnO, Mn_2O_3, &c.) are usually basic, but the higher oxides of some of these elements (SnO_2, PbO_2, &c.) are feebly acidic.

Chromium is a heavy, hard, lustrous, grey, solid. It con- **194** ducts heat and electricity fairly. This element scarcely reacts with steam ; at a high temperature a very little oxide is produced. It interacts with acids to form salts. It forms stable haloid, and also several oxyhaloid, compounds. It does not combine with hydrogen. It is positive to most of the distinctly non-metallic elements, but negative to most of the distinctly metallic elements. From what we have learned concerning basic and acidic oxides, we might expect to find the lower oxides of chromium basic, and perhaps an oxide with compara- tively much oxygen, acidic.

The elements *iron* and *manganese* resemble *chromium* in many of their properties. It is advisable now briefly to con- sider this group of elements.

	Chromium.	Manganese.	Iron.	**195**
Combining weights	52·4	55	56	
Spec. gravities (water = 1)	about 6·5	about 7	about 7·5	
Melting points	above 2000⁰	about 1500⁰	about 1500⁰	

Appearance &c. All white, lustrous, solids: chromium and manganese very hard ; iron softer ; all malleable and fairly ductile ; all fair conduc- tors of heat and electricity.

Preparation. All obtained by reducing the oxides M_2O_3 with finely powdered carbon at a very high temperature.

✓*Chemical properties.* The elements decompose steam form- ing oxides and hydrogen ; iron at about 100⁰, manganese at a higher temperature, and chromium at a still higher temperature and then only very slowly. None of these elements combines directly or indirectly with hydrogen. They all interact with oxygen to produce several oxides. The compositions of the chief oxides are expressed by the formulæ MO, M_3O_4, M_2O_3; the oxides MnO_2, CrO_2, and CrO_3 are also known.

196   The oxides MO are basic; they react with acids to form salts; thus,

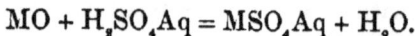

$$MO + H_2SO_4Aq = MSO_4Aq + H_2O.$$

The oxides M_3O_4, where $M = Mn$ or Fe, also react with acids to form salts; but the compositions of the salts do not correspond to that of the oxides. With dilute sulphuric acid Mn_3O_4 forms $MnSO_4$ and MnO_2; thus

$$Mn_3O_4 + 2H_2SO_4Aq = MnO_2 + 2MnSO_4Aq + 2H_2O.$$

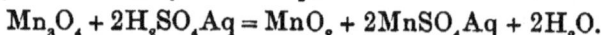

With hot concentrated sulphuric acid $MnSO_4$ is formed and oxygen is evolved; thus

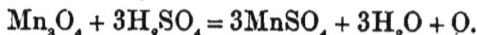

$$Mn_3O_4 + 3H_2SO_4 = 3MnSO_4 + 3H_2O + O.$$

The corresponding oxide of iron reacts with sulphuric acid to form a mixture of the two sulphates of iron; thus,

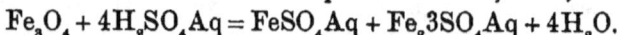

$$Fe_3O_4 + 4H_2SO_4Aq = FeSO_4Aq + Fe_23SO_4Aq + 4H_2O.$$

The existence of Cr_3O_4 is doubtful.

The oxides M_2O_3 are also basic; their interactions with sulphuric acid may be represented thus,

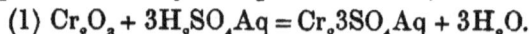

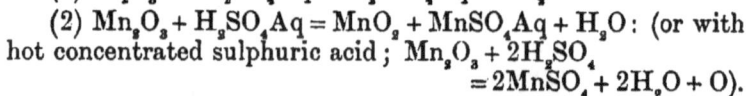

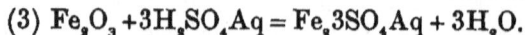

(1) $Cr_2O_3 + 3H_2SO_4Aq = Cr_23SO_4Aq + 3H_2O.$

(2) $Mn_2O_3 + H_2SO_4Aq = MnO_2 + MnSO_4Aq + H_2O$: (or with hot concentrated sulphuric acid; $Mn_2O_3 + 2H_2SO_4$
$$= 2MnSO_4 + 2H_2O + O).$$

(3) $Fe_2O_3 + 3H_2SO_4Aq = Fe_23SO_4Aq + 3H_2O.$

But the oxides M_2O_3 are also feebly acidic; salts of the form $RO.M_2O_3$ are known, where $RO = K_2O$, CaO, ZnO, &c.

Of the remaining oxides, CrO_2 has not been fully examined; it seems to react with acids to form the same salts as are obtained from Cr_2O_3, and at the same time to evolve oxygen.

MnO_2 with concentrated acids gives salts of the form MnX (where $X = SO_4$, $2NO_3$, CO_3, &c.), and at the same time oxygen is produced; thus,
$MnO_2 + H_2SO_4Aq = MnSO_4Aq + H_2O + O$ (the amount of water must be small).

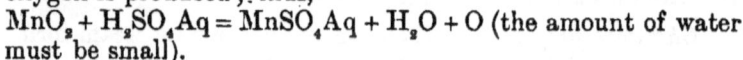

This oxide also reacts with alkalis to form salts the negative part of which is formed of the MnO_2 and oxygen ; thus when this oxide is fused with solid potash the reaction is

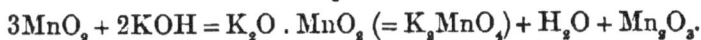

$$3MnO_2 + 2KOH = K_2O \cdot MnO_2 (= K_2MnO_4) + H_2O + Mn_2O_3.$$

The salt potassium manganate (K_2MnO_4) dissolves in water, and from the solution other salts are obtained ; thus,

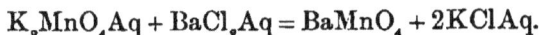

$$K_2MnO_4Aq + BaCl_2Aq = BaMnO_4 + 2KClAq.$$

These salts, the composition of which is expressed by the formula $MMnO_4$ where $M = Ba$, Ca, K_2, Na_2, &c., are called *manganates*. Their composition is similar to that of sulphates (MSO_4).

An aqueous solution of a manganate reacts with dilute acids to form a permanganate; thus,

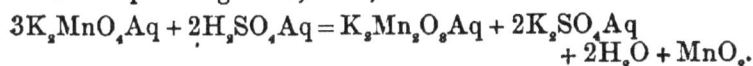

$$3K_2MnO_4Aq + 2H_2SO_4Aq = K_2Mn_2O_8Aq + 2K_2SO_4Aq$$
$$+ 2H_2O + MnO_2.$$

A series of *permanganates* is known,

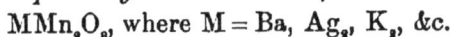

$$MMn_2O_8, \text{ where } M = Ba, Ag_2, K_2, \&c.$$

The oxide CrO_3 dissolves in water to form an acid liquid from which crystals of the acid H_2CrO_4 have been obtained. This acid is called *chromic acid*; it reacts with alkalis and basic oxides to give a series of *chromates* $MCrO_4$ $(M = K_2,$ Na_2, Ag_2, Ba, Ca, &c.). The composition of the chromates is similar to that of the manganates $(MMnO_4)$ and sulphates (MSO_4).

An aqueous solution of a chromate reacts with dilute acids to form a dichromate ; thus

$$2K_2CrO_4Aq + H_2SO_4Aq = K_2Cr_2O_7Aq + K_2SO_4Aq + H_2O.$$

The *dichromates*, MCr_2O_7, are not similar in composition to the permanganates (MMn_2O_8), but they agree in composition with a series of sulphur salts known as *di-* (or *pyro-*) *sulphates*, MS_2O_7.

The oxide CrO_3 also interacts with hot concentrated acids to form salts and oxygen ; thus,

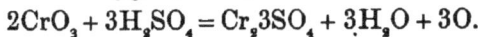

$$2CrO_3 + 3H_2SO_4 = Cr_23SO_4 + 3H_2O + 3O.$$

The manganates and permanganates react with acids to form manganese salts and oxygen; thus,

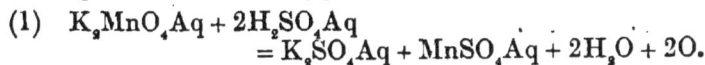

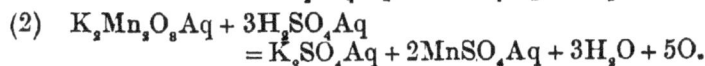

(1) $K_2MnO_4Aq + 2H_2SO_4Aq$
$$= K_2SO_4Aq + MnSO_4Aq + 2H_2O + 2O.$$

(2) $K_2Mn_2O_8Aq + 3H_2SO_4Aq$
$$= K_2SO_4Aq + 2MnSO_4Aq + 3H_2O + 5O.$$

Similarly, the chromates and dichromates react with concentrated solutions of acids to form chromium salts and oxygen; e.g.

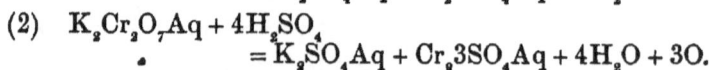

(1) $2K_2CrO_4Aq + 5H_2SO_4$
$$= 2K_2SO_4Aq + Cr_23SO_4Aq + 5H_2O + 3O.$$

(2) $K_2Cr_2O_7Aq + 4H_2SO_4$
$$= K_2SO_4Aq + Cr_23SO_4Aq + 4H_2O + 3O.$$

197   The salts of chromium, manganese, and iron,—*i.e.* compounds derived from acids by replacing hydrogen by chromium, manganese, or iron,—form two series the compositions of which are represented by the general formulæ MX and, M_23X, respectively, where $M = Cr$, Mn, or Fe, and $X = 2NO_3$, $2ClO_3$, SO_4, SO_3, CO_3, $\frac{2}{3}PO_4$, $\frac{2}{3}AsO_4$, &c.

The salts MX are called *chromous, manganous,* and *ferrous, salts*; those of the composition M_23X are called *chromic, manganic,* and *ferric salts*. A few examples of each class of salts are given :—

-ous salts.		*-ic salts.*	
Ferrous sulphate	$FeSO_4$	Ferric sulphate	Fe_23SO_4
Manganous sulphate	$MnSO_4$	Manganic sulphate	Mn_23SO_4
Chromous sulphate	$CrSO_4$	Chromic sulphate	Cr_23SO_4
Ferrous nitrate	$Fe2NO_3$	Ferric arsenate	Fe_22AsO_4
Manganous chlorate	$Mn2ClO_3$	Manganic phosphate	Mn_22PO_4
Chromous oxalate	CrC_2O_4	Chromic selenite	Cr_23SeO_3.

Many iron salts of both series are known; the ferrous salts are all fairly readily oxidised to ferric salts. Most of the known manganese salts belong to the manganous class; the manganic salts are all readily reduced to manganous salts. Very few chromous salts have been prepared; they are all easily oxidised to chromic salts.

198   Chromium and manganese resemble the halogen elements and the elements of the sulphur group in that each forms at least one acidic oxide. The resemblance between chromium and manganese and the sulphur group of elements is further shewn by the compositions of the salts obtained by the interactions of these acidic oxides with basic, or alkali-forming, oxides. Thus $(M = Ba, Pb, Ca, K_2, Na_2, Ag_2, \&c.)$;—

Chromates.		**Manganates.**		**Sulphates.**	
Oxide.	CrO_3	*Oxide.*	not isolated	*Oxide.*	SO_3
Acid.	H_2CrO_4	*Acid.*	not isolated	*Acid.*	H_2SO_4
Salts.	$\begin{cases} MCrO_4 \\ MCr_2O_7 \end{cases}$	*Salts.*	$MMnO_4$	*Salts.*	$\begin{cases} MSO_4 \\ MS_2O_7 \end{cases}$

The resemblance between manganese and the halogens is well shewn by comparing the compositions of permanganates and perchlorates. Thus ;—

Permanganates.		Perchlorates.	
Oxide.	not isolated	*Oxide.*	not isolated
Acid.	$H_2Mn_2O_8Aq$: known only in aqueous solution	*Acid.*	$H_2Cl_2O_8$
Salts.	MMn_2O_8	*Salts.*	MCl_2O_8

But chromium and manganese resemble the alkali-metals in that each forms at least one basic oxide.

If we tabulate the compositions, and indicate the pro- **199** perties, of several oxides which have now been examined, we shall find a distinct connexion between these compositions and properties.

The name *peroxide* is here used to indicate an oxide which reacts with acids to evolve oxygen, and at the same time to form salts which correspond in composition with an oxide with less oxygen than the specified peroxide. (*s.* reactions of Mn_3O_2 and MnO_2 with sulphuric acid ; par. 196.)

A. Basic Oxides.		B. Acidic Oxides.	
1.	2.	1.	2.
Alkali-forming.	*Do not form alkalis.*	*Form acids which have been isolated, or are known in aqueous solutions.*	*Do not form isolable acids; but interact with oxides to form salts.*
K_2O	$CrO . Cr_2O_3$..................CrO_3		
Na_2O	$FeO . Fe_2O_3 . Fe_3O_4$		
	MnO ....MnO_2		
	$Cl_2O . ClO_2 . I_2O_5$		

C.
Peroxides.

1.	2.	3.
Form alkalis, and oxygen, with water.	*Do not interact with water.*	*Form acids with water.*
K_2O_2		
K_2O_4		
Na_2O_3	$Mn_3O_4 . Mn_2O_3 . MnO_2 . CrO_3.$	

200   The elements potassium and sodium are very positive; they are soft, light, solids. They interact with cold water to form hydroxides and hydrogen. The elements chlorine and iodine are very negative : one is a gas, the other a lustrous, fairly heavy, solid. They interact with steam at high temperatures to form hydrides and oxygen. The elements chromium, manganese, and iron are neither very positive nor very negative; they are hard, heavy, malleable, solids. They interact slowly with steam at fairly high temperatures to produce oxides and hydrogen. The lower oxides of the three elements whose properties are intermediate between the very positive and the very negative groups are basic ; the highest oxide of one of these elements is distinctly acidic, and the oxide MnO_2 is also acidic although less distinctly so than CrO_3. But although CrO_3 and MnO_2 are acidic, yet they are not wholly acidic ; in their interactions with concentrated acids they exhibit basic properties; although neither forms a corresponding salt, yet both produce salts when they react with acids.

201   The oxide of a very positive element, then, appears to be always basic, even when it is composed of relatively much oxygen with relatively little of the positive element. The oxide of a very negative element appears to be always acidic, even when it is composed of relatively little oxygen with relatively much of the negative element. The oxide of an element which is neither very positive nor very negative appears to be only basic when it is composed of relatively little oxygen, but acidic, with basic tendencies, when it is composed of much oxygen combined with a relatively small quantity of the other element.

202   Of the members of the chromium group of elements, chromium forms the most markedly acidic oxide. Chromium has the smallest combining weight of the three elements. Manganese however also forms well marked manganates and permanganates. Considering that the differences between the combining weights of the three elements are very small, we might expect that *ferrates*, salts analogous in composition to chromates and manganates, would be produced if the proper conditions could be realised. Could we oxidise Fe_2O_3 in contact with a large quantity of a strong alkali, we might expect a higher oxide of iron to be formed and simultaneously to react with the alkali and produce a salt.

If ferric oxide (Fe_2O_3) is suspended in very strong warm potash solution and chlorine is passed into the liquid, a portion of the ferric oxide dissolves and a red coloured liquid is obtained. Similarly if ferric oxide is mixed with a large quantity of solid potash and a little potassium nitrate, the mixture is melted, kept molten for a time, cooled, and dissolved in water, a red coloured liquid is produced. These red liquids evolve oxygen, and precipitate ferric oxide, when heated. By · measuring the oxygen evolved and determining the ferric oxide precipitated, conclusions can be drawn as to the composition of the compound in the original red solution. Such measurements have been made; they are in keeping with the hypothesis that the red liquids contain potassium ferrate, K_2FeO_4. This salt appears to exist only in solution in much potash. The change which occurs when this solution is heated is probably as follows ;—

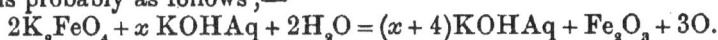

$$2K_2FeO_4 + x\,KOHAq + 2H_2O = (x + 4)KOHAq + Fe_2O_3 + 3O.$$

Although no acidic oxide of iron has been isolated, yet we may say that the formation of potassium ferrate shews that a compound of iron with much oxygen would be an acid-forming oxide.

Chromium and manganese must then be classed both with **203** the metallic and with the non-metallic elements. A consideration of their physical properties alone would lead us to place them in the class metals; a consideration of the chemical properties of their lower oxides would confirm this conclusion ; but a consideration of the chemical properties of their higher oxides shews that the elements in question are fairly closely related to the undoubtedly non-metallic elements sulphur and chlorine.

The properties of oxides are evidently conditioned by the **204** chemical characters of the elements with which oxygen is combined, and also by the ratio of the numbers of combining weights of oxygen and the other element which are united in a reacting weight of each oxide. The properties of hydroxides are also conditioned by the chemical characters, and by the relative masses, of the elements which are combined in a reacting weight of each hydroxide.

As we advance in our study of classes of elements and compounds we shall find that a similar statement holds good for each class of compounds.

Our examination of the properties and compositions of **205** classes of compounds has shewn that such terms as *basic oxide* or *acidic oxide* are relative. The oxide CrO_3 is acidic in its reactions with water and alkalis, but it is basic in its reactions with concentrated acids: the oxide MnO_2 is acidic in its reactions with alkalis, but not with water; it is basic in its reactions towards concentrated acids. It is only the oxides of the very positive elements which are basic, and the oxides of the very negative elements which are acidic, in all their reactions. But even in these cases, the terms basic and acidic imply that the interactions of the oxides with other compounds (acids and alkalis) have been examined.

Chemistry is not the study of elements and compounds alone, but it is the study of the interactions of elements and compounds.

206   The importance of chemical classification is so great that an examination of another group of elements will be made before we pass to other parts of our subject.

The elements *nitrogen, phosphorus, arsenic, antimony,* and *bismuth* are placed in the same class.

	Nitrogen.	Phos-phorus.	Arsenic.	Anti-mony.	Bismuth.
Combining weights.	14	31	75	120	208
Specific gravities (water=1).	·97 if air=1	1·9	5·7	6·7	9·9
Melting points.		45°	about 600° (data un-certain)	about 450°	about 270°
Appearance &c.	Colourless, odourless, gas: lique-fied at very low temp. and great pressure.	Soft, wax-like, solid: crystalline.	Hard, grey, crystalline, brittle, solid.	Hard, grey, very crystalline, brittle, solid.	Hard, grey, brittle, crystalline, solid.

Nitrogen and bismuth are the only elements of the class which occur uncombined with others in nature. Oxides and sulphides of the other elements, or compounds of these with other compounds, are found in several rocks ; none of the elements except phosphorus and nitrogen occurs very widely distributed or in very large quantities. The elements arsenic, antimony, and bismuth are obtained by reducing their oxides by heating with finely powdered charcoal. Phosphorus is obtained by heating calcium phosphate ($Ca2PO_3$) with char-coal.

207   These elements all combine directly with oxygen and the halogens; and all except nitrogen combine directly with sulphur. Hydrides of all, except bismuth, are known. Several oxyacids of each element, except bismuth, are known. The oxides of bismuth are basic, one is feebly acidic; the oxides of arsenic and antimony are acidic, but also slightly basic; the oxides of nitrogen and phosphorus are acidic. Each element forms at least one oxychloride. Several salts— *i.e.* derivatives of acids obtained by replacing hydrogen of the acids by positive elements—of bismuth, and a few salts of arsenic and antimony, are known. Neither nitrogen nor phosphorus, nor any oxide of these elements, interacts with acids to form salts.

The compositions of the more important compounds of the **208** elements of the nitrogen group may be expressed by the following formulae, where $M = N$, P, As, Sb, or Bi.

Hydrides. MH_3 (none known when $M = Bi$).

Oxides. M_2O, M_2O_2, M_2O_3, M_2O_4, M_2O_5.

Sulphides. M_2S_3, M_2S_5.

Haloid compounds. MX_3, MX_5 ($X = Cl$, Br, I, F.).

Oxyhaloid compounds. MOX, MOX_3, and many complex compounds $M_2O_yX_z$.

Oxyacids. HMO_2, HMO_3, H_4MO_4, $H_4M_2O_7$, &c.

Salts of commoner acids. M_26NO_3, M_23SO_4, M_23CO_3: $MONO_3$, $MO(SO_4)_2$, &c.; chiefly known when $M = Bi$.

The *hydrides*, MH_3, when $M = N$, P, As, or Sb, are colour- **209** less gases at ordinary temperatures and pressures.

Hydride of arsenic or antimony is formed when hydrogen is produced in a solution of a compound of arsenic or antimony; thus with zinc and dilute sulphuric acid,—

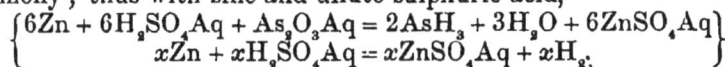

$$\left\{ \begin{array}{l} 6Zn + 6H_2SO_4Aq + As_2O_3Aq = 2AsH_3 + 3H_2O + 6ZnSO_4Aq \\ xZn + xH_2SO_4Aq = xZnSO_4Aq + xH_2. \end{array} \right\}$$

Part of the hydrogen is evolved and part reacts with the arsenious oxide.

Nitrogen hydride (ammonia) is generally prepared by heating solid salammoniac (ammonium chloride) with lime; thus

$$2NH_4Cl + CaO = CaCl_2 + H_2O + 2NH_3.$$

Phosphorus hydride may be obtained by a similar process from phosphonium iodide; thus

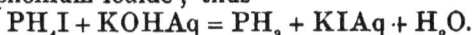

$$PH_4I + KOHAq = PH_3 + KIAq + H_2O.$$

The hydrides all dissolve in water; NH_3 and PH_3 combine with acids to form compounds similar in properties to the salts of the alkali metals; thus

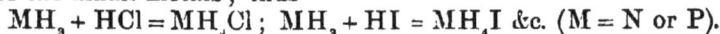

$$MH_3 + HCl = MH_4Cl; \quad MH_3 + HI = MH_4I \text{ &c. } (M = N \text{ or } P).$$

210   An aqueous solution of ammonia (NH_3) is markedly alkaline ; it turns red litmus blue, combines with carbon
dioxide to form ammonium carbonate, precipitates hydroxides of many metals from solutions of the salts of these
metals, and neutralises acids to form salts. The salts thus
produced are similar to, and generally identical or nearly
identical in crystalline form with, the salts of the alkali metals.
Compounds which react very similarly under similar conditions
have usually similar compositions. If we represent the
interactions between aqueous solutions of ammonia and potash
with acids as follows we fail to trace the similarities of
composition which we should expect to find :—

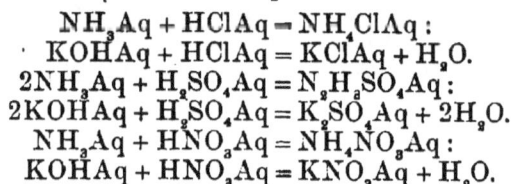

$$NH_3Aq + HClAq = NH_4ClAq :$$
$$KOHAq + HClAq = KClAq + H_2O.$$
$$2NH_3Aq + H_2SO_4Aq = N_2H_8SO_4Aq :$$
$$2KOHAq + H_2SO_4Aq = K_2SO_4Aq + 2H_2O.$$
$$NH_3Aq + HNO_3Aq = NH_4NO_3Aq :$$
$$KOHAq + HNO_3Aq = KNO_3Aq + H_2O.$$

The salts NH_4Cl and KCl, $N_2H_8SO_4$ and K_2SO_4, NH_4NO_3
and KNO_3, crystallise in the same forms. Identity, or very
close similarity, of crystalline form usually accompanies similarity of chemical composition. The alkalis KOH, $NaOH$, &c.
are metallic hydroxides. Let us assume that an aqueous
solution of ammonia contains the hydroxide $(NH_4)OH$, that
is a compound of the group of elements NH_4 with oxygen and
hydrogen, and that in the reactions of this compound with
acids the group of elements NH_4 remains undecomposed, then
we may represent the foregoing reactions thus,—

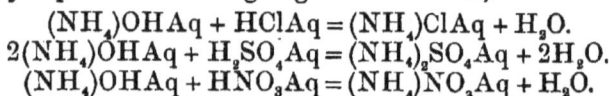

$$(NH_4)OHAq + HClAq = (NH_4)ClAq + H_2O.$$
$$2(NH_4)OHAq + H_2SO_4Aq = (NH_4)_2SO_4Aq + 2H_2O.$$
$$(NH_4)OHAq + HNO_3Aq = (NH_4)NO_3Aq + H_2O.$$

The formulae $(NH_4)Cl$ and KCl, $(NH_4)_2SO_4$ and K_2SO_4,
$(NH_4)NO_3$ and KNO_3, $(NH_4)OH$ and KOH, are evidently
similar. In order to exhibit these similarities of composition
we have assumed that a compound of one combining weight of
nitrogen and 4 c.ws. of hydrogen takes the place of one c.w. of
potassium in the reacting weights of the foregoing compounds ;
we have assumed that the compounds obtained by the interactions of an aqueous solution of ammonia (NH_3) with acids
are compounds of negative elements $(Cl, NO_3, SO_4, \&c.)$ with
the group of elements NH_4, which group behaves in these
compounds as if it were an elementary substance. The reasons

for making this assumption are the facts of similarity of properties between the compounds obtained by the interactions of the alkalis with acids and the compounds obtained by the interactions of an aqueous solution of ammonia with acids. We might represent the reaction between hydrocloric acid and ammonia solution in several ways; thus

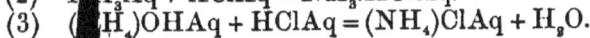

(1) $NH_3.H_2OAq + HClAq = NH_3.HClAq + H_2O.$
(2) $NH_3Aq + HClAq = NH_3.HClAq.$
(3) $(NH_4)OHAq + HClAq = (NH_4)ClAq + H_2O.$

We choose (3) because this representation of the chemical change, more clearly ▮▮▮▮▮▮▮▮, suggests the analogies between this change ▮▮▮▮▮▮ which occurs when caustic potash solution interacts with hydrochloric acid.

A study of the properties of a class of compounds has thus led to a special ▮▮▮ of the composition of these compounds.

Such a hypothetical group of elements as NH_4 is called a **211** *compound radicle*; ▮▮▮ especial compound radicle NH_4 is called *ammonium*. ▮▮▮ salts of ammonium are analogous to the salts of potassium, sodium, lithium, rubidium, and caesium. Therefore, it may be argued, if ammonium were isolated it would probably be similar in properties to the alkali metals; and if ammonium hydroxide $(NH_4)OH$ were isolated its properties would be similar to those of the hydroxides of the alkali metals.

Neither ammonium nor ammonium hydroxide has been isolated. A solid compound is known having the composition $NC_4H_{13}O$, this compound resembles potassium hydroxide; it dissolves in water forming a strongly alkaline solution which reacts with acids to form salts, &c. The methods whereby this compound is formed indicate that it is a hydroxide of the radicle $N(CH_3)_4$: the formula $N(CH_3)_4OH$ represents the compound in question as derived from $(NH_4)OH$ by replacing H_4 by $(CH_3)_4$. We have here carried the conception of compound radicle a step further. Facts seem to warrant the supposition that the four combining weights of hydrogen in the group NH_4 may be replaced by the radicle CH_3, and that the more complex group—$N(CH_3)_4$,—so obtained may react like an element.

The conception of compound radicles will be more fully **212** developed later (s. chap. XVII.). Meanwhile it is important to notice that what we have learned emphasises the importance of regarding the reacting weight of a compound as a definite mass of that compound composed of definite numbers of smaller parts (combining weights) of two or more elements, which smaller parts are arranged in some definite way relatively to each other. The reactions of ammonium compounds, for instance, almost oblige us to think of the reacting weights of these compounds as each composed of one combining weight of nitrogen closely united, in some way, with 4 c.ws. of hydrogen, and this group of combining weights as less closely united with the other elements which form a part of the reacting weight of the compound.

213 Let us return to the consideration of the properties of the hydrides MH_3. These compounds may all be oxidised : phosphorus hydride is very easily changed to phosphorus pentoxide (P_2O_5) and water by mixing with oxygen and heating ; arsenic and antimony hydrides are oxidised to oxide of arsenic or antimony (M_2O_3) and water, by burning in contact with a large quantity of oxygen ; ammonia is oxidised with difficulty, it is necessary to mix ammonia with a large quantity of oxygen and raise the temperature considerably, the products are water, nitrogen oxides (especially NO and N_2O_3), and nitrogen.

214 The *oxides* of the nitrogen group of elements are numerous ; the following table presents the compositions of the best studied of these oxides.

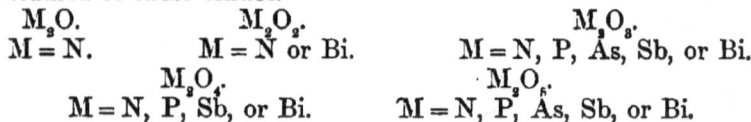

M_2O.	M_2O_2.	M_2O_3.
$M = N$.	$M = N$ or Bi.	$M = N, P, As, Sb$, or Bi.
	M_2O_4.	M_2O_5.
	$M = N, P, Sb$, or Bi.	$M = N, P, As, Sb$, or Bi.

When any one of the elements, except nitrogen, is heated in oxygen the oxide M_2O_3 is formed ; in the case of phosphorus, P_2O_5, and in the case of antimony, Sb_2O_4, is also produced. Nitrogen and oxygen combine when electric sparks are passed through a mixture of the gases ; N_2O_3 and N_2O_4 are produced. The oxides M_2O_5 are usually produced from M_2O_3 by an interaction between M_2O_3 and some compound (e.g. nitric acid), or compounds (e.g. caustic potash solution and chlorine), from which oxygen is produced. The lower oxides M_2O_2 and M_2O are formed by reducing the higher oxides ; the methods of reduction employed are very indirect. The oxides M_2O_3 are changed to M_2O_3, by the direct action of oxygen. The oxides M_2O_3, when $M = N$ or Sb, are changed to M_2O_4 by the direct action of oxygen ; when $M = P$ the oxide is easily changed to M_2O_5 by the direct action of oxygen ; when $M = As$

or Bi the oxides are unchanged by heating in oxygen. The oxides M_2O_3, except P_2O_3, are decomposed by heat, giving off oxygen, and forming either M_2O_4 ($M = N$ or Sb), or M_2O_3 ($M = As$ or Bi).

Nitrous oxide (N_2O) dissolves in water without forming an acid; but as the oxide can be obtained by heating an aqueous solution of hyponitrous acid (HNOAq), the oxide may be regarded as the anhydride of this acid.

The oxides M_2O_2 cannot be classed either as distinctly acidic or basic.

The oxides M_2O_3, except Bi_2O_3, are acidic; N_2O_3 and P_2O_3 dissolve in water to form nitrous acid HNO_2, and phosphorous acid H_3PO_3, respectively ($N_2O_3 + H_2O + Aq = 2HNO_2Aq$; $P_2O_3 + 3H_2O + Aq = 2H_3PO_3Aq$); arsenious acid ($?H_3AsO_3$) has not been isolated, but salts, e.g. K_3AsO_3, are obtained by the reaction of As_2O_3 with alkalis and basic oxides; antimonious acid H_3SbO_3 is known, although it is not obtained by the reaction of water with the oxide Sb_2O_3. The oxide Bi_2O_3 is basic; it interacts with acids to form salts and water, thus $Bi_2O_3 + 6HNO_3Aq = 2Bi3NO_3Aq + 3H_2O$. Besides being acidic towards alkalis and the more distinctly basic oxides, the oxides As_2O_3 and Sb_2O_3 are basic towards many acids; thus each interacts with hydrochloric acid to form a chloride and water; $M_2O_3 + 6HClAq = 2MCl_3Aq + 3H_2O$. These oxides also appear to interact with concentrated sulphuric acid to form sulphates M_23SO_4, but there is some doubt as to the compositions of the products of these reactions.

The oxides M_2O_5 are distinctly acidic, except Bi_2O_5. When $M = N$ or P, the oxides dissolve in water to form the corresponding acids; the acids corresponding to the oxides when $M = As$ or Sb are not obtained directly from these oxides by interaction with water. Bismuthic oxide, or bismuth pentoxide, (Bi_2O_5) interacts with acids to form the same salts as are produced when Bi_2O_3 interacts, and oxygen is simultaneously evolved; bismuthic oxide is therefore a peroxide. From what we have learned concerning the properties of oxides, and from considering the properties of the highest oxides of the elements classed with bismuth, we might expect bismuthic oxide to exhibit some acidic functions. No salts are obtained by interactions between this oxide and alkalis; but when the oxide in question is prepared by passing chlorine into very concentrated potash solution holding bismuthous oxide (Bi_2O_3) in suspension, the properties of the substances obtained render it very probable that unstable compounds of bismuthic and potassium oxides are formed at certain stages of the preparation. The composition of these unstable salts may be expressed by the formula $x\mathrm{Bi_2O_5}.y\mathrm{K_2O}$; x and y probably vary according to the relative masses of $\mathrm{Bi_2O_3}$ and KOH used in the preparation of $\mathrm{Bi_2O_5}$, and according to the temperature, &c. If bismuthic oxide exhibits any acidic functions they are certainly extremely feeble.

215 The following table presents the compositions of the more important *oxyacids* of the members of the nitrogen group; to each oxyacid there generally corresponds a certain oxide; that is, the oxide is obtained from the acid by removing hydrogen and oxygen (generally by heating the acid), or the acid is produced by the interaction of water and the oxide.

Nitrogen
{
Oxide. $\quad$ $\mathrm{N_2O}$. $\qquad$ $\mathrm{N_2O_3}$. $\qquad$ $\mathrm{N_2O_5}$.
Oxyacid. $\quad$ $\mathrm{HNOAq.}^*$ $\quad$ $\mathrm{HNO_2Aq.}^*$ $\quad$ $\mathrm{HNO_3}$.
}

Phosphorus
{
Oxide. $\qquad$. — $\qquad$ $\mathrm{P_2O_3}$. $\qquad\qquad$ $\mathrm{P_2O_5}$.
Oxyacid. $\quad$ $\mathrm{H_3PO_2}$. $\quad$ $\mathrm{H_3PO_3}$. $\quad$ $\overbrace{\mathrm{HPO_3}\,.\,\mathrm{H_3PO_4}\,.\,\mathrm{H_4P_2O_7}.}$
}

Arsenic
{
Oxide. $\qquad\qquad\qquad$ $\mathrm{As_2O_3}$
Oxyacid. salts known of forms $\mathrm{M_3AsO_3}$ and $\mathrm{MAsO_2}$ $(\mathrm{M=K},$ &c.)
}

"
{
Oxide. $\qquad\qquad\quad$ $\mathrm{As_2O_5}$
Oxyacid. $\overbrace{\mathrm{HAsO_3}\,.\,\mathrm{H_3AsO_4}\,.\,\mathrm{H_4As_2O_7}.}$
}

Antimony
{
Oxide. $\qquad\qquad$. $\qquad$ $\mathrm{Sb_2O_3}$.
Oxyacid. $\qquad$. $\quad$ $\mathrm{H_3SbO_3}$
$\qquad\qquad\qquad$ $\overbrace{\text{salts of form } \mathrm{MSbO_2} \text{ are also known.}}$
}

"
{
Oxide. $\qquad\qquad$ $\mathrm{Sb_2O_5}$
Oxyacid. $\overbrace{\mathrm{HSbO_3}\,.\,\mathrm{H_3SbO_4}\,.\,\mathrm{H_4Sb_2O_7}.}$
}

It will be noticed that the pentoxides of phosphorus, arsenic, and antimony, are represented as being each the anhydride of three acids.

Phosphorus pentoxide interacts with water to produce one or other of the three acids according to the relative masses of water and oxide used, and the temperature. The following equations represent the reactions;—

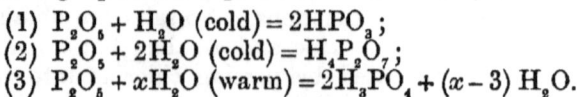

(1) $\mathrm{P_2O_5 + H_2O}$ (cold) $= 2\mathrm{HPO_3}$;
(2) $\mathrm{P_2O_5 + 2H_2O}$ (cold) $= \mathrm{H_4P_2O_7}$;
(3) $\mathrm{P_2O_5} + x\mathrm{H_2O}$ (warm) $= 2\mathrm{H_3PO_4} + (x-3)\,\mathrm{H_2O}$.

* The symbol Aq is here used to signify that the acids after which it is placed are known only in aqueous solutions.

When the acid H_3PO_4 is heated to about $200°$, it loses water and forms the acid $H_4P_2O_7$, when this is heated to about $400°$ it loses water and forms the acid HPO_3; thus,

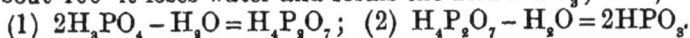

(1) $2H_3PO_4 - H_2O = H_4P_2O_7$; (2) $H_4P_2O_7 - H_2O = 2HPO_3$.

None of the acids corresponding to arsenic or antimony pentoxide is obtained by the interaction of the oxide with water. The acid H_3AsO_4 is formed by oxidising arsenious oxide in presence of water (thus $As_2O_3Aq + 5H_2O + 4Cl = 2H_3AsO_4Aq + 4HClAq$); when this acid is heated it yields $H_4As_2O_7$, and then $HAsO_3$. The acid H_3SbO_4 is produced by the interaction of antimonic chloride with a little cold water; thus $SbCl_5 + 4H_2O = H_3SbO_4 + 5HCl$; the acid $H_4Sb_2O_7$ is obtained by heating H_3SbO_4 to about $100°$, and the acid $HSbO_3$ by heating $H_4Sb_2O_7$ to about $200°$.

Solutions of either HPO_3 or $H_4P_2O_7$ when heated give a solution of H_3PO_4; solutions of $HAsO_3$ and $H_4As_2O_7$ even at the ordinary temperature change rapidly to a solution of H_3AsO_4; but solutions of $HSbO_3$ and $H_4Sb_2O_7$ seem to be more stable than a solution of H_3SbO_4.

To the acids corresponding to the oxides M_2O_5 are given names ending in -*ic*; phosphoric, arsenic, antimonic, acid. The prefix *ortho*- is employed to distinguish the acid of the form H_3MO_4, the acid of the form HMO_3 is called *meta*-, and the remaining acid, $H_4M_2O_7$, is called *pyro*-. The acids, especially the phosphoric acids, are also distinguished as *tribasic* phosphoric acid H_3PO_4, *monobasic* HPO_3, and *tetrabasic* phosphoric acid $H_4P_2O_7$. Nitrogen forms a *meta*-acid only.

All the oxyacids of nitrogen are more or less easily decomposed, by heat, or by reactions with other substances; hyponitrous acid ($HNOAq$), and nitrous acid (HNO_2Aq), combine with oxygen at ordinary temperatures to form nitric acid (HNO_3); these acids therefore act as reducing agents. Nitric acid we know is an energetic oxidising agent. The lower acids of phosphorus, H_3PO_2 and H_3PO_3, are reducing agents; but they do not combine with oxygen so rapidly or at such low temperatures as the lower acids of nitrogen do. An aqueous solution of arsenious oxide may possibly contain the acid H_3AsO_3; this solution, like that of antimonious oxide, is a weak reducing agent. The highest acids of phosphorus can scarcely be classed as oxidising agents; they are fully oxidised, but they do not easily part with oxygen: these acids are not separated into oxide and water by heat alone. The highest oxides of arsenic and antimony (M_2O_5) are reduced by heat

216

217

to lower oxides (M_2O_3) and oxygen; these oxides therefore sometimes react as oxidising agents: inasmuch as the highest acids of arsenic and antimony can be separated by heat alone into oxide (M_2O_5) and water, it follows that these acids will sometimes react as oxidisers.

218   The preceding sketch of the oxides and oxyacids of the elements of the nitrogen groups shews how closely related these elements are to each other; but it also shews a gradation of properties from nitrogen to bismuth. Bismuth is evidently more widely separated from the other members of the group than these are from each other. Nitrogen and phosphorus are distinctly non-metallic, negative, elements; bismuth is metallic; arsenic and antimony stand midway between the metals and the non-metals; these elements are sometimes classed, with one or two others, as *metalloids*.

219   Our examination of some metals and non-metals shewed that non-metals sometimes exhibit allotropy. We might reasonably expect to find nitrogen and phosphorus existing each in more than one form, and it would certainly be incumbent on us to inquire whether arsenic and antimony exhibit allotropy or not. We should scarcely expect to find more than a single form of bismuth.

The existence of more than one form of nitrogen has not been proved; experimental results are however on record which point to the possibility of nitrogen undergoing allotropic change.

220   At least two distinct modifications of phosphorus are known. Ordinary phosphorus is a yellowish-white, semitransparent, crystalline, solid; spec. gravity $= 1\cdot8$; melting point $= 44^0$; easily soluble in carbon disulphide, ether, and various oils. It combines with oxygen, the halogens, and sulphur, very rapidly and at low temperatures. It is extremely poisonous. When ordinary phosphorus is heated to about 240^0 in an atmosphere of carbon dioxide it is changed into a red amorphous solid. This change is more quickly and completely accomplished by heating ordinary phosphorus in a closed vessel to about 300^0: a portion of the phosphorus is oxidised, and the rest is transformed into red phosphorus. Red, or amorphous, phosphorus is heavier than ordinary phosphorus; spec. gravity $= 2\cdot1$; it is insoluble in carbon disulphide and ether; it combines with oxygen, the halogens, and sulphur only at fairly high temperatures. Red phosphorus is not poisonous. When heated in a stream of carbon dioxide

to about 260° red phosphorus is changed to ordinary phosphorus; but it may be heated in closed tubes considerably
above 300° without undergoing this change. To prove that
ordinary and red phosphorus are forms of the same element
it suffices to heat a weighed quantity of ordinary phosphorus
in an atmosphere of dry nitrogen or carbon dioxide to
230°—240°, until the change into red phosphorus is completed, and to weigh the product; no change of mass has
occurred. The red phosphorus is then heated to 260°—280°
in a very slow current of dry nitrogen, when ordinary
phosphorus is produced; the mass of this is the same as the
mass of the phosphorus at the beginning of the experiment.
To make the proof more complete, equal masses of ordinary
and red phosphorus are burnt in oxygen so that none of the
products escape; the product of each combination is dissolved
in water, the solutions are heated for some time, and the
quantity of phosphoric acid (H_3PO_4) in each is determined : it is
found that each gram of either phosphorus has produced 3·16
grams of phosphoric acid (H_3PO_4). If equal masses of the
two varieties of phosphorus interact with chlorine, the same
mass of the same compound is obtained in each case.

If however the quantity of heat produced during the burning **221**
of equal masses of the two varieties of phosphorus is measured, different values are obtained. Let 1 gram of ordinary
phosphorus be wholly burnt to phosphorus pentoxide, about
5900 gram-units of heat are produced; let 1 gram of red
phosphorus be wholly burnt to phosphorus pentoxide, about
5500 gram-units of heat are produced. In each case the
change which proceeds is represented thus :

$$2P + 5O = P_2O_5.$$

If the symbol P^α represents 31 grams of ordinary, and
P^β 31 grams of red, phosphorus (combining weight of phosphorus = 31), then we have these statements :—

$2P^\alpha + 5O = P_2O_5$; 365,800 gram-units of heat produced.

$2P^\beta + 5O = P_2O_5$; 341,000 „ „ „

Hence it follows that in the change from 62 grams of
ordinary, to 62 grams of red, phosphorus, 24,800 gram-units of
heat are produced. In other words energy is degraded in the
change from ordinary to red phosphorus; the quantity of
energy degraded, or rendered less available for doing work, is
equal to nearly 25,000 gram-units of heat-energy per 62 grams

of phosphorus changed from ordinary to red phosphorus. Equal masses of the two forms of phosphorus do not contain equal quantities of energy: red phosphorus contains less energy than an equal mass of ordinary phosphorus. The red variety of phosphorus is more stable, and less easily undergoes chemical change, than the ordinary variety.

222   Arsenic is known in two, or perhaps, three modifications. The change from one to the other form is attended by change from crystals to an amorphous powder, or *vice versa*, change of specific gravity and some other physical properties, and production or disappearance of heat. It is not quite certain whether antimony does or does not undergo allotropic change. Only one form of bismuth is known.

223   The *haloid compounds* of the elements of the nitrogen group have the compositions MX_3 and MX_5; P_2I_4 also exists.

The haloid compounds of nitrogen have not been fully studied; they are extremely explosive, and it is therefore very dangerous to work with them.

Phosphorus forms compounds of both compositions PX_3 and PX_5; e.g. PCl_3, PCl_5, PBr_3, PF_5; the known haloid compounds of arsenic and bismuth all belong to the type, or general form, MX_3; antimony combines with halogens in both ratios $M : X_3$ and $M : X_5$. All the compound MX_5 except PF_5, are decomposed by heat into MX_3 and X_2; most of the compounds MX_3 can be gasified without decomposition. By the interactions of the haloid compounds MX_3 with water either oxyacids (or oxides) and haloid acids, or oxyhaloid compounds, are produced. Thus

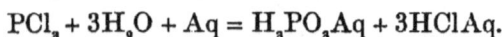

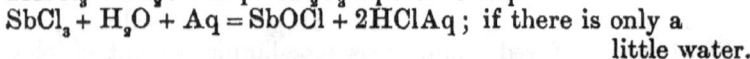

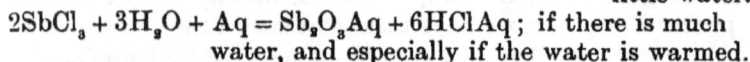

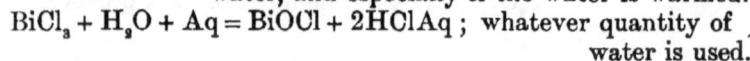

$PCl_3 + 3H_2O + Aq = H_3PO_3Aq + 3HClAq.$

$2AsCl_3 + 3H_2O + Aq = As_2O_3Aq + 6HClAq.$

$SbCl_3 + H_2O + Aq = SbOCl + 2HClAq$; if there is only a
little water.

$2SbCl_3 + 3H_2O + Aq = Sb_2O_3Aq + 6HClAq$; if there is much
water, and especially if the water is warmed.

$BiCl_3 + H_2O + Aq = BiOCl + 2HClAq$; whatever quantity of
water is used.

An oxychloride of phosphorus, $POCl_3$, is obtained by the interaction of ortho-phosphoric acid (H_3PO_4) with phosphoric chloride (PCl_5). An oxychloride of arsenic, $AsOCl$, can be obtained by heating together arsenious oxide and chloride ($As_2O_3 + AsCl_3 = 3AsOCl$).

We have already learned (par. 156) that the haloid compounds of non-metallic elements, as a class, interact with water; that the compounds of the more distinctly negative elements generally produce oxides (or oxyacids) and haloid acids ; and that the compounds of the less negative non-metals generally produce oxyhaloid compounds and haloid acids. The oxyhaloid compounds of metals are usually very complex, and are produced either by heating the oxides with haloid compounds, or by passing halogen over a heated mixture of oxide and carbon.

224 A classification of the elements of the nitrogen group into metals and non-metals based on the behaviour of their haloid compounds with water would place phosphorus and arsenic with the non-metals, antimony with the metalloids, and bismuth either with the metalloids or the metals. A classification of the same elements based on the properties of their oxides would lead to the same, or about the same, arrangement; bismuth would certainly be placed with the metals. A classification based on the existence and properties of hydrides would result in all the elements except bismuth being placed among the non-metals. A classification based on the occurrence or non-occurrence of allotropy would result in phosphorus and arsenic being placed with the non-metals, antimony with the metalloids, and bismuth with the metals.

225 The properties of the sulphides of a group of elements sometimes afford class-marks by using which the elements may be subdivided into families.

+All the elements of the nitrogen group form *sulphides* : the following are certainly known :—N_2S_3, P_2S_3, P_2S_5, and several other compounds of phosphorus and sulphur ; As_2S_3, As_2S_5 ; Sb_2S_3, Sb_2S_5 ; Bi_2S_3, and probably other sulphides of bismuth. Nitrogen sulphide differs considerably from the others : it is very explosive ; it is decomposed by water or aqueous alkalis giving ammonia and ammonium salts of sulphur oxyacids. The sulphides of phosphorus react with aqueous solutions of potash to form potassium salts of sulpho- (or *thio-*) acids of phosphorus ; but these salts are unstable and easily decomposed. The sulphides of arsenic and antimony dissolve in aqueous solutions of potash or potassium sulphide (K_2S) to form potassium sulpho- or thio-arsenite and antimonite, respectively (K_3AsS_3 ; $KSbS_2$) ; these solutions are decomposed by boiling, or by interacting with dilute acids, giving arsenious, or antimonious, sulphide, and hydrogen sulphide. Bismuth sulphide does not interact either with caustic potash or potassium sulphide.

The sulphides of phosphorus, arsenic, and antimony, are therefore acidic in their reactions with alkalis and alkaline sulphides; sulphide of bismuth is not acidic, it interacts with acids to form salts and hydrogen sulphide, just as oxide of bismuth interacts with acids to form salts and hydrogen oxide. The following formulae exhibit the relations of composition between (1) acidic oxides and their oxyacids, (2) basic oxides and their salts, (3) acidic sulphides and their sulpho-acids, and (4) basic sulphides and the salts obtained by their reactions both with acids and with acidic sulphides.

Acidic Oxides.	Oxyacids.	Basic oxides.	Salts, *obtained by interactions of basic oxides with acids.*
P_2O_3.	H_3PO_3.	K_2O.	K_2SO_4; KNO_3; K_2CO_3.　(+H_2O)
N_2O_5.	HNO_3.	CaO.	$CaSO_4$; $Ca2PO_4$; $Ca2NO_3$. (+H_2O)
P_2O_5.	H_3PO_4.	FeO.	$FeSO_4$; Fe_32PO_4; $FeCO_3$. (+H_2O)
SO_3.	H_2SO_4.	Cr_2O_3.	Cr_23SO_4; $CrPO_4$; Cr_26NO_3. (+H_2O)

Acidic Sulphides.	Sulpho-acids: *in many cases only salts of these acids are known.*	Basic Sulphides.	Salts *obtained by reactions,* (1) *of* acidic sulphides;	(2) *of* oxyacids.
As_2S_3.	H_3AsS_3.	K_2S	K_3AsS_3.	K_2SO_4 (+H_2S)
Sb_2S_3.	$HSbS_2$.	CaS.	$Ca(SbS_2)_2$.	$Ca2NO_3$ (+H_2S)
SnS_2.	H_2SnS_3.	Bi_2S_3.	none	$Bi3NO_3$ (+H_2S)
Au_2S_3.	$HAuS_2$.	FeS.	none	$FeSO_4$ (+H_2S)

A classification of the elements of the nitrogen group based on the properties of their sulphides would place phosphorus, arsenic, and antimony, with the non-metals, and bismuth with the metals. Nitrogen would be placed with the non-metals, but to some extent apart from the other members of the group.

226 We have now learned something of the methods used in chemistry for classifying elements and compounds. We cannot classify elements without at the same time classifying compounds; we cannot arrange compounds in groups without at the same time arranging elements in groups.

As class-marks we have used, the properties and the composition of oxides; the properties and the composition of sulphides; the formation, composition, and interaction with water, of haloid compounds; the formation, or non-formation, the properties, and the composition, of hydrides; the production of compounds with oxygen and hydrogen which are either acids, alkalis, or intermediate between these; the occurrence or non-occurrence of allotropic change; the position of elements in the electrical series; &c., &c.

We have traced the change from one class of compounds to another accompanying the increase or decrease in the relative mass of one of the elements in a series of binary or ternary compounds. We have seen basic oxides combining with oxygen and forming acidic oxides; we have compared basic hydroxides with acidic hydroxides of the same element. We have also traced the change from one class of compounds to another accompanying the substitution of a positive by a negative element in compounds of similar composition. We have compared acidic hydroxides of negative elements with alkaline hydroxides of positive elements, and both with neutral hydroxides of elements which were neither markedly positive nor strongly negative.

We have also learned something of the conditions under which classes of compounds are formed. We have seen some elements combining directly at ordinary temperatures with oxygen; others only at high temperatures; others only when they interact with compounds under such conditions that oxygen is produced in the interactions.

Of each compound we have asked; of what elements, and of how much of each, is a reacting weight of this compound composed? What can this compound do, and under what conditions does it perform its chemical functions? Composition, and properties, these have been our guides.

In examining the classification of elements and compounds we have travelled far from the starting point of our inquiries. We have not looked much to the general conditions of chemical change, or to the circumstances which modify the course and the final goal of chemical processes. We have been content to know that acids, for instance, are compounds of hydrogen with one or more negative elements, generally with oxygen and another more or less negative element, and that these compounds interact with metals, with basic oxides, and with alkalis, to produce salts. We have not asked; under what conditions does this or that acid interact? if two acids react with one alkali will one combine with the whole of the alkali, or will the alkali be divided between the acids? But we must now come back to such inquiries as these.

CHAPTER XII.

CONDITIONS WHICH MODIFY CHEMICAL CHANGE.

227 BODIES interact chemically only when brought into very close contact. The contact may be effected by dissolving, gasefying, or melting, one or more of the reacting substances; or, in some cases, by submitting the substances to very great pressure.

Barium chloride and sodium sulphate, for example, remain unchanged when mixed; but when aqueous solutions of these compounds are mixed, barium sulphate and sodium chloride are at once produced. Ammonia and hydrogen chloride gases combine to produce ammonium chloride. A mixture of iron and sulphur remains unchanged for an indefinite time; but when iron filings are added to molten sulphur chemical change occurs and iron sulphide is formed. The same product is formed by exposing the mixture of finely divided iron and sulphur to a pressure of several thousand tons on the square inch.

228 As a general rule chemical change proceeds more rapidly between a solid and a liquid, or a solid and a gas, the more finely divided the solid is, that is, the greater the surface exposed by a given quantity of the solid. Thus a piece of iron oxidises slowly in ordinary air; but very finely divided iron—obtained by strongly heating iron tartrate in a glass tube and closing the tube while hot—oxidises so rapidly in ordinary air that the particles of the oxidising iron glow and emit light. Granulated zinc, that is zinc in thin irregular-shaped pieces, dissolves in dilute sulphuric acid much more quickly than a compact mass of zinc.

229 If one or more of the possible products of a chemical inter-action is gaseous under the experimental conditions, that chemical interaction usually occurs readily. Thus calcium carbonate $(CaCO_3)$ rapidly interacts with dilute hydrochloric acid solution, to produce calcium chloride $(CaCl_2)$ which remains in solution, and carbon dioxide (CO_2) which escapes as a gas;

$$CaCO_3 + 2HClAq = CaCl_2Aq + H_2O + CO_2.$$

Calcium carbonate is decomposed by heat to solid calcium oxide and gaseous carbon dioxide : $CaCO_3$ (heated) $= CaO + CO_2$. Zinc and dilute sulphuric acid interact at ordinary temperatures to produce zinc sulphate and gaseous hydrogen :

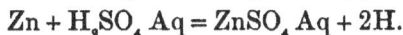

$$Zn + H_2SO_4 Aq = ZnSO_4 Aq + 2H.$$

If one or more of the possible products of a chemical inter- **230** action between solutions is a solid under the experimental conditions, that chemical change usually occurs readily. Thus aqueous solutions of barium chloride and sodium sulphate interact when mixed to produce solid barium sulphate, and sodium chloride which remains in solution :

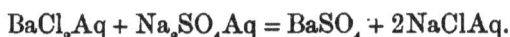

$$BaCl_2Aq + Na_2SO_4Aq = BaSO_4 + 2NaClAq.$$

A solution of antimony chloride in hydrochloric acid interacts with water to produce solid antimony oxychloride, and hydrochloric acid which remains in solution :

$$SbCl_3 \text{ (in HClAq)} + H_2O + xH_2O = SbOCl + 2HClAq + xH_2O.$$

Potassium acetate is soluble in alcohol, potassium carbonate is insoluble in alcohol; if carbon dioxide is passed into an alcoholic solution of potassium acetate, the formation of an insoluble compound becomes possible; this compound, potassium carbonate, is formed and precipitated. On the other hand, if acetic acid is added to an aqueous solution of potassium carbonate, the formation of a gas becomes possible; this gas, carbon dioxide, is formed and the carbonate is decomposed. These reactions furnish an instance of the reversal of a chemical change by alterations in the experimental conditions such that in one case the formation of a solid, and in the other the formation of a gas, becomes possible.

The conditions which chiefly modify chemical changes, besides those already mentioned, are temperature, and the relative masses of the interacting substances.

Influence of temperature on chemical change. A **231** chemical change occurs within a definite range of temperature. Some changes take place readily and completely at ordinary temperatures; other changes begin only at higher temperatures.

Thus sodium or potassium oxidises rapidly in ordinary air; iroñ filings are oxidised rapidly only by heating in oxygen. Iron and dilute sulphuric acid readily react to produce iron sulphate and hydrogen; copper and sulphuric acid do not interact until the temperature is raised to 100° or more. Hydrogen and oxygen do not combine until the temperature is raised very considerably.

232　　　The products of a chemical interaction sometimes vary according to the temperature at which the substances are caused to interact. ·Thus sodium chloride and sulphuric acid react at ordinary temperatures to produce sodium-hydrogen sulphate and hydrogen chloride, but at higher temperatures the chief product, besides hydrogen chloride, is sodium sulphate; the two reactions may be represented thus:

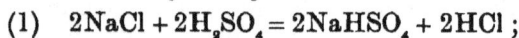

(1)　$2NaCl + 2H_2SO_4 = 2NaHSO_4 + 2HCl$;

(2)　$2NaCl + H_2SO_4 = Na_2SO_4 + 2HCl$.

A solution of bismuth iodide in hydriodic acid interacts with cold water to precipitate bismuth iodide; BiI_3 (in $HIAq) + xH_2O$ (cold) $= BiI_3 + HIAq + xH_2O$. But the same solution interacts with hot water to precipitate bismuth oxyiodide;

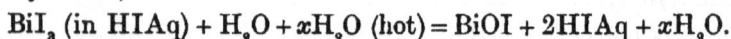

BiI_3 (in $HIAq) + H_2O + xH_2O$ (hot) $= BiOI + 2HIAq + xH_2O$.

A cold aqueous solution of copper sulphate reacts with cold caustic potash solution to precipitate copper hydroxide; a hot aqueous solution of copper sulphate reacts with hot caustic potash solution to precipitate copper oxide:

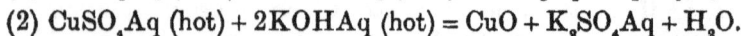

(1)　$CuSO_4Aq$ (cold) $+ 2KOHAq$ (cold) $= CuO_2H_2 + K_2SO_4Aq$.

(2)　$CuSO_4Aq$ (hot) $+ 2KOHAq$ (hot) $= CuO + K_2SO_4Aq + H_2O$.

233　　　Sometimes a certain chemical change occurs within a defined range of temperature, and the reverse change takes place within another defined range of temperature. Thus lime and carbon dioxide combine at ordinary temperatures to form calcium carbonate; but calcium carbonate can be wholly changed to lime and carbon dioxide by raising the temperature;

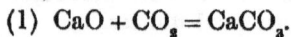

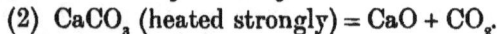

(1)　$CaO + CO_2 = CaCO_3$.

(2)　$CaCO_3$ (heated strongly) $= CaO + CO_2$.

Ammonia and hydrogen chloride gases combine when mixed to form ammonium chloride; but ammonium chloride is completely resolved into ammonia and hydrogen chloride at a moderately high temperature: (1) $NH_3 + HCl = NH_4Cl$; (2) NH_4Cl (heated) $= NH_3 + HCl$. Sulphur trioxide and water combine to form sulphuric acid; but sulphuric acid is decomposed by heating into sulphur trioxide and water-vapour: (1) $SO_3 + H_2O = H_2SO_4$; (2) H_2SO_4 (heated) $= SO_3 + H_2O$. Hydrogen and iodine combine when heated to 400°—500° to form hydrogen iodide; but hydrogen iodide is separated at a higher temperature into hydrogen and iodine:

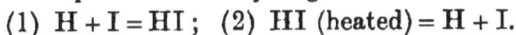

(1) $H + I = HI$; (2) HI (heated) $= H + I$.

When carbon dioxide is passed over moist sodium carbonate, sodium-hydrogen carbonate is formed; but this salt, when heated, is changed to water, carbon dioxide, and sodium carbonate:

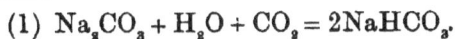

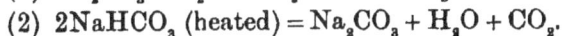

(1) $Na_2CO_3 + H_2O + CO_2 = 2NaHCO_3$.

(2) $2NaHCO_3$ (heated) $= Na_2CO_3 + H_2O + CO_2$.

If amylic bromide, $C_5H_{11}Br$, is heated in an enclosed space to about 270° it is changed into amylene, C_5H_{10}, and hydrogen bromide, HBr; if the temperature is now allowed to fall the amylene and hydrogen bromide recombine to form amylic bromide.

When phosphorus pentachloride, PCl_5, is heated the vapour produced is a mixture of phosphorus trichloride, PCl_3, and chlorine; when this mixture of gases is allowed to cool solid phosphorus pentachloride is re-formed.

The amount of chemical change produced, in many of these **234** cases, is conditioned not only by the temperature but also by the pressure. Thus when nitrogen tetroxide, N_2O_4, is heated, it is changed to nitrogen dioxide, NO_2; the amount of change at 16° under a pressure of 229 mm. of mercury is the same (20 p. c.) as that at 27° under a pressure of 755 mm.

The change of calcium carbonate into carbon dioxide and **235** calcium oxide, brought about by the action of heat, and the reverse change of calcium oxide and carbon dioxide into calcium carbonate, may be regarded as a completed cycle of change. Changes such as this are classed together under the name *dissociation*.

The following changes, already referred to, which are brought about by altering conditions of temperature, are instances of *dissociation*; (1) the change of hydrogen iodide into hydrogen and iodine, and of hydrogen and iodine into hydrogen iodide; (2) the change of ammonium chloride into ammonia and hydrogen chloride, and of a mixture of these gases into ammonium chloride, (3) the change of amylic bromide into amylene and hydrogen bromide, and of a mixture of these gases into amylic bromide, (4) the change of phosphorus pentachloride into phosphorus trichloride and chlorine, and of a mixture of these gases into phosphorus pentachloride, &c. &c.

236   Every dissociation is brought about by the action of heat alone; in every case there is (1) the production of less complex from more complex substances, one at least of the less complex being a gas; (2) the possibility of reversing the process by cooling the products of the change in contact with each other.

When calcium carbonate is heated to a specified temperature in a closed vacuous vessel, to which is attached an apparatus for measuring the pressure inside the vessel, dissociation into solid calcium oxide and gaseous carbon dioxide proceeds until the pressure of the carbon dioxide reaches a certain amount, when the dissociation ceases, and the system, consisting of calcium carbonate, calcium oxide, and carbon dioxide, remains in equilibrium. If temperature is now raised, more carbonate is decomposed, more carbon dioxide accumulates, the pressure increases, and at last the dissociation stops. If temperature is lowered, some of the carbon dioxide and calcium oxide combine, and pressure falls until a new state of equilibrium is attained. If temperature is kept constant when a certain quantity of carbonate has been dissociated and the system is in equilibrium, and pressure is suddenly lowered by removing some of the carbon dioxide, dissociation proceeds until the accumulation of carbon dioxide brings the pressure to its former value; when this pressure is reached dissociation stops, and the system remains in equilibrium.

The relations of temperature and pressure to chemical change are important, not only in cases of dissociation, but also in cases of ordinary double decomposition in which gases are produced and the original bodies and the products of the change remain in contact and are capable of chemically interacting. Thus, when steam is passed over hot iron, iron oxide and hydrogen are produced; if the hydrogen is removed as it is formed and more steam is supplied, the whole of the iron is changed to oxide; but if the hydrogen is caused to accumulate in contact with the other members of the system, pressure increases, the rate of change decreases, and at last a state of equilibrium is reached, the system is composed of definite relative masses of steam, hydrogen, iron, and iron oxide, and no more change occurs. If hydrogen is now removed, other conditions remaining constant, pressure falls, and change proceeds with formation of more iron oxide and more hydrogen; if hydrogen is forced into the apparatus, other conditions being constant, change occurs with production of more iron and steam. If all the members of the changing system are allowed to remain in contact with each other, then a state of equilibrium is sooner or later attained.

The conception of a chemical system as being in equili- **237** brium under certain conditions, and as having this equilibrium overthrown by altering these conditions, is of much importance. We shall have to return to this conception again, and more fully to examine the conditions which affect the equilibrium of such systems. (s. Chaps. XIII. and XIV.)

Influence of the relative masses of the interacting **238** substances on chemical change. We have already had many examples of the necessity of using an *excess* of one or other of two interacting substances in order to complete a chemical reaction. Thus, to change a specified mass of lead monoxide (PbO) into lead dioxide (PbO_2), the lead monoxide is treated with a very large excess of potassium hypochlorite (Chap. XI. par. 158); a given mass of manganese dioxide is changed to potassium manganate by fusion with caustic potash, but very much more potash must be used than is actually converted into the new salt (Chap. XI. par. 196).

Sulphuric acid and sodium nitrate interact as shewn in the equation $2NaNO_3 + H_2SO_4 = Na_2SO_4 + 2HNO_3$; but if a large quantity of nitric acid is heated with sodium sulphate, sodium nitrate and sulphuric acid are produced

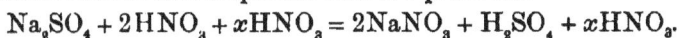

$$Na_2SO_4 + 2HNO_3 + xHNO_3 = 2NaNO_3 + H_2SO_4 + xHNO_3.$$

If silver chloride is heated with a considerable quantity of an aqueous solution of hydriodic acid some silver iodide is produced; but if silver iodide is warmed with hydrochloric acid silver chloride is formed. The two changes may be represented thus;

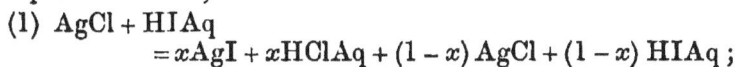

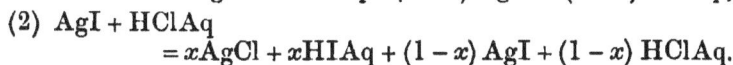

(1) $AgCl + HIAq$
$$= xAgI + xHClAq + (1 - x) AgCl + (1 - x) HIAq;$$
(2) $AgI + HClAq$
$$= xAgCl + xHIAq + (1 - x) AgI + (1 - x) HClAq.$$

In the first reaction little AgI is formed unless a large excess of HI is used; in the second reaction a considerable quantity of AgCl is produced even when a small excess of HCl is used.

If aqueous solutions of potassium sulphocyanide (KCNS) and ferric chloride are mixed, ferric sulphocyanide [$Fe_2(CNS)_6$] and potassium chloride are formed. In this case the four substances all remain in solution. The change may be represented thus ;

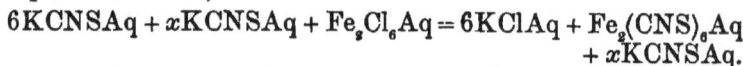

$$6KCNSAq + xKCNSAq + Fe_2Cl_6Aq = 6KClAq + Fe_2(CNS)_6Aq + xKCNSAq.$$

If the salts are allowed to react in the ratio $6KCNS : Fe_2Cl_6$ very little of either is decomposed; in order to change the mass of ferric chloride represented by the formula Fe_2Cl_6 almost completely into ferric sulphocyanide the salts must be mixed in about the ratio $500KCNS : Fe_2Cl_6$. But at tho oloso of the reaction only $6KCNS$ has been decomposed for every Fe_2Cl_6 changed; the rest of the potassium sulphocyanide remains chemically unchanged.

When hydrogen and iodine are mixed at 440° and under a pressure of 340 mm. a portion of the elements combine to form hydrogen iodide; the amount of hydrogen iodide formed is increased by increasing the mass of either hydrogen or iodine relatively to the other gas. The following numbers shew the influence of increasing the mass of hydrogen :—

Ratio of H to I.	Ratio of HI formed to total possible HI.
I : H	·74
I : 2H	·84
I : 3H	·87
I : 4H	·88

239   The influence of the relative masses of the interacting substances on the course of a chemical change is most marked when all the interacting substances and all the products of the change are in solution, or are liquids. Ethylic alcohol and acetic acid for instance react to produce ethylic acetate and water, thus $C_2H_5.OH + C_2H_3O.OH = C_2H_3O.OC_2H_5 + H_2O$; but much more acetic acid than is represented in this equation must be used to complete the change. Such changes as this, or as that which occurs between solutions of potassium sulphocyanide and ferric chloride (*v. ante*), may be divided into two parts ; the direct, and the reverse, change. The direct change in the case of alcohol and acetic acid is that shewn in the equation, the reverse change is that from ethylic acetate and water to ethylic alcohol and acetic acid ($C_2H_3O.OC_2H_5 + H_2O = C_2H_3O.OH + C_2H_5.OH$). The direct and reverse changes in the case of potassium sulphocyanide and ferric chloride solution may be represented thus,

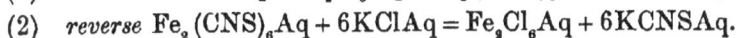

(1) *direct* $6KCNSAq + Fe_2Cl_6Aq = Fe_2(CNS)_6Aq + 6KClAq$;

(2) *reverse* $Fe_2(CNS)_6Aq + 6KClAq = Fe_2Cl_6Aq + 6KCNSAq$.

Neither change is completed under ordinary conditions; the changing system reaches a state of equilibrium, which may be overthrown by altering the relative masses of the members of the system, or by raising or lowering the temperature.

When one or more of the products of a chemical change **240** between liquids or bodies in solution is a solid or a gas, that product is removed from the sphere of interaction of the members of the changing system as quickly, or almost as quickly, as it is formed; hence the change proceeds more or less rapidly to a conclusion, and the direct change is only slightly retarded by interactions among the products tending to reproduce the original substances. Thus the change $BaCl_2Aq + Na_2SO_4Aq = BaSO_4$ (solid) $+ 2NaClAq$ is realised when barium chloride and sodium sulphate are mixed in solution in the ratio $BaCl_2 : Na_2SO_4$. The change $CaCO_3$ (solid) $+ 2HClAq = CaCl_2 + H_2O + CO_2$ (gas) is completed by using the masses of the interacting compounds shewn in the equation. But comparatively few chemical changes are completed unless an excess of one of the interacting substances is used. It is certain that in many cases, and it is probable that in most cases, the products of the direct change interact to produce the substances originally present; the whole system swings in two directions, and the result is a state of equilibrium, which is generally more or less easily overthrown by adding more of one of the interacting substances, or by removing one of the products of the direct change, or by altering the temperature.

Whether a chemical change will or will not occur, and if it **241** occurs, what will be the extent and the products of the change, depend chiefly (1) on the chemical characters of the reacting bodies, (2) on the relative masses of these bodies allowed to interact, and (3) on the temperature, and (4) in some cases, especially where gases are produced, on the pressure.

We have already learnt something of the meaning of the expression *chemical character* of a substance; we know, for instance, that certain compounds are acids, others are alkalis, and others salts; these names imply, in each case, a certain composition and the property of interacting in a certain definite way.

But why is it that acids and alkalis interact to produce salts? Does a specified acid always produce the same mass of a salt when it interacts with a specified alkali? The formula of an acid tells us the composition and something about the reactions of the acid, it expresses the composition of the reacting weight of the acid; but would it be possible to find a number for each acid which should express how much of a specified chemical change—say interaction with alkali to form a salt—this acid is capable of performing? Or are the conditions of chemical action so complex, and chemical actions themselves so much modified by slight changes in the conditions under which they occur, as to render hopeless all attempts to determine constants of chemical change? We must now attempt to find answers to such questions as these.

CHAPTER XIII.

CHEMICAL AFFINITY.

THAT property of elements and compounds by virtue of **242** which they interact to produce new combinations is called *chemical affinity.*

A chemical change may always be regarded as one of the **243** results of the mutual action and reaction between two or more substances or systems of substances. There is always a change in the configuration of each part of the chemical system. One member, or one part, of the system may be said to exert force on the other members, or on the other parts, of the system. Thus, when solutions of potash and sulphuric acid interact to produce potassium sulphate and water, the potash may be said to exert force on the acid, and the acid may be said to exert force on the alkali.

Of chemical forces, that is the forces which come into play in chemical changes, we know very little. It is probably better, at any rate for the present, to make no attempt to consider chemical change from the strictly dynamical point of view.

Let us rather look to the manifestations of that property **244** of bodies which is called chemical affinity.

If mercury is placed in a solution of silver in sulphuric acid, the silver is slowly precipitated and some of the mercury is dissolved; if a piece of copper is placed in a solution of mercury in sulphuric acid, the mercury is slowly precipitated and some of the copper is dissolved; if a piece of iron is brought into a solution of copper in sulphuric acid the copper is slowly precipitated and some of the iron is dissolved. The older chemists said, mercury has a greater affinity for sulphuric acid than silver has, but the affinity for sulphuric acid of copper is greater than that of mercury, and the affinity for the same acid of iron is greater than the affinity of copper.

Similarly they said that the affinity for potash of hydrochloric acid is greater than that of phosphoric acid because potassium phosphate is decomposed by hydrochloric acid, that the affinity of nitric acid for potash is greater than that of hydrochloric acid, and the affinity of sulphuric acid for potash is greater than that of nitric acid, because potassium chloride is decomposed by nitric acid, and potassium nitrate is de-composed by sulphuric acid.

Chemical affinity was regarded by the older chemists as analogous to a mechanical force. As two mechanical forces opposite in direction and unequal in magnitude produce motion in the direction of the greater, so it was supposed that a stronger affinity always overcame a weaker and produced chemical change. The affinity of nitric acid for potash, for instance, was supposed to be stronger under all circumstances than that of hydrochloric acid for potash. According to this conception, chemical affinity acts exclusively in one direction.

245   The facts that potassium chloride is decomposed by heating with a considerable quantity of nitric acid, but that potassium nitrate is decomposed by heating with much hydrochloric acid, shew that a complete account of chemical change cannot be given by regarding only the affinities of the interacting substances. It is necessary to pay attention also to the relative masses of these substances.

246   In the early years of this century Berthollet formulated the statement "Every substance which tends to enter into chemical combination with others reacts by reason of its affinity and its mass*." Berthollet taught that a chemical change between substances in solution, wherein neither solids nor gases are formed, results in the production of a system in equilibrium; that each member of the complete system interacts with each other in proportion to its affinity and its mass; and that therefore the equilibrium of the system may be overthrown by changes in the relative masses of one or more of the members of the system.

Berthollet further taught that changes in which solid or gaseous substances are formed are not suitable for the study of chemical affinity, because in these changes all the members of the chemical system are not free to interact, inasmuch as some

* "Toute substance qui tend à entrer en combinaison, agit en raison de son affinité et de sa quantité."

of them are removed from the sphere of interaction almost as quickly as they are formed. The typical normal case of chemical change, according to Berthollet's view, is one wherein every member of the system is free to interact with all the other members throughout the whole of the change; those changes wherein a final distribution of the interacting substances is quickly established by the formation of solid precipitates or the evolution of gases are special limiting cases.

Berthollet's *law of mass* has been developed in recent years **247** chiefly by the researches of Guldberg and Waage and of Ostwald. Guldberg and Waage formulate the *law of mass* thus *chemical action is proportional to the active mass of each substance taking part in the change.* By *active mass* is meant that quantity of a substance measured in equivalent weights which is present in unit volume of the chemical system.

The expression *equivalent weights* will be explained more fully hereafter (*s.* Chap. XVII.). We know that the amounts of potash and soda which severally neutralise 36·5 parts by weight of hydrochloric acid ($HCl = 36.5$) are those expressed by the formulae KOH (56) and NaOH (40), respectively. We also know that to neutralise a reacting weight of sulphuric acid ($H_2SO_4 = 98$) 112 parts by weight of potash (2KOH) or 80 of soda (2NaOH) are required. So far as neutralising by alkali is concerned, the quantities expressed by the formulae 2HCl (or H_2Cl_2) and H_2SO_4 are *equivalent*; so far as neutralising by acid is concerned the quantities KOH and NaOH (or 2KOH and 2NaOH) are *equivalent.*

Suppose that a solution of 112 parts by weight of potash (2KOH), 73 parts by weight of hydrochloric acid (2HCl), and 98 parts by weight of sulphuric acid (H_2SO_4), is diluted with water to a specified volume; then the *active masses* of potash, hydrochloric acid, and sulphuric acid, respectively, in this solution are one equivalent of each, provided that by one equivalent is meant the quantity expressed by the formulae 2KOH (or $K_2O_2H_2$), 2NaOH (or $Na_2O_2H_2$), and H_2SO_4, respectively.

The law of mass-action has been experimentally proved in many different reactions; it probably holds good in all chemical changes.

The principle of the *coexistence of reactions* states that **248** when several reactions occur simultaneously, each proceeds as if it alone took place. No direct experimental investigation of this principle has been made; but it has been largely applied in work on chemical affinity, and numerous results have been obtained in keeping with the principle.

249   Assuming the law of mass-action, and the principle of the coexistence of reactions, let us briefly examine a fairly simple chemical change. Let equivalent quantities of the alkali caustic potash, and the acids hydrochloric and sulphuric, be mixed in dilute aqueous solution; let the substances be present in the ratio $K_2O_2H_2 : H_2Cl_2 : H_2SO_4$. The possible products of the interactions are potassium sulphate (K_2SO_4), potassium-hydrogen sulphate ($KHSO_4$), potassium chloride (KCl), and water (H_2O). But these substances may interact to reproduce the original substances. We have then certain direct changes and certain reverse changes possible. Chemical equilibrium will result when the velocities of the opposite reactions have become equal, that is, when the quantities of the substances formed in the direct change are equal to the quantities of the substances formed in the reverse change, in unit of time. But we say that each change, the direct and the reverse, is proportional to the affinities, and the masses, of the reacting substances. Now we can measure the mass of each substance present at the beginning of the change, and we can also measure the mass of each substance present when equilibrium is established; hence we can deduce numerical values for the affinities of the reacting substances. Guldberg and Waage, Ostwald, van 't Hoff, and others, have deduced the necessary equations from the fundamental statements already made.

But there is another method by which values for the relative affinities of the substances taking part in a chemical change may be deduced from experimental data. The change may be allowed to proceed to a certain extent only, but not until the system has settled down into equilibrium; the quantity of each substance present in the system may then be measured, and the velocity of the change may thus be determined. Then, assuming that the change which has occurred is proportional to the affinities and the masses of the interacting substances, we may deduce relative values for these affinities from our measurements of the masses. The necessary equations have been deduced by Guldberg and Waage, Ostwald, and others.

250   One of the great difficulties in applying these methods is to find reactions which are sufficiently simple. Very many chemical changes which appear to be simple are complicated by the occurrence of secondary reactions among the products of the primary change.

Another difficulty is to measure the masses of the substances present when equilibrium is established. Suppose, for instance, that potash and sulphuric and nitric acids have been mixed in the ratio $K_2O_2H_2 : H_2N_2O_6 : H_2SO_4$, in dilute aqueous solution; how are we to determine how much potassium nitrate, and how much potassium sulphate, is actually present in the solution? The ordinary methods of analysis are useless here, because they are based on the use of reagents other than the substances in the solution; but the addition of any reagent is forbidden because the equilibrium of the system would thereby be destroyed. We cannot here go into the methods adopted; many of them consist in measuring some definite physical change and using this as an index of the chemical change which has occurred.

In the case of the two acids, nitric and sulphuric, reacting **251** with potash in equivalent quantities it has been shewn, with a very high degree of probability, that about 10 parts of potash combine with the nitric acid to form potassium nitrate, for each 7 parts which combine with the sulphuric acid to form potassium sulphate, when the system is in equilibrium. Hence it is concluded that the ratio of the affinities for potash of nitric and sulphuric acids is approximately 10 : 7. When the acids are nitric and hydrochloric, and the base is potash, the potash divides itself almost equally between the two acids; that is one half of the potash reacts with the nitric acid to produce potassium nitrate, and one half with the hydrochloric acid to produce potassium chloride. Hence the affinities of hydrochloric and nitric acids for potash are approximately equal. When the acids are hydrochloric and acetic almost the whole of the potash reacts with the hydrochloric acid; only about ·4 parts of potash react to produce potassium acetate for each 100 parts which react to produce potassium chloride. Hence the ratio of the affinities for potash of hydrochloric and acetic acids is approximately 100 : ·4.

It is very important to observe what is the exact meaning we are now giving to such a statement as this 'the relative affinity for potash of nitric acid is to that of hydrochloric acid as 1 : 1,' or as this 'the ratio of the affinities for potash of hydrochloric and acetic acids is 100 : ·4.' When these statements are amplified they assert, (1) that if equivalent masses of caustic potash (KOH), hydrochloric acid (HCl), and nitric acid (HNO_3), are allowed to interact freely in dilute aqueous solution, one half of the potash combines with each acid to produce potassium chloride and potassium nitrate, respectively; (2) that if equivalent masses of caustic potash, hydrochloric acid, and acetic acid ($C_2H_4O_2$), are allowed to interact freely in dilute aqueous solution, then for every 100 parts of potash which are changed to potassium chloride only ·4 parts of potash are changed to potassium acetate.

If these statements are correct, it is evident that when equilibrium results the solution contains in the first case, potassium chloride and nitrate and also hydrochloric and nitric acids; and in the second case, much potassium chloride, a little potassium acetate, a little hydrochloric acid, and much acetic acid.

252 The experiments of Ostwald and Thomsen have shewn that the relative affinities of acids are almost, if not quite, independent of the nature of the base; in other words that if equivalent masses of, say, hydrochloric and nitric acids, are mixed in dilute aqueous solution with an equivalent mass of caustic potash (KOH), or soda ($NaOH$), or ammonia (NH_4OH), or caustic lime (CaO_2H_2), or caustic baryta (BaO_2H_2), &c. one half of the base combines with each acid.

Ostwald's experiments have also rendered it very probable that the ratio of the affinities for bases of any two acids is independent of the temperature, at least within such a range as 0° to 60°.

253 There are many chemical changes brought about by acids other than those which take place between acids and bases. Some of these changes have been examined with the object of determining whether they are quantitatively conditioned by the same values as have been found to condition the reactions between acids and bases. Thus Ostwald made a number of measurements of the effects of different acids on the velocity of the change of acetamide into ammonium acetate. This change may be represented thus

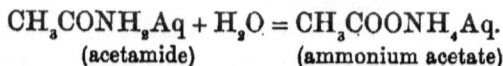

$$CH_3CONH_2Aq + H_2O = CH_3COONH_4Aq.$$
$$\text{(acetamide)} \qquad\qquad \text{(ammonium acetate)}$$

The change occurs more or less quickly in the presence of acids; each acid increases the amount of change in unit time to a certain definite extent. The equations deduced from the fundamental statement, that chemical action is proportional to the relative affinities and the active masses of the interacting substances, shew that the ratio of the affinities in such a change as that under consideration is equal to that of the square roots of the velocities of the reactions. Ostwald measured the velocities of the reaction for various acids, and hence deduced the relative affinities of these acids. The numbers obtained agree well with those found by the examination of the reactions between the same acids and potash, soda, and other bases. Among other chemical changes examined were, the change of methylic acetate in presence of water and an acid into methylic alcohol and acetic acid

$$(CH_3COOCH_3Aq + H_2O = CH_3OHAq + CH_3COOHAq);$$

and the change of cane sugar in presence of water and an acid into glucose $(C_{12}H_{22}O_{11}Aq + H_2O = 2C_6H_{12}O_6Aq)$. In each case the velocity of the change was determined for various acids; the ratios of the square roots of the velocities were taken (as indicated by theory) as the ratios of the relative affinities of the acids. The numbers agree as well as could be expected among themselves, and also with the numbers found by the study of the reactions between acids and bases.

The electrical conductivities of solutions of acids are proportional to the velocities of the chemical changes produced by these acids. Hence measurements of the electrical conductivities of acids in aqueous solutions of various concentrations give data from which the relative affinities of these acids may be deduced. Many measurements have been made of the electrical conductivities of acids in aqueous solutions, chiefly by Ostwald; the values of the relative affinities deduced from these results agree very well with those found by more strictly chemical methods.   **254**

The outcome of the work which has been done in recent years on the subject of the affinities of acids is to establish the conclusion that it is possible to determine for each acid a *specific affinity-constant* which quantitatively conditions all the reactions brought about by this acid.   **255**

Of course when it is said that this or that reaction is brought about by an acid, the reaction is regarded as being more simple than it really is. The reaction is brought about by all the substances which form the chemically changing system. But it seems that we may regard the complete change as made up of various parts each of which occurs in accordance with its own laws. The more completely a specified change is dependent on the character of the acids which take

part in that change the more suitable is it for deducing values
for the relative affinities of these acids.

The interactions of acids and bases in dilute aqueous solutions are conditioned only by the characters of the acids and
the bases.   The results of measurements of these changes
render it very probable that each acid has a specific affinity-
constant which is independent of the nature of the base interacting with the acid, and that each base has a specific affinity-
constant which is independent of the nature of the acid interacting with the base.

The results of measurements of many other reactions which
occur only in the presence of acids, and which may justly be
said to be caused by these acids, render it very probable that
the amount of change occurring in a specified time, or the
amount of change which has occurred when the system has
settled down into equilibrium, is conditioned by the values of
the same specific affinity-constants which condition the interactions between these acids and bases in dilute aqueous
solutions.

256   If these conclusions are granted—and they rest on a large
body of carefully verified facts—it follows that measurements
of the specific affinity-constants of the acids are of the utmost
importance.

Of the various methods hitherto employed for making
these measurements the most promising seems to be that
based on the proportionality between the electrical conductivities and the velocities of the chemical reactions brought
about by acids.   This method presents no great experimental
difficulties, and it is free, or nearly free, from the disturbing
influence of secondary reactions.   Most, if not all, purely
·chemical methods are open to the objection that the primary
change to be measured is complicated and modified by the
occurrence of other changes, and that the influence of these
secondary changes can scarcely be eliminated by any experimental arrangements.   Many measurements of the electrical
conductivities of aqueous solutions of acids have been made,
and data have thus been accumulated for comparing the
relative affinities of many acids.   We shall shortly consider
some of these data when we have learned more about the
composition of acids (s. Chap. XVII.)

Acids with large affinity-values are called *strong* acids;
those with small affinity-values are called *weak* acids.

257   Answers can then be given to the questions propounded at the close of Chap. XII.; we have learnt that a number can be found for each acid, and each base, which expresses the amount of chemical change which this acid, or base, is capable of producing under defined conditions.

It is probable that, as investigation proceeds, specific affinity-constants will be determined for the members of other classes of compounds besides acids and bases.

We have now learnt something about chemical composition **258** and chemical classification. We have also found that many, and probably all, chemical reactions brought about by the compounds classed together as acids are quantitatively conditioned by the affinity-constants of these acids.

We ought now to inquire into the connexions between the compositions of acids and the values of their affinity-constants. But we are not yet ready for this inquiry; we must learn more regarding chemical composition. (*s.* Chap. XVII.)

No attempt has been made in this chapter to analyse the **259** meaning of the term affinity; we have not asked why this body chemically interacts with that; we have not inquired as to the nature of chemical affinity. We have been content to call affinity that property of elements and compounds by virtue of which they interact to produce new combinations. We have found it possible to assign quantitative values to this property in the cases of acids and bases.

Before proceeding to consider in some detail the generally accepted theory regarding the mechanism of chemical change, we shall briefly glance at the relations between chemical changes and the changes of energy which invariably accompany them.

CHAPTER XIV.

RELATIONS BETWEEN CHEMICAL CHANGES AND CHANGES OF ENERGY *. .

260 EVERY chemical change consists of two parts, a change in the form of combination of the matter of the system, and a change in the total quantity, or in the form, or in both the quantity and form, of the energy of the system.

Energy is the power of doing work. *Work* is the "act of producing a change of configuration in a system in opposition to a force which resists that change."

If one system does work on another system, one loses and the other gains energy ; and the energy lost by one is equal to the energy gained by the other. If both systems are included in a larger, the total energy of this system is unchanged. If one part of a system does work on another part, the total energy of the system is unchanged, although one part has gained and another part has lost energy.

The principle of the *conservation of energy* affirms that ;—

"The total energy of any material system is a quantity which can neither be increased nor diminished by any action between the parts of the system, though it may be transformed into any of the forms of which energy is susceptible." (Clerk Maxwell.)

261 The energies of actually existing material systems depend upon the states of these systems at any moment. The state of a system is conditioned by many variables; among the more important are chemical composition, pressure, temperature, and volume.

If we wish to connect changes of energy with changes of chemical composition we must start with chemical systems of

* The subject of energy is treated very shortly. The student should refer to a book on Physical principles, *e.g.* to Clerk Maxwell's *Matter and Motion.*

definite and defined composition, in definite and defined states, and we must cause these to change to other definite and defined states; we must then determine the compositions of the resulting systems, and we must measure the changes of energy which have accompanied these changes of composition and of state.

Of two equal quantities of energy one may be more **262** available for doing work than the other. Thus, in order to cause thermal energy to do work it is necessary to allow it to pass from a body at a higher to a body at a lower temperature. A certain body may be at a very low temperature and yet contain thermal energy; but it may be impossible to cause this energy to do work, because of the impossibility of framing an engine consisting of the cold body and another system at a lower temperature than the cold body. A quantity of heat as it exists in a hot body is more available for doing work than the same quantity of heat as it exists in a colder body.

When energy passes from a more available, or higher, to a **263** less available, or lower, form it is said to be *degraded*. All forms of energy can be directly or indirectly transformed into heat. A given quantity of heat-energy cannot be wholly transformed into one of the higher forms of energy. Every transformation of energy involves the degradation of a portion of the energy. But every chemical change is accompanied by a transformation of energy from the form of chemical energy to other forms of which thermal energy is usually one; every chemical change therefore is accompanied by a degradation of energy. It is not asserted that the whole of the energy which changes form during a chemical change is necessarily degraded.

The chemical system represented by the symbols $2H + O$ **264** contains more energy than the system represented by the symbol H_2O. In the passage from one of these systems to the other energy is lost by the changing system; the energy so lost by the system is gained by neighbouring systems, by the vessel in which the change is accomplished, the surrounding air, &c. But although there is no destruction, there is degradation, of energy. If $2H$ represents 2 grams of hydrogen, O represents 16 grams of oxygen, and H_2O represents 18 grams of liquid water, all measured at normal pressure and at about $15°—16°$, then the change $2H + O = H_2O$ is accompanied by the production of 68,360 gram-units of heat. If we assume that the whole of the energy which changes form during the chemical change $2H + O = H_2O$ appears as heat, then 68,360 gram-units of heat represents the difference between the energies of the two systems $2H + O$ and H_2O. Whether this quantity of heat does or does not measure the total difference of energy between the two systems, it is certain that the change from the one system to the other is always accompanied by the production of the same quantity of heat. And what is true of this chemical change is true of others also. Each definite chemical change from one system of defined composition, under defined conditions, to another system of defined composition, under defined conditions, is accompanied by the production or disappearance of a fixed quantity of heat. The following examples illustrate this point. In each case the original and final systems are under the normal pressure (760 mm.) and the temperature of each is about 16°. The symbols represent the combining, or reacting, weights taken in grams.

Original System.	New System formed.	Gram-units of heat which are produced or disappear. [The sign + signifies produced, the sign −, disappears.]
$Cl + H$	HCl	$22,000 +$
$Br + H$	HBr	$8,440 +$
$I + H$	HI	$6,040 -$
$S + 2H$	H_2S	$4,740 +$
$2C + 4H$	C_2H_4	$2,710 -$
$C + 2O$	CO_2	$96,960 +$
$K + Cl + 3O$	$KClO_3$	$95,860 +$
$KCl + 3O$	$KClO_3$	$9,750 -$
$KClO_3 + Aq$	$KClO_3Aq$	$10,040 -$
$K + Cl + 3O + Aq$	$KClO_3Aq$	$85,820 +$
$H_2O + 2Cl + Aq$	$2HClAq + O$	$10,270 +$
$HClOAq$	$HClAq + O$	$9,380 +$
$S + 3O$	SO_3	$103,240 +$
$S + 3O + H_2O$	H_2SO_4	$124,560 +$
$SO_3 + H_2O$	H_2SO_4	$21,320 +$
$H_2SO_4Aq + H_2S$	$H_2S_2O_3Aq + H_2O$	$9,320 -$

265 In some of these changes heat disappears from the system, that is to say, energy is raised from the form of heat-energy to some other more available, or higher, form. Yet it has been asserted (par. 263) that every chemical change is accompanied by a degradation of energy. Take, for example, the change H + I = HI ; in this change 6040 gram-units of heat disappear. But in order to bring about the formation of HI from H + I it is necessary to heat the system H + I to 300°—400°; that is to say it is necessary to add energy from without the system. This added energy is employed in bringing the system H + I into a condition such that chemical action becomes possible; chemical action results and this action is attended with a degradation of energy.

Suppose a stone to rest at the bottom of an inclined plane AB (Fig. 19). Let the stone be moved from A to B ; to perform

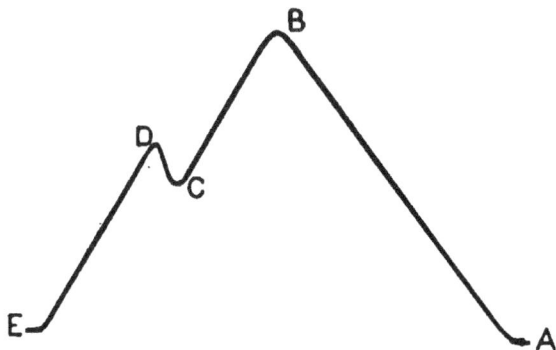

Fig. 19.

this work a certain amount of energy must be used. The stone now possesses more energy, by virtue of its position, than it did when it was at A. Let the stone be moved a very little way over the crest of the plane AB, it will now roll down the inclined plane BC until it comes to rest at C. In the passage from B to C energy has been degraded; but the stone at C possesses more energy than it did at A because the level of C is higher than that of A. If the stone is rolled up the short incline CD, by the expenditure of energy, it will be in a position to descend from D to E ; in this descent energy will be degraded. When the stone reaches E it will possess the same energy as when it was at rest at A.

The system H + I corresponds to the stone at A ; to bring the system into such a condition that chemical change can occur, energy must be expended. The system ready to undergo chemical change corresponds to the stone at B. Chemical change occurs; the system passes from B to C, from H + I to

HI, and this passage is attended with degradation of energy. But the system HI possesses more energy than the system H + I did before energy was expended upon it in causing it to pass into that condition in which chemical change could occur; therefore although the *chemical* change from H + I to HI has been attended with degradation of energy, yet the *whole* change from 1 gram of gaseous hydrogen and 127 grams of solid iodine, at say 15° or 16°, to 128 grams of gaseous hydrogen iodide at the same temperature, has been attended by a raising of energy from a lower to a higher form. The system HI is readily separated into the elements H + I. To effect this separation the temperature of the gas is raised; this corresponds to the rolling of the stone up the small incline CD. The separation of HI into H + I is attended with the degradation of a considerable quantity of energy, just as the descent of the stone from D to E is attended with the degradation of energy.

In this example the energy required to bring the original system into that state in which chemical action could occur was furnished in the form of thermal energy from without the system.

But the necessary energy for bringing a chemical system into that state in which chemical change can take place, and energy can be degraded, is frequently supplied by chemical actions occurring within a larger system of which the special system under consideration forms a part. Thus the change from 71 grams of gaseous chlorine and 16 grams of gaseous oxygen to 87 grams of gaseous chlorine monoxide ($2Cl + O = Cl_2O$) is accompanied with the disappearance of 17,900 gram-units of heat. Chlorine monoxide is produced by passing chlorine over mercuric oxide; the equation representing the chemical change is $2HgO + 4Cl = Cl_2O + Hg_2OCl_2$. This change consists of various parts; (1) formation of $HgCl_2$ and O, thus $HgO + 2Cl = HgCl_2 + O$, this is accompanied by the production of 16,250 gram-units of heat; (2) production of the oxychloride Hg_2OCl_2, thus $HgCl_2 + HgO = Hg_2OCl_2$, this is accompanied by the production of 8,900 gram-units of heat; (3) formation of Cl_2O, thus $2Cl + O = Cl_2O$, this is accompanied by the disappearance of 17,900 gram-units of heat. Now $16,250 + 8,900 = 25,150$; and $25,150 - 17,900 = 7,250$. The complete change from $2HgO + 4Cl$ to $Hg_2OCl_2 + Cl_2O$ is accompanied by the production of 7,250 gram-units of heat.

To bring the system $2Cl + O$ into that state in which

chemical union can take place, work must be done upon the system, energy must be expended; this energy is supplied by the other chemical changes $HgO + 2Cl = HgCl_2 + O$ and $HgO + HgCl_2 = Hg_2OCl_2$, in each of which energy is degraded.

Every chemical change is attended with the degradation of energy; but to bring a given system into that state in which chemical change can take place it may be necessary to expend energy upon the system; this energy is sometimes supplied by processes altogether outside the system, sometimes by making the required chemical change one part of a cycle of changes in some of which energy is set free under such conditions as to be directly available for bringing the special part of the whole system into that state in which the wished for chemical change can occur.

We have had examples of oxidation accomplished by arranging a series of chemical changes so that oxygen shall be produced in contact with the substance to be oxidised. Thus lead monoxide (PbO) was oxidised to lead dioxide (PbO_2) by suspending the monoxide in concentrated warm potash solution and passing chlorine into the liquid: potassium hypochlorite (KClO) was thus produced, but was at once decomposed to potassium chloride (KCl), and oxygen which combined with the lead monoxide to produce lead dioxide. If lead monoxide is suspended in water or potash solution and oxygen is passed into the liquid no lead dioxide is formed. Now to bring the system $PbO + O$ into that state in which the system PbO_2 can be produced, energy must be expended: but the interaction of potash solution with chlorine, to produce potassium chloride, water, and oxygen, is attended with the setting free of a large quantity of energy; a portion of this energy is employed in bringing the part of the whole changing system represented by the symbols $PbO + O$ into that condition in which chemical action is possible (s. Chap. XI. par. 158.)

When zinc and dilute sulphuric acid interact a solution of zinc sulphate, and hydrogen gas, are produced; if a solution of sodium sulphite (Na_2SO_3) is added to this changing system, hydrogen sulphide (H_2S) is evolved. But if hydrogen gas is passed into a solution of sodium sulphite, hydrogen sulphide is not produced (s. Chap. XI. par. 175.) To effect the change $Na_2SO_3Aq + 6H = Na_2OAq + 2H_2O + H_2S$ energy must be expended: when Na_2SO_3Aq is added to dilute sulphuric acid in contact with zinc, both the material and the energy needed for effecting the chemical change are provided; and

moreover the energy is provided exactly in the form in which
it is required for accomplishing the chemical work to be done.

267   When energy is degraded in a chemical change, a
portion, or perhaps in some cases the whole, of the energy is
degraded to the form of heat. But the quantity of heat
produced seldom if ever affords a direct measure of the
chemical energy degraded. Chemical changes are always
accompanied by more or less marked physical changes ; the
production of the energy which appears as heat is generally
due partly to the chemical, and partly to the physical, portion
of the complete change. Thus when gaseous hydrogen and
oxygen combine to form liquid water a large quantity of
energy is degraded, and much heat is produced [$2H + O = H_2O$
$= 68,360$ gram-units of heat $+$]; some of this heat is due to the
chemical change from the system $2H + O$ to the system H_2O,
some of it is due to the physical change from gaseous water to
liquid water.

Some chemical changes, with their accompanying physical
changes, occur with the disappearance of heat. Thus selenion
dioxide dissolves in water to form a solution of selenious acid,
$SeO_2 + H_2O + Aq = H_2SeO_3Aq$; this change is accompanied by
the disappearance of 920 gram-units of heat. Similarly iodine
pentoxide dissolves in water to form a solution of iodic acid,
$I_2O_5 + H_2O + Aq = 2HIO_3Aq$; this change is accompanied by
the disappearance of 1790 gram-units of heat. [The formulae
represent reacting weights taken in grams ; SeO_2 e.g. means
here 111 grams of selenion dioxide.] In such cases we may
suppose that the whole of the energy degraded in the chemical
part of the complete change, and some of the energy degraded
in portions of the physical change, are degraded to heat, but
that this quantity of heat is wholly used in effecting other
portions of the physical change. Or we may suppose that
only a portion of the chemical energy is degraded to heat, and
that this heat is used, along with other forms of energy
produced by the degradation of the chemical energy, in
effecting the physical portion of the complete change. Either
supposition is in keeping with the fact that the total change
occurs with the disappearance of heat, and with the generalisation that every chemical change is accompanied by degradation of energy.

268   It is important to note once again that the statement that
every chemical change is accompanied by degradation of
energy does not assert or imply that the whole of the energy

which changes form during a chemical change is degraded : some of it is degraded, but some of it may be raised to a more available form.

In chapters XII. and XIII. we saw that many chemical **269** changes may be justly regarded as proceeding in two directions simultaneously, and that equilibrium results when the velocities of the direct and reverse changes become equal. We saw also that such chemical equilibrium may generally be overthrown by changing the‧temperature, or sometimes the ‚pressure, of the system, or the relative masses of the interacting substances. The considerations concerning the relations of energy-changes and chemical changes shortly developed in the present chapter may be applied to the conception of chemical equilibrium gained in chaps. XII. and XIII. Suppose that the masses of ferric chloride and potassium sulphocyanide shewn by the formulae Fe_2Cl_6 and $K_6C_6N_6S_6$ ($= 6KCNS$) are mixed in dilute aqueous solution; the system is in a condition in which chemical change can occur; chemical change occurs, and a system is produced the composition of which may be represented by the equation $Fe_2Cl_6Aq + 6KCNSAq + x Fe_2Cl_6Aq$
$+ x'$ KCNSAq $= Fe_2 (CNS)_6Aq + 6KClAq + x Fe_2Cl_6Aq$
$+ x'$ KCNSAq (*comp.* Chap. XII. par. 238). Some energy is degraded in this change ; a portion of this energy appears as heat, a portion is probably employed in effecting some of the physical changes (contraction or expansion of volume &c.) which accompany the chemical change; a portion of the energy degraded in one part of the change is also probably employed in bringing the products of the change into a state in which they can interact to reproduce the original substances. After a very short time the system settles down into a state in which there is equilibrium of energy and of chemical distribution of the interacting substances. A little more potassium sulphocyanide is now added; chemical change again occurs; energy is degraded ; and after a short time equilibrium is established. Potassium sulphocyanide is added little by little until the whole of the ferric chloride originally present has been changed; the addition of more sulphocyanide cannot now cause chemical change; no more energy can be degraded by chemical processes ; the system has reached its state of final equilibrium. Each addition of potassium sulphocyanide disturbed the equilibrium of the energies of the system, and this disturbance was attended by chemical change ; but a disturbance of the equilibrium

of the energies of the system would also be produced by raising the temperature of the system; hence raising the temperature of the system would also alter the distribution of the elements forming the members of the system, ·i.e. would cause chemical change to occur.

Now consider a chemical change, one of the products of which is a solid under the conditions of the experiment. Suppose aqueous solutions of barium chloride and sodium sulphate to be mixed in the ratio $BaCl_2 : Na_2SO_4$. Chemical change occurs; energy is degraded; there is change from liquid to solid, and more energy is degraded. As one of the products of the chemical change (barium sulphate) is removed from the sphere of action, by precipitation in the solid form, none of the energy degraded in the direct change can be used to bring the products of this change into a state in which they can chemically react to reproduce the original substances; hence the whole, or at any rate nearly the whole, of the energy degraded to the form of heat passes out of the system. The system is, so to speak, rapidly rolling down hill. Chemical change proceeds until the whole of the energy which can be degraded to heat has been degraded. The system is now in its final state of equilibrium. And this final state has been reached without adding an excess of either of the interacting substances.

270   We have now gained some fairly clear conceptions regarding chemical change.

Elements and compounds interact to produce other elements and compounds. Numbers are given to the elements expressing the masses of them which combine or interact with unit mass of one element chosen as a standard, and which also interact or combine with each other. These numbers we have called the combining weights of the elements. Numbers are also given to compounds which express the smallest masses of them which chemically interact with each other. These numbers we have called the reacting weights of the compounds. But it is necessary in chemistry to have regard not only to composition but also to properties. Elements are classified in accordance with their properties into metallic or positive, and non-metallic or negative, elements. They are also classified in groups in accordance with the properties and compositions of their oxides, hydrides, haloid and oxyhaloid compounds, &c. This classification of elements carries with it a classification of compounds also. Compounds are classified in accordance with their properties into acids, bases, salts, &c. But with the properties connoted by each of these terms there is associated a certain composition. The term *acid*, for instance, implies certain common properties, and a certain common composition.

But chemistry is not content with finding an answer to the question—What is produced in this process and how much of it is produced? it seeks to find an answer to this question also—How is it produced? Chemistry therefore examines the conditions and general laws of the interactions of elements and compounds. One substance interacts with another to produce new substances; but the new substances also interact, unless they are prevented by the removal of one or more of them from the sphere of action, and tend to reproduce the original substances. Chemical change results in chemical equilibrium. Each substance taking part in a chemical change, wherein all the substances are free to act and react, probably produces a certain definite and measurable effect on the change, which effect is independent of the interactions of the other substances. In certain classes of changes at any rate it is possible to assign to each of the two primarily interacting substances a definite number, a knowledge of which enables us to predict the amount of change which will occur under defined conditions. Chemical change is accompanied by change of energy; there is a redistribution of the matter which undergoes change and also of the energies of the parts of the changing system.

Although we have thus gained some fairly clear conception **271** of the general character of chemical change, and of the kind of phenomena studied in chemistry, yet we stand greatly in want of a general theory which shall bring the facts together and bind them into a connected whole. There is a theory which to a great extent does this. This theory we must now endeavour to understand.

CHAPTER XV.

THE MOLECULAR AND ATOMIC THEORY.

272   THE Greek philosophers Leucippus and Democritus (about 440—400 B.C.) were among the first to give definite shape to the conception that "the bodies which we see and handle, which we can set in motion or leave at rest, which we can break in pieces and destroy, are composed of smaller bodies, which we cannot see or handle, which are always in motion, and which can neither be stopped, nor broken in pieces, nor in any way destroyed or deprived of the least of their properties" (Clerk Maxwell). This doctrine was developed by Epicurus (340—270 B.C.). In the poem *De Rerum Natura*, Lucretius gives what purports to be an account of the teaching of Epicurus on the subject. The conception of atoms is fully elucidated in this poem, and on it is based a theory of the physical universe, and to some extent also a theory of things moral and spiritual. Lucretius says that nothing exists except atoms and empty space, that the atoms are of different forms and different weights, and that the number of atoms of each form is infinite; that the atoms are in constant motion, and that all change consists in the separation and combination of atoms. According to Lucretius, every atom is indestructible, and its motion is indestructible likewise. Atoms unite to form different kinds of substances; the properties of the substances so formed depend on the mutual relations of the atoms—"it matters much with what others and in what positions the atoms of things are held in union, and what motions they mutually impart and receive."*

Nearly complete as it was in many respects, the Lucretian theory failed as a scientific conception; it did not work. It

* Lucretius, *De Rerum Natura*, II. 1007—9 (Munro's translation).

did not admit of accurate applications to the facts of nature. It was not a science-producing theory, but rather a speculation about the possible causes of natural events.

The teachings of the atomists were opposed by the follow- **273** ers of Aristotle, for whom the names of things were as real or more real than the things themselves. As Aristotelianism prevailed during the middle ages, atomism declined. The atomic theory of the Greek philosophers was revived towards the end of the 16th century by Gassendi. Boyle and Newton were upholders of this theory. Newton's demonstration of the action of the force of gravitation made a science of atomic physics possible; but the great difficulty was, and still is, to form a clear image of the action of gravitation in terms of the atomic conception of the structure of matter.

Not much was done to advance the applications of the **274** atomic theory, after Newton, until in the early years of the present century Dalton made a serious attempt to determine the conditions under which the atoms of elementary bodies unite to form the atoms of compound bodies. Dalton said that it is possible to find the relative weights of the atoms of elements and compounds, and he indicated the method by which this could be done.

Dalton found that the mass of hydrogen which combined **275** with carbon to form a certain compound of these elements was twice as great as the mass of hydrogen which combined with the same quantity of carbon to form another compound of these elements. He found also that a specified mass of carbon combined with a certain mass of oxygen to form one oxide of carbon, and with twice that mass of oxygen to form another oxide of carbon. He noticed similar regularities in the masses of oxygen which combined with a fixed mass of nitrogen. Meanwhile Dalton, led thereto by his physical experiments on the absorption of different gases by water, had been thinking a great deal about the ultimate structure of matter. He pictured to himself a quantity of carbonic acid gas as built up of innumerable minute particles, or atoms, each of which was itself composed of one atom of carbon and two atoms of oxygen; a quantity of nitrous oxide gas as built up of a vast number of atoms, each of which was itself composed of yet smaller portions of matter, viz. of one atom of nitrogen and one atom of oxygen; and a quantity of hydrogen gas as built up of minute particles, which were single atoms each composed only of hydrogen. Fig. 20, copied from the original in the

13—2

New System of Chemical Philosophy, gives Dalton's pictorial presentment of his conception of the atoms of these three gases.

276

Dalton's application of the term atom to compounds, such as water, carbonic acid, &c., shews that he did not use the word *atom* in its strict etymological meaning as 'that which cannot be cut' but rather as signifying the smallest portion of a body which could exhibit the properties of the body. The atom of water, for instance, could be separated into atoms of hydrogen and oxygen, but this act of separation into parts produced kinds of matter wholly different from the water. The properties of the parts of the atom of water were quite unlike those of the atom itself; whereas the properties of a quantity of water were regarded by Dalton as the same as those of the atom of water.

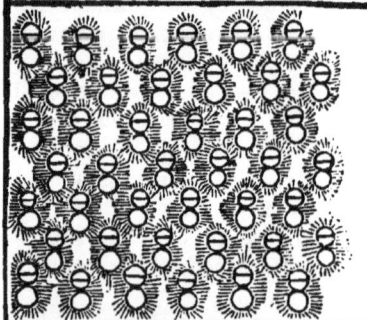

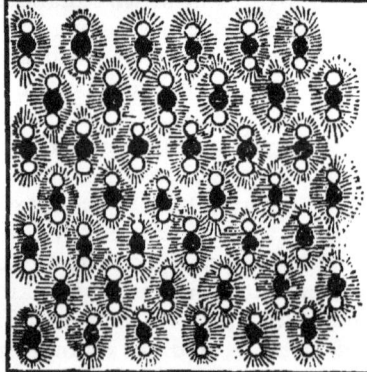

277

The reason of the regularities in the compositions of the two oxides of carbon, or the two compounds of carbon and hydrogen, examined by Dalton, was to be found, according to him, in the nature of the atoms of carbon, hydrogen, and oxygen. It is only

Fig. 20.

necessary to assume that the atoms of these elements do not separate into parts in chemical reactions, then the facts find a simple explanation. One atom of hydrogen combines with one atom of carbon to form one atom of a certain hydride of carbon; if another compound of these elements is formed containing more hydrogen, relatively to the same mass of carbon, the next smallest quantity of hydrogen which can combine with the atom of carbon is two atoms. Similarly with the oxides of carbon. One atom of carbon combines with one atom of oxygen to form an atom of an oxide of carbon: this is the simplest possible compound of the two elements. The next compound which can be formed by combining more oxygen with the same mass of carbon, must be that the atom of which is composed of one atom of carbon united with two atoms of oxygen.

Chemical action was thus conceived by Dalton to be an **278** action between atoms. A mass of any element, or compound, was regarded as constituted of a vast number of very small particles of matter, all alike, all having the same weight, but all unlike the atoms of any other element, or compound. The atoms of elements were supposed to be capable of combining together to form atoms of compounds. In some cases the atoms of compounds might combine together to form atoms of more complex compounds. An atom of one element might combine with one, two, three, &c. atoms of another element to form one, two, three, or more, distinct compounds; but the atoms of elements could not separate into parts. The atoms of compounds on the other hand separated into parts when compounds interacted to produce new kinds of matter, and these parts, which were elementary atoms, rearranged themselves to form atoms of the new compounds produced in the interactions. Dalton pictured to himself an atom of a compound as a structure, or building, formed of a definite number of elementary atoms arranged in a more or less definite manner. He used symbols to represent elementary atoms, and he grouped these symbols together to represent compound atoms. Thus the symbol ○ represented an atom of *oxygen*; ⊙, an atom of *hydrogen*; ⊕, an atom of *nitrogen*; ⊕, an atom of *sulphur*; ⊙, an atom of *aluminium*; ⊕, an atom of *potassium*; and so on. In Fig. 21 are given a few of Dalton's symbols for the atoms of compounds. *A* represents an atom of *potash alum*; *B* an atom of *aluminium nitrate*; *C* an atom of *barium chloride*; and *D* an atom of *barium nitrate*.

Thus did Dalton's conception of the atom throw light on the laws of fixity of composition, multiple proportions, and reciprocal proportions. .

"It is one great object of this work" says Dalton in his **279** *New System of Chemical Philosophy*, "to shew the importance and advantage of ascertaining the relative weights of the ultimate particles both of simple and compound bodies, the number of simple elementary particles which constitute one

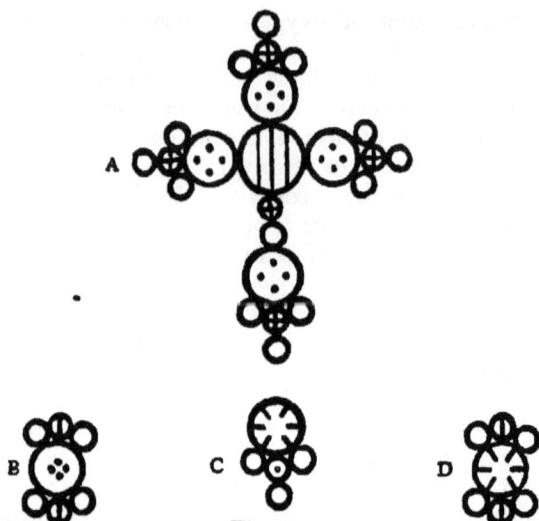

Fig. 21.

compound particle, and the number of less compound particles which enter into the formation of one more compound particle." How then did he determine the relative weights of the ultimate particles of simple bodies?

Let us take the case of oxygen. The atom of oxygen, said Dalton, is 8 times heavier than the atom of hydrogen: let us call the weight of the atom of hydrogen, or the *atomic weight* of hydrogen, one; then the atomic weight of oxygen is asserted to be 8. Masses of hydrogen and oxygen combine in the ratio 1 : 8, to form water; but an atom of water is formed by the union of atoms of hydrogen and oxygen; *if it is assumed that an atom of water is formed by the union of one atom of hydrogen with one atom of oxygen*, then the atomic weight of oxygen must be 8. In saying that the atomic weight of oxygen was 8, Dalton implicitly made this assumption. But it might be assumed that an atom of water is composed of one atom of oxygen united with two atoms of hydrogen; as an atom of hydrogen is the unit in terms of which the weights of the atoms of other elements are stated, it follows from this assumption that the atomic weight of oxygen is 16; because 2 : 16 = 1 : 8. Or it might be assumed that an atom of water is composed of three atoms of hydrogen united with one atom of oxygen; in this case the atomic weight of oxygen must be 24. Or it might be assumed that an atom of water is composed of two atoms of oxygen united with one atom of hydrogen; in this case the atomic weight of oxygen is 4.

Before the atomic weight of oxygen could be determined **280** from the data of the composition of water, it was necessary to determine the number of atoms of oxygen and hydrogen which united to form an atom of water. Dalton's conception of the atom supplied no method whereby this could be done. To get over this difficulty if possible, Dalton framed several empirical rules regarding the compositions of the atoms of binary compounds.

He classified compound atoms formed by the union of two elements into binary, ternary, quaternary, &c., atoms. Calling the two elements *A* and *B*, he said that a binary atom is formed by the union of one atom of *A* with one atom of *B*; a ternary atom, by the union of, either one atom of *A* with two of *B*, or two atoms of *A* with one of *B*; a quaternary atom, by the union of, either one atom of *A* with three of *B*, or three of *A* with one of *B*. He then laid down the following rules:

" I.   When only one combination of two bodies [elements] can be obtained, it must be presumed to be a *binary* one, unless some cause appears to the contrary.

II.   When two combinations are observed, they must be presumed to be a *binary* and a *ternary*.

III.   When three combinations are obtained, we may expect one to be a *binary*, and the other two *ternary*.

IV.   When four combinations are observed, we should expect one *binary*, two *ternary*, and one *quaternary*, &c. &c."

By applying these rules to water, which was the only compound of hydrogen and oxygen known at the time, Dalton concluded that the atom of water was a binary atom; but his own analyses had convinced him that hydrogen and oxygen combine nearly in the ratio 1 : 7 to produce water [we now know that the ratio is 1 : 8]; therefore he concluded that the atomic weight of oxygen was approximately 7.

Dalton's rules for determining the compositions of compound atoms were not based on any general principle deduced from his fundamental conception of the atom; this conception could not indeed supply such a principle. If only one com-

pound of two specified elements was known, the simplest
assumption to make was certainly that embodied in Dalton's
first rule. But the simplest assumption is not always the
best.

281   The law of combination by volumes of gaseous elements
enunciated by Gay-Lussac in 1809 (*s.* Chap. vi. par. 87) may
be stated as follows. *The gaseous elements combine in the ratios
of their combining volumes, or in ratios which bear a simple
relation to these.*

By *combining volume* is here meant the smallest volume of
a gaseous element which combines with unit volume of hydrogen ; and unit volume of hydrogen is defined to be the volume
occupied by unit mass of hydrogen. All measurements of
volumes are assumed to be made at the same temperature and
pressure.

Gay-Lussac interpreted his law, in the light of the Daltonian
theory, to mean that the ratios of the masses of the combining
volumes of gaseous elements are also the ratios of the masses
of the atoms of these elements. Thus ; 2 volumes of hydrogen
combine with 1 volume of oxygen to form water ; but a volume
of oxygen weighs 16 times as much as an equal volume
of hydrogen ; therefore an atom of oxygen is 16 times heavier
than an atom of hydrogen ; but the atomic weight of hydrogen
is 1, therefore the atomic weight of oxygen is 16.

Again, 1 volume of chlorine combines with 1 volume of
hydrogen to form hydrogen chloride ; but chlorine is 35·5
times heavier than hydrogen, bulk for bulk ; therefore, the
atomic weight of chlorine is 35·5.

282   If this interpretation of Gay-Lussac's law is admitted, the
law supplies a means for determining the atomic weights of
gaseous elements. But Gay-Lussac ventured on the further
generalisation that *equal volumes of gases* (measured at the
same temperature and pressure) *contain equal numbers of
atoms.*

Dalton shewed that this generalisation was inadmissible.
Thus, consider the combination of hydrogen and chlorine.
One volume of hydrogen combines with one volume of chlorine
to form 2 vols. of hydrogen chloride ; therefore, by Gay-
Lussac's generalisation, x atoms of hydrogen combine with
x atoms of chlorine, to form $2x$ atoms of hydrogen chloride.
To make the statement more definite, let us assume that $x = 1$;
then, a single atom of hydrogen by combining with a single
atom of chlorine has produced 2 atoms of hydrogen chloride.

Hence, each atom of hydrogen chloride is composed of half an atom of hydrogen united with half an atom of chlorine. But, by definition, an atom of an element is not separated into parts when it interacts chemically with atoms of other elements or compounds. Hence either Gay-Lussac's generalisation is wrong, or the Daltonian definition of the elementary atom must be modified.

In 1811 the Italian naturalist Avogadro modified the **283** Daltonian atomic theory by introducing the conception of two orders of small particles, the molecule and the atom. The *molecule* of an element or compound, said Avogadro, is the smallest mass of it which exhibits the characteristic properties of that element or compound. The molecule, he said, is formed of smaller parts; these are *atoms*. The atoms which form the molecule of an element are all of one kind; the atoms which form the molecule of a compound are of two, or more, different kinds. Avogadro's conception of the structure of matter applied to the case of water asserted that, if the separation of a quantity of water could be carried far enough, we should at last come to very minute particles each of which would exhibit the properties of water; but if these particles were separated into parts we should no longer have particles of water, but particles some of which would exhibit the properties of hydrogen and some the properties of oxygen. Similarly, if the separation into parts of a quantity of hydrogen could be carried far enough, the hypothesis asserted that we should at last come to very minute particles each of which would exhibit the characteristic properties of hydrogen; but if these particles were separated into parts we should no longer have particles of what we know as hydrogen, but particles more or less unlike hydrogen, yet each the same as all the others.

In other words, the Avogadrean conception of the structure of elements and compounds asserts, (1) that a quantity of a compound, or of an element, consists of a vast multitude of minute particles each of which possesses the characteristic properties of the compound, or of the element; these particles are called molecules; (2) that each of these molecules itself consists of a fixed number of yet smaller particles; these smaller particles are called atoms; (3) that the properties of the atoms which form the molecule of a compound are very different from the properties of the molecule itself; (4) that the properties of the atoms which form the molecule of an element are also different from the properties of the molecule of that

element, but that inasmuch as the atoms which form the molecule of an element are all of one kind, and are only more
minute portions of the same kind of matter as the molecule
itself, there is not so marked a difference between the properties of these atoms and the properties of the molecule formed
by their union, as there is between the properties of the atoms
of the different elements which form a compound and the properties of the molecule of that compound.

284   Avogadro modified the generalisation of Gay-Lussac, and
gave it the following form :—

*Equal volumes of gaseous elements and compounds, measured
at the same temperature and pressure, contain equal numbers of
molecules.*

285   Let us apply this generalisation to the combination (1) of
hydrogen and chlorine to form hydrogen chloride, (2) of hydrogen and bromine to form hydrogen bromide.   In each case we
shall suppose that a certain volume of hydrogen, which we
shall call 1 volume, is caused to combine with the other
element, and that the volume of the gaseous compound is
measured, the temperature and pressure at which all measurements are made being the same.   The data are these ;—

> 1 volume of hydrogen combines with 1 volume of chlorine to form
>   2 volumes of hydrogen chloride.
> 1 volume of hydrogen combines with 1 volume of bromine to form
>   2 volumes of hydrogen bromide.

Let there be x molecules of hydrogen in the 1 vol. used,
then the data translated into the language of the Avogadrean
hypothesis read thus ;—

> x molecules of hydrogen combine with x molecules of chlorine and produce
> $2x$ molecules of hydrogen chloride.
> x molecules of hydrogen combine with x molecules of bromine and produce
> $2x$ molecules of hydrogen bromide.

Now as every molecule of hydrogen chloride is composed
of both hydrogen and chlorine, and as every molecule of hydrogen bromide is composed of both hydrogen and bromine,
the necessary conclusion—if we grant Avogadro's hypothesis
—is that one molecule of hydrogen chloride (or bromide) is
composed of half a molecule of hydrogen and half a molecule
of chlorine (or bromine) ; in other words, that each molecule
of hydrogen, and each molecule of chlorine and bromine, has
separated into at least two parts, and that these parts of

molecules have combined to produce the molecules of the compounds formed in the reactions.

If we now tabulate the data for the reverse chemical changes, and translate these data into the language of Avogadro's hypothesis, we have the following statements :—

{ 2 volumes of hydrogen chloride produce 1 volume of hydrogen and 1 volume of chlorine.
$2x$ molecules of hydrogen chloride produce x molecules of hydrogen and x molecules of chlorine.

{ 2 volumes of hydrogen bromide produce 1 volume of hydrogen and 1 volume of bromine.
$2x$ molecules of hydrogen bromide produce x molecules of hydrogen and x molecules of bromine.

As we concluded from the former data that a single molecule of hydrogen reacting with a single molecule of chlorine (or bromine) produces 2 molecules of hydrogen chloride (or bromide), so now we conclude that 2 molecules of hydrogen chloride (or bromide), when decomposed produce 1 molecule of hydrogen and 1 molecule of chlorine (or bromine).

The outcome of Avogadro's conception of the structure of **286** matter is given in the statement already enunciated ; *equal volumes of gases contain equal numbers of molecules.* The application of this generalisation to the interactions between hydrogen and chlorine, and hydrogen and bromine, has led to the conclusion that the molecules of these elementary gases are composed each of at least two parts, and that these parts part company when the gases interact to form hydrogen chloride and bromide, respectively.

Since the time of Avogadro the physical conception of the **287** molecule, as a minute portion of matter, has been much advanced. Every attempt to gain a consistent notion of the mechanism of physical changes has led to the recognition of the grained structure of matter. The hypothesis which asserts that a mass of apparently homogeneous matter is really homogeneous, that however small are the parts into which the body is divided each part exhibits all the properties of the body, has failed to explain any large class of physical facts. Physicists have fully adopted the view that a quantity of any kind of matter consists of a vast number of very minute particles in constant motion. These minute portions of matter they call molecules. The molecules of a gas are supposed to be continually moving about, frequently colliding against each other and rebounding again, but yet remaining intact during these collisions. The physical definition of the molecule of a gas is given in the following words of Clerk Maxwell.

"*A gaseous molecule is that minute portion of a substance which moves about as a whole, so that its parts, if it has any, do not part company during the motion of agitation of the gas.*"

288 The theory that every portion of a body we can see or handle is composed of a great number of very minute particles, in constant motion, each of which is possessed of the properties which characterise the body in question, does not assert or deny the infinite divisibility of matter. What this theory asserts, to use the words of Clerk Maxwell, is "that after we have divided a body into a certain finite number of constituent parts called molecules, then any further division of these molecules will deprive them of the properties which give rise to the phenomena observed in the substance."

289 The relations between the motions and the space occupied by a number of molecules which are mutually independent have been investigated by mathematical analysis. The equations arrived at, after making a justifiable assumption as to the dynamical meaning of temperature, express with considerable accuracy the observed relations between the volume, temperature, and pressure, of gases considerably removed from their liquefaction-points; that is to say the equations agree well with the laws of Boyle and Charles.

The properties of a system of molecules moving about freely, and acting on each other only when they come into contact, have been investigated mathematically. One of the deductions arrived at is the generalisation which was stated by Avogadro in 1811; 'Equal volumes of gases contain equal numbers of molecules.' This generalisation is thus raised from a merely empirical statement to the rank of a deduction, made by dynamical reasoning, from a simple hypothesis regarding the structure of matter, which is itself justified by many classes of experimentally established facts.

290 The generalisation of Avogadro is of fundamental importance in chemistry. It is essential that the student should understand that this statement rests on physical evidence and dynamical reasoning; and also that he should understand that the statement presupposes the physical definition of the molecule of a gas (s. par. 287). When this generalisation is applied to many chemical changes taking place between gaseous elements, it leads to the necessary conclusion that the molecules of most gaseous elements are composed of parts, and that these do part company when the molecules chemically interact. Hence in chemistry we must recognise two orders of small particles; molecules, and the parts of molecules or atoms.

Avogadro's generalisation, or *Avogadro's law** as it is **291** usually called, furnishes a means for determining the relative weights of gaseous molecules. For, if the number of molecules in equal volumes of two gases (at the same temperature and pressure) is the same, it follows that *the ratio of the densities of the gases is also the ratio of the masses of the two kinds of molecules.*

Now a specified volume of oxygen is 16 times heavier than an equal volume of hydrogen; therefore a molecule of oxygen weighs 16 times as much as a molecule of hydrogen. Therefore if the weight of the molecule of hydrogen is taken as unity, the molecular weight of oxygen must be 16.

But ought the molecular weight of hydrogen to be taken **292** as unity? We have already found that the application of Avogadro's law to the interactions which occur between hydrogen and chlorine, and hydrogen and bromine, requires us to assert that each molecule of hydrogen separates, in these reactions, into at least two parts. A similar examination of other reactions between hydrogen and various gaseous elements confirms this conclusion.

A molecule of hydrogen then is composed of at least two parts, or atoms. But we agree to call the atomic weight of hydrogen one, and to make this the standard in terms of which the relative weights of the atoms of other elements are to be stated. Hence the smallest value which we can give to the molecular weight of hydrogen is two.

Of course we may assume that when hydrogen and chlorine react, each molecule of either element separates into 4, 6, 8, 10, &c. parts or atoms; we must assert that each separates into *at least two* parts. Suppose the assumption is made that each molecule separates into 4 atoms; then, as there are twice as many molecules of hydrogen chloride formed as the number of molecules of hydrogen or chlorine taking part in the reaction, it follows that each molecule of hydrogen chloride is composed of 2 atoms of hydrogen and 2 atoms of chlorine. But no chemical reactions of hydrogen chloride are in keeping with this conclusion. When this compound is decomposed with separation of

---

* The student should observe that the term *law* is used here in a sense different from that in which the same term is applied to the facts of chemical combination : *s.* note at end of Chap. XVI.

hydrogen or chlorine, the whole of the hydrogen or of the chlorine is removed. But the chemical reactions of a gaseous compound are regarded by the molecular theory as the re-actions of the molecules of that compound; therefore, when a molecule of hydrogen chloride reacts chemically with other molecules, the whole of the hydrogen, or the whole of the chlorine, is removed. The conclusion is that, most probably, a molecule of hydrogen chloride is composed of one atom of hydrogen and one atom of chlorine.

293   For reasons such as these, we conclude that the molecular weight of hydrogen is almost certainly two; that is, that the molecule of hydrogen is composed of two atoms, the weight of each of which we have agreed to call one.

Now oxygen is 16 times heavier than hydrogen; but the molecular weight of hydrogen is 2; therefore the molecular weight of oxygen is 32. Similarly, chlorine gas is 35·5 times heavier than hydrogen; therefore the molecular weight of chlorine is 71. Mercury-gas is 100 times heavier than hydro-gen; therefore the molecular weight of mercury-gas is 200: and so on, for all the elements which can be obtained as gases.

Similarly with gaseous compounds. Water-gas is 9 times heavier than hydrogen; therefore the molecular weight of water-gas is 18. Ammonia is 8·5 times heavier than hydrogen; therefore the molecular weight of ammonia is 17. Alcohol-gas is 23 times heavier than hydrogen; therefore the molecular weight of alcohol-gas is 46; and so on, for all compounds which can be obtained as gases.

294   We can now define the term molecular weight of a gas. The definition may be stated in various forms of words.

The molecular weight of a gaseous element or compound is twice the specific gravity of the gas referred to hydrogen.

Or; *The molecular weight of a gaseous element or compound is a number which tells the weight of two volumes of the gas, that is, the weight of that volume of the gas which is equal to the volume occupied (under the same conditions of temperature and pressure) by two parts by weight of hydrogen.*

Or, inasmuch as air is 14·435 times heavier than hydrogen, we may say that; *The molecular weight of a gaseous element or compound is the product obtained by multiplying the specific gravity of the gas referred to air, by 28·87.*

295   ∨ The application of Avogadro's law to chemical interactions leads to the recognition of the atom as a particle of matter

weighing less than the molecule; it also gives a means for determining the maximum weights of the atoms of those elements which form gaseous compounds.

The atom of an element is, by definition, the ultimate particle of the element of which cognisance is to be taken in chemistry; hence, it is evident that the molecule of a compound gas formed by the union of (say) two elements, A and B, must be formed by the union of at least one atom of A with at least one atom of B. Or, in general terms, a molecule of a compound gas must be composed of at least one atom of each of the elements which unite to produce the compound. This is equivalent to saying, *the atom of an element is the smallest mass of that element which combines with other atoms to form a gaseous molecule.*

As we have agreed to call the mass of one atom of hydrogen unity, and to state the weights of all other atoms in terms of that of the atom of hydrogen, we arrive at a definition of the maximum value to be given to the atomic weight of an element.

The smallest mass of an element, in terms of hydrogen as unity, which is found to combine with other elements to form a gaseous molecule represents the maximum value to be given to the atomic weight of the element in question.

Or; *The number which expresses how many times heavier than the smallest mass of hydrogen which combines with other elements to form gaseous molecules, is the smallest mass of a specified element which combines with other elements to form gaseous molecules, also expresses the maximum value which can be given to the atomic weight of the element in question.*

The greater the number of gaseous compounds of the **296** specified element which have been examined, the greater is the probability that the maximum value deduced for the atomic weight of the element represents the true value.

Let it be required to determine the atomic weight of **297** oxygen. The definition of atomic weight tells; (1) that several gasifiable compounds of oxygen must be prepared; (2) that these compounds must be gasified and the specific gravity of each, and hence the molecular weight of each, determined; (3) that each compound must be analysed, and the results stated as parts of each element per molecule of the compound. Then the smallest mass of oxygen in any one of these molecules is taken as the atomic weight of oxygen.

Here are some of the data.

Data for determining the atomic weight of oxygen.

Gaseous compound.	Sp. gravity; air = 1.	Sp. Gr. × 28·87 corrected (s. par. 300) i.e. molecular weight.	Composition of one molecular weight.
Carbon dioxide	1·53	43·89	31·92 oxygen + 11·97 carbon.
Sulphur dioxide	2·25	63·90	31·92 ,, + 31·98 sulphur.
Sulphur trioxide	2·85	79·86	47·88 ,, + 31·98 ,,

If no other gaseous compounds of oxygen were known, we should select the number 31·92 for the atomic weight of this element. But the following data shew that this conclusion would be incorrect.

Data for determining the atomic weight of oxygen.

Gaseous compound.	Sp. gravity; air = 1.	Sp. Gr. × 28·87 corrected (s. par. 300) i.e. molecular weight.	Composition of one molecular weight.
Carbon monoxide	·97	27·93	15·96 oxygen + 11·97 carbon.
Water	·63	17·96	15·96 ,, + 2 hydrogen.
Nitric oxide	1·04	29·97	15·96 ,, + 14·01 nitrogen.

298   As a great many gaseous compounds of oxygen are known, and as a molecule of none of them contains less than 15·96 parts by weight of oxygen, this number is taken to be the atomic weight of oxygen.

Some of the data from which the value 35·37 is deduced for the atomic weight of chlorine are presented in the following table.

Data for determining the atomic weight of chlorine.

Gaseous compound.	Sp. Gr. air = 1.	Sp. Gr.×28·87 corrected i.e. molecular weight.	Composition of one molecular weight.
Carbon tetrachloride	5·83	153·45	141·48 chlorine + 11·97 carbon.
Silicon tetrachloride	5·94	169·48	141·48 ,, + 28 silicon.
Phosphorus trichloride	4·85	137·07	106·11 ,, + 30·96 phosphorus.
Antimony trichloride	7·8	226·11	106·11 ,, + 120 antimony.
Zinc chloride	4·7	135·64	70·74 ,, + 64·9 zinc.
Tungsten hexachloride	13·4	395·82	212·22 ,, + 183·6 tungsten.
Sulphuryl chloride	4·67	134·64	70·74 ,, + 31·92 oxygen + 31·98 sulphur.
Chloroform	4·1	119·08	106·11 ,, + 1 hydrogen + 11·97 carbon.
Thallium chloride	8·3	289·01	35·37 ,, + 208·64 thallium.
Methyl chloride	1·74	50·34	35·37 ,, + 11·97 carbon + 3 hydrogen.
Nitrosyl chloride	2·3	65·84	35·37 ,, + 15·96 oxygen + 14·01 nitrogen.
Hydrogen chloride	1·25	36·37	35·37 ,, + 1 hydrogen.

When only a few compounds of a specified element have **299** been gasified and analysed, the value thence deduced for the atomic weight of that element may be, and very possibly is, too large; it cannot be too small. Thus only three compounds of aluminium have been gasified &c.; a molecule of each is composed of 54·04 parts by weight of aluminium, combined with 212·22 parts by weight of chlorine, 478·5 of bromine, and 759·18 of iodine, respectively. Hence the atomic weight of aluminium is not greater, but may be less, than 54·04. From other data we know that the atomic weight of this metal is

$$\frac{54 \cdot 04}{2} = 27 \cdot 02. \quad (s. \text{ par. } 305.)$$

Determinations of the specific gravities of gases are subject **300** to several sources of error. But the mass of an element which combines with one part by weight of hydrogen, or eight parts by weight of oxygen, or 35·5 parts by weight of chlorine, or 16 parts by weight of sulphur, i.e. the smallest value of the combining weight of the element (v. ante, Chap. v. pars. 73 to 75), can be determined with great accuracy. It is evident that the molecular weight of an element must be either equal to, or a whole multiple of, the combining weight of the element; and that the molecular weight of a compound must be either equal to, or a whole multiple of, the sum of the combining weights of the constituent elements. Hence the data required for an accurate determination of the molecular weight of an element are (1) an accurate determination of the combining weight of the element, and (2) a fairly accurate determination of the specific gravity of the element in the gaseous state. Similarly, the data required for an accurate determination of the molecular weight of a compound are (1) accurate determinations of the combining weights of the elements which form the compound, and (2) a fairly accurate determination of the specific gravity of the compound in the gaseous state.

Thus 35·37 parts by weight of chlorine combine with 1 part by weight of hydrogen; therefore the molecular weight of chlorine is n35·37, where n is a whole number. A determination of the specific gravity of chlorine shews that this gas is approximately 35½ times heavier than hydrogen; therefore the molecular weight of chlorine is approximately 35·5 × 2 = 71. But 35·37 × 2 = 70·74; therefore the molecular weight of chlorine is 70·74. Again, 10·32 parts by weight of phosphorus combine with 1 part by weight of hydrogen to produce

M. E. C.   14

phosphorus hydride; therefore the molecular weight of this compound is n 11·32 ($n =$ a whole number). Gaseous phosphorus hydride is found to be about 17 times heavier than hydrogen; therefore the molecular weight of this gas is about $17 \times 2 = 34$. But $11\cdot32 \times 3 = 33\cdot96$; therefore the molecular weight of gaseous phosphorus hydride is 33·96.

The meaning of the heading of col. III. in the tables in pars. 297 and 298,—'$Sp.$ $Gr.$ ($air =$ 1) $\times 28\cdot87$ corrected, i.e. molecular weight'—will now be apparent.

301   By applying Avogadro's law, values have been obtained for the atomic weights of rather more than half the elements; gaseous compounds of the remaining elements have not yet been obtained, and hence the atomic weights of these elements have not been determined by the method based on the law of Avogadro.

The following table gives the results of the application of Avogadro's law to determining the atomic weights of elements.

Maximum Atomic Weights of Elements. (*Avogadro's law.*)

Element	Max. Atom. Weight	Element	Max. Atom. Weight	Element	Max. Atom. Weight
Hydrogen	1	Chromium	52·4	Antimony	120
Beryllium	9·1	[Aluminium*	54·04]	Tellurium	125
Boron	10·95	Zinc	64·9	Iodine	126·53
Carbon	11·97	Germanium	72·3	[Copper*	126·8]
Nitrogen	14·01	Arsenic	74·9	[Gallium*	138]
Oxygen	15·96	Selenion	78·8	Tantalum	182
Fluorine	19·1	Bromine	79·75	Tungsten	183·6
Silicon	28	Zirconium	90	Osmium	19Ꙩ(?)
Phosphorus	30·96	Niobium	94	Mercury	199·8
Sulphur	31·98	Molybdenum	95·9	Thallium	203·6
Chlorine	35·37	[Iron*	111·8]	Lead	206·4
Potassium	39·04	Cadmium	112	Bismuth	208
Titanium	48	Indium	113·4	Thorium	232
Vanadium	51·2	Tin	117·8	Uranium	240

302   Avogadro's law—equal volumes of gases contain equal numbers of molecules—furnishes chemists with a method whereby they may determine the relative weights of the molecules of all gaseous compounds and elements, and the maximum values to be given to the relative weights of the atoms of all elements which form gaseous compounds. But at present only 14 or 15 elements have been gasified, and gaseous

* The atomic weights of these four elements are almost certainly 27·02, 55·9, 63·4, and 69, respectively (*s.* par. 305).

compounds of only 42 elements have been prepared and analysed. Hence the application of the method based on the law of Avogadro is limited. This method is at present the only general method for determining the relative weights of the gaseous molecules of elements and compounds. But there is another general method whereby values may be found for the atomic weights of elements. This method is contained in the statement ;—

The products of the specific heats of solid elements, determined in each case at the temperature-interval for which specific heat is nearly constant, into the atomic weights of these elements, approach a constant, the mean value of which is 6·4.

This statement is a modification of the so-called law of Dulong and Petit. From their study of the specific heats of 13 solid elements in the year 1819, these naturalists announced that "the atoms of all simple bodies have exactly the same capacity for heat." Investigation has shewn that this statement was too absolute. The specific heats of some solid elements, e.g. carbon, boron, silicon, beryllium, vary much with variations of temperature, and become approximately constant only at high temperatures. The specific heat of a solid also varies to some extent with variations in the greater or less compactness of the specimen.

The product *specific heat of solid element* × *atomic weight* is usually called the *atomic heat* of the element.

The specific heats of a few elements have not yet been **303** determined. Values which may be approximately correct, have been indirectly obtained for some of these ; but too great stress must not be laid on these values. The indirect method in question is based on the assumption, to some extent verified by facts, that the '*molecular heat*' of a solid compound, i.e. the product of the specific heat into the mass of the compound expressed by its formula, is equal to the sum of the atomic heats of the elements in the compound; therefore if the '*molecular heat*' (as thus defined) of a solid compound is known, and the atomic heats of all the elements in the compound except one are known, the atomic heat of the remaining one element can be calculated.

The following statements summarise the present state of **304** knowledge with regard to the atomic heats of the 42 elements maximum values for the atomic weights of which have been determined by applying the law of Avogadro (par. 301).

I. *Solid elements*, 28 *in number, the specific heats of which*

14—2

have been directly determined, and the atomic heats of which are approximately equal to 6·4 :—P, S, K, Ti, Cr, Al, Zn, As, Se, Br, Zr, Mo, Fe, Cd, In, Sn, Sb, Te, I, Cu, W, Os, Hg, Tl, Pb, Bi, Th, U.

II.   *Solid elements, 6 in number, the specific heats of which have been directly determined, and the atomic heats of which are approximately equal to* 5·5 :—Be, B, C, Si, Ga, Ge.

III.   *One solid element, the atomic heat of which has been indirectly determined and is probably equal to* 6·4:—Vanadium.

IV.   *Five gaseous elements, the specific heats of which in solid form have only been determined indirectly and are extremely doubtful :—*H, N, O, F, Cl.

V.   *Two solid elements the specific heats of which have not been determined directly or indirectly :—*Nb, Ta.

These data establish a very fair probability in favour of the statement made in par. 302 regarding the constant value of the atomic heat of the solid elements.   If this statement is granted, then an approximate value may be found for the atomic weight of an element by dividing 6·4 by the specific heat of that element in the solid form.

305   The maximum values found for the atomic weights of aluminium, iron, copper, and gallium, by the use of Avogadro's law were 54·04, 111·8, 126·8, and 138, respectively (*s.* Table, par. 301).   Now the spec. heats of these elements are ·225, ·114, ·097, and ·08, respectively; dividing 6·4 by each of these numbers we get the quotients, 28·5, 56·1, 65·9, and 80. Therefore we conclude that the maximum values found for the atomic weights of these elements by applying Avogadro's law must be halved, and we adopt the numbers 27·02, 55·9, 63·4 and 69, as very probably the true atomic weights of aluminium, iron, copper, and gallium, respectively.

306   There is another physical method which has sometimes been found useful in determinations of atomic weights, but which can only be used as a guide to point the way to experimental inquiries.   This method is founded on the generalisation, that similarity of chemical composition is usually associated with close similarity of crystalline form.   In some cases marked similarity of composition is accompanied by identity of crystalline form ;   e.g. the oxides of arsenic and antimony, As_2O_3 and Sb_2O_3, crystallise in identical forms ; they are *isomorphous.*

The difficulties in applying this method—generally known as the *method of isomorphism*—lie in the vagueness of the expressions 'similarity of chemical composition' and 'similarity of crystalline form.' The following example will indicate how the so-called *law of isomorphism* has been used as an aid in determining the atomic weight of gallium.

Gallium sulphate was found to form a double salt with ammonium sulphate; the crystalline form of this double sulphate was identical with that of ordinary, ammonia-alum. Therefore the double sulphate in question doubtless belonged to the class of alums. Now the composition of the alums is expressed by the general formula $M_2 3SO_4 . N_2 SO_4 . 24H_2O$ where M = Al, Fe, Cr, or Mn, and N = Na, K, Cs, Rb, or NH$_4$. In the case of common ammonia-alum $M_2 = Al_2 = 2 \times 27 \cdot 02$ parts by weight of aluminium; in the double sulphate of gallium and ammonium M$_2$ was found to represent 138 parts by weight of gallium. Hence, as 2 atoms of aluminium have been replaced by 138 parts by weight of gallium without altering the crystalline form or the general chemical type of the compound, it was concluded that the atomic weight of gallium was $\frac{138}{2} = 69$. This number was afterwards verified by the application of the law of Avogadro, and also by the specific heat method*.

There are then two generally applicable methods whereby **307** values may be found for the atomic weights of the elements; the method founded on the law of Avogadro; and the method based on the specific heats of solid elements. Besides these, there is another method, arising out of the relations between the chemical composition and crystalline forms of similar compounds, which is useful as a guide in determinations of atomic weights. The first method is applicable to determinations of the atomic and molecular weights of elements, and the molecular weights of compounds, but it is restricted to bodies which are gasifiable without decomposition. The second and third methods can be strictly applied only to find values for the atomic weights of solid elements, and to some extent of elements which form solid compounds. All the methods are essentially physical; they are based on physical conceptions, and they are to a great extent developed by physical reasoning. Thus the image of the molecule which is called up in the mind by the statement "equal volumes of gases contain equal numbers of molecules" is that of a very small, definite,

---

* We do not propose to go more fully into the method of isomorphism here. The study of this subject is more suited to the advanced student of chemistry.

portion of matter, moving about without separation into parts, colliding with other like particles of matter, and rebounding after collision. The application of this conception to chemical changes obliges us to admit that in many of these changes the molecule is shattered into parts. Thus we are led to the chemical conception of the atom, as a portion of matter smaller than the molecule, and either itself without parts, or else composed of parts which, so far as we know at present, do not part company during any of the changes which the atom undergoes. The study of the properties of atoms leads to the generalisation that the atoms of all solid elements, at certain temperatures, have equal capacities for heat.

The molecular and atomic theory regards the molecule of a gas as the smallest portion of it in which the properties of the gas inhere. Chemical change, it looks on as an interaction between molecules; in most cases of chemical change the interacting molecules are separated into parts and these parts are rearranged to form new molecules; but in some cases it is probable that one kind of molecules combines with other kinds to form more complex molecules.

The Daltonian atomic theory applied the term atom to elements and compounds alike; but the atom of an element was supposed to have no parts, whereas the atom of a compound was separable into unlike parts. The molecular and atomic theory applies the term molecule to elements and compounds alike; but the molecule whether of an element or a compound is regarded as built up of parts which may be either all of one kind, or of different kinds.

308   The atomic weights of most of the elements have been determined by one or other of the physical methods arising out of the molecular and atomic theory. But there are a few elements no compounds of which have yet been gasified, and the specific heats of which have not yet been determined. The values assigned to the atomic weights of these elements have been gained by studying the chemical analogies between these elements and others to which the methods of the molecular and atomic theory are directly applicable.

The metal rubidium is a case in point. No compound of this metal has been gasified; hence the molecular weights of rubidium compounds are not known; and hence the atomic weight of the element has not been determined by the application of the law of Avogadro. Nor has the specific heat of rubidium been determined. The value given to the atomic weight of rubidium is 85·2 ; how has this number been obtained ?

There can be no doubt that rubidium belongs to the class of elements which comprises sodium and potassium (for details of the properties of this group, s. Chap. XI. pars. 160—168). The atomic weights of sodium and potassium are 23 and 39 (in round numbers) respectively ; 23 parts by weight of sodium and 39 parts by weight of potassium severally combine with (a) 8 parts by weight of oxygen, and (b) 35·5 parts by weight of chlorine ; the specific heats of these metals are, for sodium ·293, for potassium ·166 ; now ·293 × 23 = 6·7, and ·166 × 39 = 6·5. But if 23 and 39 are the atomic weights of sodium and potassium, respectively, and if 16 is the atomic weight of oxygen, then analyses of the oxide, chloride, &c. &c. of these metals shew that the formulæ of these compounds must be M_2O, MCl, M_2SO_4, M_2CO_3, &c. &c. where M = one atomic weight of sodium or potassium. Now the compounds of rubidium are very similar in their properties to the compounds of potassium and sodium, hence the oxide, chloride, &c. &c. of rubidium ought to be represented by the formulæ Rb_2O, $RbCl$, Rb_2SO_4, $RbCO_3$, &c. &c. where Rb = one atomic weight of rubidium. But in order to do this, the number 85·2 must be assigned to the atomic weight of rubidium.

The method based on a study of the analogies between the chemical properties of a specified element and those of other elements is also frequently used to check the results of the determinations of atomic weights gained by applying the two physical methods. But a fuller examination of this chemical method will be better made when we come to consider the *periodic law* (s. Chaps. XVIII. and XXVI.).

In the sketch which has been given of the molecular theory **309** of the structure of matter, the conception of the molecule has been applied only to gases. The theory regards liquids and solids also as built up of minute particles. It asserts that the minute particles of a liquid have less freedom of motion than the molecules of a gas, and that they are so frequently in collision with each other that the paths which they describe are far removed from being straight lines. The minute particles of a solid are supposed to oscillate about positions of equilibrium, and never to travel far from these positions. The particles of both liquids and solids, moreover, are probably aggregations of smaller particles ; and the complexity of the

particles of a specified liquid or solid is probably not the same
for all the particles, nor even for the same particle at different
times.

310   It is customary to apply the term *molecule* to the particles
of liquids and solids, as well as to the much more rigidly
defined particles of gases. But no generalisations can at
present be made, in terms of the molecular theory, regarding
the properties of liquids and solids comparable with those
which have been made for gases, and which are known as the
laws of Boyle, Charles, and Avogadro. We cannot define the
term molecule as applied to a liquid or solid body; we can
define the term when applied to a gas.

When therefore we speak of the molecular weight of an
element or a compound we ought to mean the relative weight
of the molecule of the gaseous element or compound. The
expression molecular weight is not always used in chemistry
in this strict meaning; it is frequently applied to what in
former chapters we have called the *reacting weight* of a body.

To take an example. No compounds of sodium have been
gasified; therefore we do not know the molecular weights of
any compounds of this element. But the atomic weight of
this metal has been determined by applying the method of
specific heat. The simplest formulæ which can be given to
the compounds of sodium, when we know that $Na = 23$, are
Na_2O, $NaCl$, $NaBr$, $NaNO_3$, Na_2CO_3, Na_2SO_4, &c. Moreover
these formulæ enable us to express the chemical reactions of
the compounds in a consistent and satisfactory manner.
These formulæ are therefore adopted. But we must carefully observe that the formulæ do not necessarily express
the atomic compositions of *molecules* of the compounds:
indeed we cannot in strict accuracy speak of a molecule of
sodium oxide, or sodium chloride, because these compounds
are only known as solids and the term molecule corresponds
to an accurately defined conception only when it is applied to
gases.

311   We formerly used the term *reacting weight* of a compound;
thus Na_2O represents the composition of a reacting weight of
sodium oxide. We may now, in the light thrown on chemical
interactions by the molecular and atomic theory, widen the
meaning we give to this term *reacting weight*. Although
Na_2O does not certainly represent the atomic composition of a
molecule of sodium oxide, yet it almost certainly represents
the ratio of the number of atoms which constitute the reacting

weight of this oxide. The reacting weight of a compound may now mean for us a group or collocation of atoms which interacts chemically with other groups of atoms. The formula of a solid or liquid compound does not necessarily express the number of atoms in this chemically reacting group of atoms— indeed the number may vary under different circumstances— but in all probability it does represent the ratio between the number of atoms in this reacting group. The atomic composition of the reacting weight of sodium oxide may perhaps be better represented by one of the formulæ Na_4O_2, Na_6O_3, $Na_{10}O_5$, than by the simpler formula Na_2O; but

$$4 : 2 = 6 : 3 = 10 : 5 = 2 : 1.$$

CHAPTER XVI.

312   THE molecular and atomic theory asserts that a quantity of any gaseous element or compound is constituted of a very great number of minute particles, all having the same masses and the same properties, and all in constant motion. These particles, or molecules, are constituted of smaller particles which have a certain freedom of motion among themselves ; these smaller particles, or atoms, are of one kind and of equal masses when the molecule formed by their union is the molecule of an element ; but the atoms are of different kinds and different masses when the molecule formed by their union is the molecule of a compound.

Chemical change, according to the molecular and atomic theory, is an interaction between molecules, and it results in the formation of new molecules. In very many cases of chemical change the interacting molecules are separated into their constituent atoms, and these atoms rearrange themselves to form new molecules; but in some cases the interaction of the original molecules probably consists in the direct formation of more complex molecules. Thus the interactions of hydrogen and chlorine to produce hydrogen chloride, and of hydrogen and oxygen to produce water-gas, are represented thus by the theory of atoms and molecules :—

(1) $H_2 + Cl_2 = 2HCl$; (2) $2H_2 + O_2 = 2H_2O$. The symbols H_2, Cl_2, O_2, HCl, H_2O, each represents the atomic composition of a molecule of an element or compound. But the interactions of water and cobalt chloride, or water and copper sulphate, are probably best represented by equations which assume the

change to consist in the combination of molecules of the interacting substances to produce more complex molecules : thus,

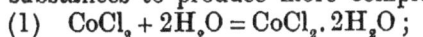

(1) $CoCl_2 + 2H_2O = CoCl_2 . 2H_2O$;
(2) $CuSO_4 + 5H_2O = CuSO_4 . 5H_2O$.

The theory regards most physical changes as changes in the   **313**
rates of motion, without changes in the atomic compositions, of
molecules. But changes usually called physical may result in the
coalescence of molecules into more or less complex aggregations which are stable under definite conditions of temperature,
pressure, &c.

The theory of molecules and atoms does not therefore give
us a means of sharply distinguishing between physical and
chemical change. The typical chemical change results in a
redistribution of the atoms of the interacting molecules so as
to form new molecules; the typical physical change results in
changes in the rates of motion of molecules without any
redistribution of the parts of these molecules. But there are
many changes which cannot be placed wholly in one or other
of these classes. Every chemical change is accompanied by
physical change : the portion of the change we call chemical
is only one part of the complete occurrence. Even if the
theory gave a sharp and clear definition of each kind of
change, it could not give a means whereby we might classify
all actually occurring changes into chemical on the one hand,
and physical on the other.

The laws of chemical combination find a simple explana-   **314**
tion in terms of the molecular and atomic theory. The atom is
the ultimate particle of matter of which we take cognisance in
chemistry. The properties of a molecule depend, among other
conditions, on the nature and number of the atoms which
form it; this is the law of fixity of composition. If two or
more different kinds of atoms combine to form several different
molecules, each molecule must be composed of x atoms of one
kind + x' atoms of another kind + x'' atoms of another kind
+ x''' atoms of another kind &c., and x, x', x'', x''' must be
whole numbers, because the atom is, by definition, indi-
visible; this is the law of multiple, and the law of reciprocal,
proportions.

The molecular and atomic theory throws light on the   **315**
conceptions of combining and reacting weights. The reacting
weight of a gas is the molecular weight of that gas. The
combining weight of an element, as the term was defined
in Chap. VI. par. 79, is the atomic weight of that element.

Thus we found (s. Chap. VI. par. 86) that the reacting weight of water is 18; and that one reacting weight of this compound is composed of two combining weights of hydrogen united with 1 c. w. of oxygen. Translated into the language of the molecular and atomic theory this statement reads as follows :—the molecular 'weight of water-gas is 18, and a molecule of water-gas is formed by the union of 2 atoms of hydrogen with 1 atom of oxygen.

We know that 36·5 parts by weight of hydrogen chloride are formed by the combination of 1 part by weight of hydrogen with 35·5 parts by weight of chlorine. In order to express this fact in terms of combining and reacting weights, we say that one c. w. of hydrogen combines with one c. w. of chlorine to produce one reacting weight of hydrogen chloride. The molecular and atomic theory expresses the same fact by saying that one molecule of hydrogen interacts (not combines) with one molecule of chlorine to produce 2 molecules of hydrogen chloride.

We formerly applied the term reacting weight to compounds only. We now apply the term molecule to elements as well as to compounds. But when we are dealing with solid bodies which have not been gasified, we cannot in strict accuracy speak of the interactions of molecules of these bodies. Thus when boron and aluminium are strongly heated together under proper conditions two compounds, AlB_2 and AlB_{12}, are produced. As neither boron nor aluminium has been gasified, and as neither of the borides of aluminium has been gasified, we do not know the molecular weights of any of the bodies taking part, or formed, in this reaction : we cannot therefore say how many molecules of each element have taken part in the change nor how many molecules of each compound have been formed. Again, when solutions of barium chloride and sodium sulphate, in aqueous solutions, are mixed in the ratio $BaCl_2 : Na_2SO_4$, the change of composition which occurs may be represented thus:—

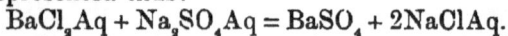

$$BaCl_2Aq + Na_2SO_4Aq = BaSO_4 + 2NaClAq.$$

We may read the equation as meaning :—one reacting weight of barium chloride interacts with one reacting weight of sodium sulphate to produce one r. w. of barium sulphate and 2 r. ws. of sodium chloride; but the equation cannot be read as certainly meaning, one molecule of barium chloride interacts with one mol. of sodium sulphate to produce one mol. of barium sulphate and 2 mols. of sodium chloride. As none of the bodies taking part in this change have been gasified we do not know the molecular weight of any of them. We might read the equation thus; atomic aggregates of barium chloride and sodium sulphate interact to produce atomic aggregates of barium sulphate and sodium chloride.

The definition of molecule is a physical definition; it is stated in terms which have an accurate meaning only when used of gaseous elements and compounds. If we choose to use the term in speaking of the phenomena of liquid and solid bodies we must not forget that the term cannot then be accurately defined. The definition of reacting weight is a chemical definition; but the term *reacting weight* is much vaguer than the term *molecule*. The reacting weight of a solid or liquid compound is doubtless an aggregation of atoms which interacts, as a whole, with other aggregations of atoms; but whether the number of atoms in this aggregation is the same in all chemical changes we do not know.

We have seen in Chap. xv. that the atomic weights of **316** most of the elements have been determined, either by the method based on the law of Avogadro, or by that founded on the generalisation 'atomic weight into spec. heat = a constant.' But the molecular weights of only a few elements have been determined. About 70 elements are known; 14 of these have been gasified; therefore the molecular weights of only 14 are known. The specific gravities in the gaseous state, and hence the molecular weights, of some of these 14 elements are constant through a wide range of temperature; the specific gravities, and hence the molecular weights, of others have very different values at different temperature-intervals. The most probable explanation of the changes in the values of the molecular weights of certain elements is that the atomic compositions of the molecules of these elements are different at different temperatures. Thus sulphur-gas from about 450° to about 600°, is 96 times heavier than an equal volume of hydrogen at the same temperature; therefore the molecular weight of sulphur-gas between 450° and about 600° is approximately $96 \times 2 = 192$. But from about 800° and upwards sulphur-gas is only 32 times heavier than hydrogen; therefore the molecular weight of sulphur-gas at temperatures above 800° is approximately $32 \times 2 = 64$. The atomic weight of sulphur is 31·98; this number is determined by applying Avogadro's law to many gaseous compounds of sulphur, and it is verified by determinations of the spec. heat of sulphur.

Now $31\cdot98 \times 6 = 191\cdot88$, and $31\cdot98 \times 2 = 63\cdot96$; therefore we conclude; (1) that sulphur-gas at 450^0 to 600^0 has the molecular weight $191\cdot88$, and at 800^0 and upwards the molecular weight $63\cdot96$; and (2) that the molecule of gaseous sulphur at 450^0 to 600^0 is composed of 6 atoms, or is *hexatomic*, and that the molecule at 800^0 and upwards is composed of 2 atoms, or is *diatomic*.

317 The expression *atomicity of a molecule* is used to denote the number of atoms which form the gaseous molecule of an element or compound. The data for classifying the molecules of elements in accordance with their atomicities are presented in the following table.

Atomicity of elementary gaseous molecules.

Monatomic	Diatomic	Triatomic	Tetratomic	Hexatomic
Zinc	Hydrogen	Oxygen	Phosphorus⎱	Sulphur
Cadmium	Chlorine	(as ozone)	Arsenic ⎰	(about 400^0 to 600^0)
Mercury	Bromine	Selenion	(at temps. below	
Iodine	Iodine	(700^0 to about 800^0)	white heat)	
(at about 1500^0 and upwards)	(200^0 to about 1000^0)			
(? Bromime at 1800^0 and upwards)	Oxygen			
	Sulphur			
Na	(800^0 and upwards)			
	Selenion			
K,	(1200^0 and upwards)			
	Tellurium			
	Nitrogen			
	Phosphorus⎱			
	Arsenic ⎰			
	(at white heat)			

318 The molecular weights of some gaseous compounds also vary with variations of temperature. Thus nitrogen tetroxide at very low temperatures is about 46 times heavier than hydrogen, but at higher temperatures it is only 23 times heavier than hydrogen; therefore this gas has two molecular weights which are approximately equal to 92 and 46 respectively. Determinations of the atomic weights of nitrogen and oxygen, and accurate analyses of the compound nitrogen tetroxide, shew that the composition of the molecule of this gas at very low temperatures is represented by the formula N_2O_4 ($N_2 = 28\cdot02$, $O_4 = 63\cdot84) = 91\cdot86$, and at higher temperatures by the formula $NO_2 = 45\cdot93$. We shall examine the relations between changes of molecular weight and changes of temperature in more detail later (s. pars. 334 to 337).

319 The conception expressed in the term *atomic weight* is now seen to be much more definite than that expressed in the term *combining weight*. The former term brings before the mind the picture of a small definite portion of matter with definite properties; the latter term merely expresses a ratio. The values of atomic weights are determined by two methods, of general applicability, which are deduced from the principles of a theory of the structure of matter which gives a fairly simple explanation of most, if not all, of the observed physical properties of matter. The value given to the combining weight of an element is a purely empirical value; it must be determined for each element by methods specially applicable to that element; it is one of several possible values, and it is selected on the ground of general convenience and expediency.

The conception expressed in the term *molecular weight* is **320** also much more definite than that underlying the expression *reacting weight*. The molecule of a gas is a perfectly definite quantity of matter with defined properties; it is a physical conception, deduced by dynamical reasoning from a physical theory of the structure of matter. This theory presents us with one generally applicable method for determining the relative weights of gaseous molecules. The reacting weight of a body is said to be 'the smallest relative mass of it which takes part in chemical interactions'; but this statement involves terms which cannot, at present at any rate, be accurately defined. The value which, under the circumstances, is the best to be given to the reacting weight of a substance must be deduced for each substance by methods which to a very great extent are empirical, and many of which are applicable only to the special case under consideration. It is true that the molecular and atomic theory has not yet enabled us to define the term molecule as applied to a liquid or solid body; but the theory has thrown light on the chemical conception of reacting weight, and in place of regarding the values of reacting weights merely as numbers, we may now look on them as expressing the relative weights of certain aggregations of atoms which interact with each other to produce new aggregations of atoms.

The molecular and atomic theory then regards the properties **321** of a gas as the properties of the molecules of that gas; and the properties of the molecules as dependent, among other conditions, on the nature and number of the atoms which form these molecules. But if this view is correct, and if it is true that the number of atoms in the molecule of a gaseous element may vary, we should expect sometimes to find differences in the properties of one and the same element.

322   The descriptions given in Chap. XI. pars. 173, 220, 221, of the properties of sulphur and phosphorus shewed that each of these elements differs very considerably in properties under different conditions.   But as both elements exhibit differences in the solid state we cannot say whether these differences are, or are not, connected with differences in the atomicities of the molecules of the elements.

323   Some of the prominent properties of oxygen were described in Chap. VIII. pars. 118 to 121.   If a quantity of pure dry oxygen is confined over sulphuric acid and a series of electric induction-sparks is passed through the oxygen, the volume of the gas diminishes until it has become about $\frac{1}{12}$ less than it was at the beginning of the experiment.   The properties of the gas after the diminution of volume has ceased are markedly different from those of oxygen; nevertheless it has been conclusively proved that nothing has combined with the oxygen during the change which has occurred.   The new gas is called *ozone* (because of its smell).   The whole of a specified quantity of oxygen cannot be changed into ozone, so that pure ozone has not yet been obtained.   But experiments have proved; (1) that the relation between the volume of that portion of a quantity of oxygen which is changed into ozone, and the volume of the ozone formed is expressed by the ratio 3 : 2; and (2) that the weight of the ozone produced is equal to the weight of the oxygen which has been changed into ozone. If the gas through which electric sparks have been sent until diminution of volume has ceased is heated to 360° or so, expansion occurs, and the original volume of oxygen is reproduced.

The outcome of the experiments on the volume- and mass-relations between oxygen and ozone is that the change of oxygen to ozone, or *vice versa*, is attended with no change of mass, but that 3 volumes of oxygen condense to 2 vols. of ozone, and 2 volumes of ozone are changed by heat to 3 volumes of oxygen.   Now as oxygen is 16 times heavier than hydrogen, it follows that ozone is 24 times heavier than hydrogen; therefore, as the molecular weight of a gas is twice its specific gravity referred to hydrogen, it follows that the molecular weight of ozone is $24 \times 2 = 48$.   As ozone is only modified oxygen, and as the atomic weight of oxygen is 16, the molecule of ozone must be composed of 3 atoms of oxygen.

Therefore the atomic composition of the molecule of ozone
is expressed by the symbol O_3, that of the molecule of oxygen
being expressed by the symbol O_2.

Oxygen and ozone are both colourless gases; oxygen is
odourless, ozone has a very pronounced odour; ozone is a very
energetic oxidiser, e. g. when passed into mercury at ordinary
temperatures it produces mercuric oxide, and it interacts with
lead sulphide (PbS) to produce lead sulphate ($PbSO_4$); oxygen
does not react with an aqueous solution of potassium iodide
(KI), ozone interacts with this salt in aqueous solution and
produces potassium oxide, iodine, and oxygen; thus
$$2KIAq + O_3 = K_2OAq + I_2 + O_2.$$
The existence of the two kinds of molecules, O_2 and O_3,
one diatomic and the other triatomic, each characterised by its
own properties yet each composed of the same kind of atoms,
and of atoms all of which are alike, shews that the properties
of some molecules at any rate are conditioned (among other
circumstances) by the number of atoms which are combined to
form these molecules.

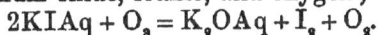

The other cases of *allotropy* (*s.* Chap. XI. par. 173) exhibited
by elements are exhibited by those elements in the solid state.
The numbers of atoms in the atomic aggregates which compose
the reacting weights of the different solid forms of phosphorus,
sulphur, arsenic, &c. may perhaps be different; but we cannot
at present decide whether this is so or not.

If the properties of a gas are dependent only on the
nature and number of the atoms which form the molecule
of that gas, then the existence of more than one gaseous
compound having a specified composition must be impossible.
But, as a matter of fact, several compounds frequently exist,
all having the same composition and the same molecular
weight. For instance 3 compounds having the composition
C_5H_{12} are known; these bodies have all been gasified; their
specific gravities as gases, and therefore their molecular
weights, are identical. But the molecule of each compound
is composed of 5 atoms of carbon united with 12 atoms of
hydrogen. Therefore we conclude that the properties of some
gaseous molecules are conditioned by other circumstances
besides the nature and number of their constituent atoms,
and that one of these other circumstances probably is the
arrangement of the parts of the molecule relatively to each
other.

The existence of more than one compound with the same

M. E. C. 15

molecular composition is called *isomerism*. We shall return
to this subject in the next chapter.

327   We now see how necessary it has become to widen our
conception of chemical composition. Restricting ourselves to
gases, we have found that there may exist more than one form
of the same element—we have found for instance that there
are two oxygens ; we have also learnt that the same quantities
of the same elements may be combined so as to produce
different compounds ; and that the same numbers of the same
atoms may be combined so as to produce several kinds of
molecules, each differing in properties from the others.

The properties of an element or compound are the properties of its molecules ; the properties of these molecules are
conditioned not only by the nature, but also by the number,
of atoms which form them ; but the properties of these molecules, it appears, are also probably conditioned by the relative
arrangement of their parts.

328   This conception of the relation between chemical properties and composition helps us to understand, more fully than
we could do before, the relations between the composition and
the properties of such classes of compounds as acids, alkalis,
and salts.

In Chaps. VIII. and IX. we learned that the compounds of
hydrogen with oxygen and another negative element, or other
negative elements, are generally acids, and that the compounds
of hydrogen with oxygen and a markedly positive element are
alkalis. We also found that the lower oxides of elements
which are neither very markedly positive or negative are
usually basic, but that the combination of more oxygen with
such oxides produces acidic oxides.

We may now translate these statements into the language
of the molecular and atomic theory, and say that molecules
formed by the union of atoms of hydrogen and oxygen with
atoms of negative elements interact, under proper conditions,
with molecules of metals, basic oxides, or alkalis, and exchange some or all of their atoms of hydrogen for atoms of
metal. We may also say that molecules composed of fairly
positive atoms united with a small number of atoms of oxygen
are ready to exchange their positive atoms for atoms of hydrogen when they interact with acids, under suitable conditions; but that molecules composed of many oxygen atoms
united with a small number of atoms of fairly positive elements
do not thus exchange their positive atoms for hydrogen, but

on the other hand are ready to interact with water to produce acids.

As few acids, alkalis, basic oxides, or salts, have been gasified, the term *molecule* is used in the foregoing paragraph to include the aggregates of atoms which form the reacting weights of solid bodies.

We know that acids may be classified in accordance with **329** their *basicity* (s. Chap. XI. pars. 188 and 189). Instead of the statement that 'an *n*-basic acid is an acid from the reacting weight of which *n* combining weights of hydrogen can be displaced by sodium or potassium under suitable conditions'; we may now say that 'the *molecule* of an *n*-basic acid contains *n* atoms of replaceable hydrogen.' It must be carefully noted that the term *molecule* is here used with a wider meaning than theory strictly justifies; it must be taken to mean that aggregate of atoms (this may or may not be a true molecule) which forms the reacting weight of an acid, &c. The account of the reactions of acids given in Chaps. X. and XI. shews what is meant by *replaceable hydrogen*. The compositions of a few acids and of some of the salts derived from them are presented in the following table; by considering the relations between the compositions of these acids and their salts, the student will be better able to grasp the meaning of the definition of the basicity of an acid which has been given. The formulae represent reacting atomic aggregates, which in some cases are probably true molecules.

Acids.   *Salts formed.*

HCl KCl, $NaCl$, $ZnCl_2$, $BiCl_3$, $SnCl_4$, &c.

H_2SO_4 $KHSO_4$, K_2SO_4, $FeSO_4$, Fe_23SO_4, $Sn2SO_4$, &c.

H_3PO_4 KH_2PO_4, K_2HPO_4, K_3PO_4, $CaHPO_4$, Ca_22PO_4, $FePO_4$, &c.

H_3PO_3 KH_2PO_3, K_2HPO_3, &c., (K_3PO_3 cannot be formed).

$H_4C_2O_2$ $KH_3C_2O_2$, $NaH_3C_2O_2$, $Pb(H_3C_2O_2)_2$, $Fe_2(H_3C_2O_2)_6$, &c.

In Chap. XII. we learned a little about the relative affinities **330** of acids. A comparison of the relative affinities of a series of acids the compositions of which differ only by the relative quantities of oxygen the acids contain brings out the fact, that the affinity-constants of the acids generally increase as the quantity of oxygen increases; thus the relative affinities of the three acids H_2SO_3, H_2SO_4, $H_2S_2O_6$, are in the ratio 66 : 150 : 178. Again the replacement of hydrogen by a strongly negative

15—2

element such as chlorine is attended with an increase in the affinity-constants; thus the relative affinities of the acids $H_4C_2O_2$, $H_3ClC_2O_2$, $H_2Cl_2C_2O_2$, $HCl_3C_2O_2$ (acetic, monochloracetic, dichloracetic, trichloracetic, acid) are in the ratio $5\frac{1}{2} : 38 : 76 : 79$.

Increase of the number of oxygen atoms relatively to that of the other atoms forming the molecule of an acid is then frequently accompanied by an increase in the affinity of the acids. And similarly, an increase in the affinity accompanies the replacement of atoms of hydrogen by atoms of more negative elements, such as chlorine or bromine.

These facts concerning the connexion between the compositions and the properties of acids establish the existence of definite relations between the acidic or non-acidic character of compounds of hydrogen with oxygen and a third element, the basicities of acids, and the values of the affinity-constants of acids, on the one hand, and the nature, number, and probably the relative arrangement, of the atoms in the *molecules* of these compounds, on the other hand.

331 If we suppose that every atom in a molecule, or reacting atomic aggregate, is directly related in some way to a limited number of other atoms, then it appears that an atom of hydrogen which is directly related to strongly negative atoms is usually easily replaceable by atoms of positive elements; or, to use a shorter form of words, we may say that this atom of hydrogen *performs an acidic function in the molecule*, or *is acidic*. It also appears that an atom of hydrogen which is directly related to strongly positive atoms, e.g. atoms of potassium or sodium, is not acidic, that is cannot be replaced by the atoms of positive elements. But we are not yet in a position to discuss this subject of the connexion of properties with molecular structure otherwise than in the most general way (*s.* Chap. XVII.).

332 In Chap. XII. we made a slight examination of some of the circumstances which condition the course and final results of a chemical change. We then arrived at the conception of chemical equilibrium as the result of direct and reverse processes of change occurring simultaneously in the changing system (*comp.* pars. 233 to 236).

The general representation which the molecular and atomic theory puts before us of a system of substances free to act and react chemically—for instance of a mixture in dilute aqueous solution of equivalent quantities of potash, soda, and nitric acid

—is that of a great many small particles moving about freely, colliding and exchanging parts so as to produce new particles, which also collide with each other and with those of the original particles which remain unchanged, and some of which in so doing are decomposed with the re-formation of the original particles. This redistribution of atoms is accompanied by a redistribution of energy; energy is degraded in some of the molecular changes, and raised to a higher form in other parts of these changes, but the net result is a degradation of a portion of the energy of the whole system. The strife of molecules proceeds until equilibrium results; this equilibrium may be overthrown by introducing a fresh number of molecules of one of the original constituents of the system, or by altering the physical conditions under which the equilibrium was attained.

In Chap. XII. we briefly examined some cases of chemical **333** change wherein less complex substances, one at least of which was a gas, were produced from a more complex substance, by the action of heat alone. These changes, which were classed together under the name *dissociation*, were found to be reversible; that is to say, the original more complex body was re-formed when the products of the change were allowed to cool in contact with each other. We learned that there was a certain distribution of the changing substances at any specified temperature and pressure, but that change of temperature or pressure was attended with chemical change either in the direct or reverse direction.

The explanation which the molecular and atomic theory gives of dissociation can only be indicated here. Dissociation is regarded by this theory as essentially a change of one kind of molecules into two or more kinds of simpler molecules, brought about by adding heat-energy to the system. The explanation is based on a deduction from the fundamental assumptions of the theory, to the effect that there must be differences in the states of motion of individual molecules in a mass of gaseous molecules of one kind. The kinetic energy of the molecules is made up of two parts; the energy of the motion of the molecules as wholes, and the energy of the rotation of the parts of the molecules. Although the sum of these must be constant as long as temperature is unchanged, yet the distribution of the two motions, and hence of the two energies, may differ much, as regards the individual molecules. The energy due to the rotation of the parts of

some of the molecules may be so great, that collision between
these molecules may cause them to separate into parts; the
energy due to the motion of other molecules as wholes may be
so much greater than the energy due to the rotation of their
parts that a considerable quantity of energy must be added to
these molecules from a source external to the system before
they separate into parts. When the system is heated, those
molecules whose kinetic energy is chiefly due to the motion of
rotation of their parts will be at once separated into parts;
they will be dissociated. Some of the heat-energy added to
the system will be used in increasing the motion of rotation of
parts of other molecules, until these molecules also are dissociated. The process of dissociation will proceed rapidly for
a time, but as the number of molecules which are not separated
into parts becomes fewer so will the rate of dissociation become less, as temperature rises. But it will be possible for
some parts of molecules to reunite and reproduce some of the
original kinds of molecules, the rotational energy of which
will not be greater than that which brings about the separation of molecules into parts. Reunion of parts of molecules
will therefore occur to some extent. If temperature is kept
constant, the processes of separation of molecules into parts,
and of recombination of parts of molecules, will proceed until
both the kinetic energy of the system, and the atoms which
form the molecules of the system, are so distributed that
equilibrium results.

334   In par. 318 we learned that the specific gravity of gaseous
nitrogen tetroxide at low temperatures is about half what it
is at high temperatures. The most probable explanation of
this fact, in terms of the molecular theory, is that gaseous
nitrogen tetroxide has two molecular weights, one half as
great as the other. The formulae N_2O_4 and NO_2 express the
atomic compositions of the two molecules. The action of
heat on the molecules N_2O_4 is to convert them into the molecules NO_2; the change is represented in an equation thus
$N_2O_4 = 2NO_2$. But at any temperature between that at which
only N_2O_4 molecules exist and that at which only NO_2. molecules exist, the gas must be composed of a mixture of both
N_2O_4 and NO_2 molecules. If we adopt this explanation of the
action of heat on nitrogen tetroxide, then it is evident that
determinations of the spec. gravity of the gas at varying
temperatures give data from which the relative number of
molecules of each kind (N_2O_4 and NO_2) at that temperature

can be calculated. The change from N_2O_4 to NO_2 is a process of *dissociation* (*s*. pars. 233 to 236).

The only other explanation of the change of spec. gravity **335** of nitrogen tetroxide is that which asserts that this gas does not even approximately obey the ordinary law of the expansion of gases by heat. This explanation obliges us to assume that the rate of expansion of gaseous nitrogen tetroxide as temperature rises differs widely from the normal rate of expansion of gases; and also that the rate of expansion of nitrogen tetroxide itself is very different at different temperatures. To make this assumption is to go against the mass of evidence concerning the relations of the volumes of gases to temperature; whereas the assumption which is made in the statement of the molecular explanation of the facts is wholly in keeping with a large mass of evidence, and at the same time brings the apparently abnormal behaviour of nitrogen tetroxide within the number of those occurrences which find a simple explanation in terms of the molecular and atomic theory.

Many other instances of so-called *abnormal vapour-densities* **336** have been observed. For instance, the gas obtained by heating sulphuric acid, H_2SO_4, is about 24·5 times heavier than hydrogen. Now the molecular weight of a gaseous compound is twice the spec. gravity of that compound referred to hydrogen as unity; therefore, reasoning only from determinations of the spec. gravity of the gas obtained by heating sulphuric acid, we should conclude that the molecular weight of gaseous sulphuric acid is approximately 49. But the simplest formula which can express the composition of a molecule of sulphuric acid is $H_2SO_4 = 98$, if the atomic weights of sulphur and oxygen are 32 and 16 respectively. The density of the vapour of sulphuric acid seems then to be abnormal. As the definition of the molecular weight of a gas is deduced from the law of Avogadro, we begin to doubt this so-called law. But our doubts are laid when experiment proves that the gas obtained by heating sulphuric acid is a mixture of equal volumes of water-gas and sulphur trioxide. The specific gravity of such a mixture is calculated thus: the formulae H_2O and SO_3 represent the compositions of molecules of gaseous water and gaseous sulphur trioxide, respectively; but $H_2O = 18$, and $SO_3 = 80$; therefore the molecular weights of the two gases are 18 and 80, respectively; but the molecular weight of a gas is the weight of 2 volumes of that gas (*comp.*

Chap. xv. par. 294); therefore a mixture of equal volumes of H_2O and SO_3 is $\dfrac{18 + 80}{4} = 24 \cdot 5$ times heavier than hydrogen. Therefore the observed specific gravity of the gas obtained by heating sulphuric acid is identical with the specific gravity calculated from the experimentally determined composition of this gas.

337   Much discussion at one time took place as to the specific gravity of phosphorus pentachloride in the gaseous state. Analyses of this compound, and determinations of the atomic weights of phosphorus (31) and chlorine (35·5), shewed clearly that the simplest formula which could be given to the solid compound was PCl_5. The specific gravity of this compound in the state of gas, referred to hydrogen as unity, must be $\dfrac{31 + (35 \cdot 5 \times 5)}{2} = 104 \cdot 25$; or, referred to air as unity, the spec. gravity must be 7·2.

The following table gives the observed spec. gravities (air = 1) at different temperatures of the gas obtained by heating phosphorus pentachloride, PCl_5. In each case the pressure was 760 mm.

Temp.	Spec. gravity	Ratio of PCl_3 to total possible PCl_3	Temp.	Spec. gravity	Ratio of PCl_3 to total possible PCl_3	Temp.	Spec. gravity	Ratio of PCl_3 to total possible PCl_3
182°	5·08	·42	230°	4·30	·68	289°	3·69	·96j
190	4·99	·45	250	3·99	·80	300	3·65	·98
200	4·85	·49	274	3·84	·98	327	3·65	·98
			288	3·67	·97	336	3·65	·98

The specific gravity, and therefore probably the molecular weight, of the gaseous compound varies; but at no temperature does the specific gravity approximate to that required by the formula PCl_5. Here again we have an instance of so-called *abnormal vapour-density*. Further investigation however shewed that the numbers are not really abnormal, but that they find a ready and simple explanation in terms of the molecular and atomic theory. Experiments proved that phosphorus pentachloride is gradually dissociated by heat into phosphorus trichloride and chlorine; $PCl_5 = PCl_3 + Cl_2$. The gas obtained by heating PCl_5 is therefore a mixture of more than one substance. The specific gravity of a mixture of phosphorus trichloride and chlorine in the ratio PCl_3 : Cl_2 must be $\dfrac{PCl_3 + Cl_2}{4} = 72 \cdot 125$ if hydrogen is taken as unity, and $3 \cdot 6$ if air is taken as unity. The specific gravity of the gas obtained by heating PCl_5 to $300°$ is practically identical with that of a mixture of PCl_3 and Cl_2; the specific gravity at different temperatures lower than that at which dissociation is complete is that of a mixture of these two gases with some gaseous PCl_5 in varying proportions. The change of PCl_5 into PCl_3 and Cl_2, brought about by the action of heat, is therefore a normal process of dissociation. A portion of the PCl_5 is volatilised without change, but at the same time some of it is dissociated into the simpler molecules PCl_3 and Cl_2; as temperature rises the quantity of unchanged PCl_5 decreases, and the quantities of PCl_3 and Cl_2 increase, until at about $300°$ the whole of the PCl_5 is changed to $PCl_3 + Cl_2$. The numbers in the third columns of the table shew the ratio, at each temperature, of PCl_3 to the total PCl_3 possible as the asumption that the whole of the PCl_5 had been dissociated.

There are other cases of so-called *abnormal vapour-densities* which are not quite so easily explained by the theory of atoms and molecules; but we cannot discuss these here.

The practical outcome of these facts as bearing on deter- **338** minations of molecular weights is, that the specific gravity of a gas must be constant through a considerable interval of temperature before we are justified in deducing an approximate value for the molecular weight, from the observed specific gravity, of this gas. If the specific gravity varies considerably with temperature-changes, then the gas is probably a mixture; but the expression *molecular weight of a mixture* has no meaning.

In Chap. XI. par. 175 we had examples of chemical changes **339** brought about by oxygen and hydrogen, respectively, when these elements were themselves products of one part of the complete cycles of change. Thus, when chlorine is passed into a warm solution of potash (KOH), potassium hypochlorite (KClO) is formed, but this compound is quickly decomposed with production of potassium chloride (KCl) and oxygen; if lead monoxide (PbO) is suspended in the warm potash solution, the oxygen, or a portion of the oxygen, produced from the decomposition of the hypochlorite combines with the lead monoxide to form dioxide (PbO_2). But if oxygen is passed into warm potash solution holding lead monoxide in suspension

no lead dioxide is produced; the oxygen must be produced by
a chemical reaction in the system of which the body to be
oxidised forms a part. Similarly, if hydrogen is produced, by
the interaction of zinc and dilute sulphuric acid, in a solution
containing sodium sulphite (Na_2SO_3), hydrogen sulphide (H_2S)
is produced; but if hydrogen is passed into a solution of
sodium sulphite hydrogen sulphide is not produced: the
hydrogen must be produced by a chemical reaction in the
system of which the sodium sulphite forms a part.

In Chap. XIV. par. 266 we briefly considered the changes
of energy which accompany such chemical changes as these.

The molecular and atomic theory throws some light on these
changes. This theory leads to the view that a system composed of atoms of a specified element, could such a system
exist, would differ from a system composed of molecules of the
same element. It also leads to the view that in very many,
if not most, chemical changes, the formation of the molecules
of the products of the change is preceded by the breaking up
into atoms, or sometimes into groups of atoms, of the molecules of the interacting substances. And, lastly, the theory
almost obliges us to believe that a system composed of atoms
of one of those elements the molecules of which are built up
of more than a single atom (s. table in par. 317), if it could
exist would be extremely unstable, and would almost at once
pass into a system composed of molecules of the same element.

The application of these conceptions to the class of changes
we are considering affords some explanation of the mechanism
of these changes. The explanation may be stated as follows.
Under ordinary conditions quantities of oxygen or hydrogen
consist of molecules of these gases. Oxygen passed into
potash holding lead monoxide in suspension does not oxidise
the lead oxide, because the affinity between molecules of
oxygen and lead monoxide is not sufficient to produce this
change, and there is not sufficient energy available in the
system for separating the molecules of oxygen into atoms: but
when potash and chlorine interact, atoms of oxygen are
produced; these atoms combine with the molecules of lead
monoxide to form lead dioxide, and in this change more energy
is degraded than would be the case if the atoms of oxygen
had combined with each other to form molecules of oxygen.

A similar explanation would be given of the interaction
between sodium sulphite and the atoms of hydrogen produced
by the interaction of zinc and sulphuric acid.

Chemical changes which are brought about by elements only when these are themselves products of a part of the complete change are sometimes classed together as *nascent actions*. The name has been useful as marking a class of reactions which have a common feature. If the view here taken of these reactions is correct, there is nothing in any way abnormal about them ; they belong to the ordinary type of chemical change.

It is evident then that the molecular and atomic theory **340** brings into one point of view, and gives fairly simple explanations of, many classes of chemical occurrences. It also indicates directions in which experiments ought to proceed with the object of discovering and explaining new classes of chemical events.

The words *law, hypothesis*, and *theory*, have been frequently used in this book.

The word *law* has sometimes been employed as synonymous with a general truth; for instance the laws of chemical combination are general truths, they summarise many facts. The same term, law, is sometimes used as meaning an abstract truth ; for instance, Newton's laws of motion are truths involved in many phenomena although actually seen in none. The statement 'equal volumes of gases contain equal numbers of molecules' has been called a *law*. This statement is really a deduction from a theory. The deduction has a definite meaning when the terms in which it is made are defined, but this can be done only by granting the fundamental assumptions of the theory. The 'law' stands or falls with the theory.

The molecular and atomic *theory*, like all scientific theories, is based on certain assumptions. The fewer, the simpler, and the more binding, the assumptions on which a theory rests the better is the theory. One of the marks of a satisfactory theory is the impossibility of escaping from discrepancies between observed facts and deductions from the theory by the invention of subsidiary hypotheses which do not follow directly from the assumptions on which the theory rests. The molecular and atomic theory, it must be confessed, has been too elastic in this respect.

An *hypothesis* is specially framed to explain a definite occurrence, or a series of occurrences. For instance, when Davy found that nitric acid was formed at the positive electrode during the electrolysis of water, he framed the hypothesis that the air surrounding the decomposing water was the source of the acid : he was able to prove by direct experiment that this hypothesis was correct. An hypothesis is sometimes stated in very general terms, and is used to explain a great many apparently unconnected facts. For instance, very many facts concerning chemical change are generalised in the hypothesis that 'the amount of a chemical change is proportional to the affinities and the active masses of the substances taking part in the change'. A direct and final experimental proof of such an hypothesis as this can scarcely be given. If the terms can be accurately defined, and if after prolonged inquiry no facts are discovered which negative the hypothesis, it is adopted as a trustworthy guide.

CHAPTER XVII.

ISOMERISM AND STRUCTURAL FORMULAE.

341   In the last chapter we had an instance of *isomerism*; namely, the existence of three different compounds all having the molecular composition expressed by the formula C_5H_{12}. The prominent fact of isomerism is, that two or more compounds sometimes exist having identical compositions, and identical specific gravities in the state of gases, and yet exhibiting different properties. The statement of this fact in the language of the molecular and atomic theory is, that two or more gaseous molecules may exist composed of the same number of the same atoms, and yet differing from each other in their properties.

342   An instance of isomerism is furnished by the existence of two compounds having the composition C_2H_6O. One of these is *ethylic alcohol*, the other is *methylic ether*. Ethylic alcohol interacts with potassium or sodium thus, $C_2H_6O + K = C_2H_5KO + H$; methylic ether and potassium (or sodium) do not interact. Phosphorus pentachloride interacts with both isomerides; the interactions are these :—

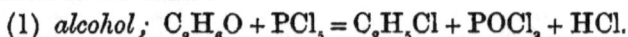

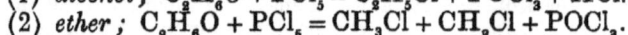

(1) *alcohol;* $C_2H_6O + PCl_5 = C_2H_5Cl + POCl_3 + HCl.$
(2) *ether;* $C_2H_6O + PCl_5 = CH_3Cl + CH_3Cl + POCl_3.$

The alcohol is easily oxidised, first to aldehyde C_2H_4O, then to acetic acid $C_2H_4O_2$; the ether is not easily oxidised. Ethylic alcohol is a colourless, volatile, liquid, boiling at $78^\circ\cdot3$; methylic ether is a colourless gas which may be condensed by cold to a liquid boiling at -21°.

Another instance of isomerism is furnished by the existence of four hydrocarbons having the molecular composition C_8H_{10}. These bodies are all liquids, boiling at 134°, 136°—137°,

$137°$—$137°·5$, and $140°$—$141°$, respectively. That which boils at $134°$ is easily oxidised to an acid having the composition $C_7H_6O_2$. The other three are oxidised to three different acids all having the composition $C_8H_6O_2$. The conditions under which these three hydrocarbons are oxidised vary somewhat.

The four hydrocarbons C_8H_{10} are evidently very similar in **343** their chemical properties; the two compounds C_9H_6O are less closely related to each other. Compounds which have the same composition and the same molecular weight, but which shew differences in their chemical properties so decided as to require them to be placed in different classes, are sometimes said to be *metameric*. *Metamerism* is included in the wider term *isomerism*.

The molecular and atomic theory endeavours to explain **344** isomerism by saying that the properties of molecules depend, among other conditions, on the arrangement of their parts. Is this assertion justified by facts?

Before we can profitably attempt an answer to this question we must understand what is meant by the phrase 'arrangement of the parts of a molecule'.

In Chap. XIII. par. 247 a very brief account was given of the **345** use of the expression 'equivalent weights of two alkalis'. We must now look more fully at the notion of *chemical equivalency*.

$88·8$ parts by weight of potash (KOH), $63·5$ parts by weight of soda (NaOH), and $38·1$ parts by weight of lithia (LiOH), severally neutralise 100 parts by weight of nitric acid. These masses, $88·8$, $63·5$, $38·1$, of the three alkalis are therefore equivalent as regards power of neutralising a specified mass of nitric acid, inasmuch as these masses are of equal value in exchange. 100 parts by weight of sulphuric acid are neutralised by $114·3$ parts by weight of potash, or by $81·6$ parts by weight of soda, or by 49 parts by weight of lithia. These numbers, $114·3$, $81·6$, and 49, represent masses of the three alkalis which are equivalent as regards power of neutralising a specified mass of sulphuric acid. Now the ratio $88·8 : 63·5 : 38·1$ is the same as the ratio $114·3 : 81·6 : 49$. If the masses of these three alkalis required to neutralise 100 parts by weight of each of several acids are determined, it is found that these masses always bear the same ratio to one another. Hence it is possible to assign values to these three alkalis representing those masses of them which are equivalent as regards power of neutralising one and the same

mass of any specified acid. Similarly, it is possible to assign
values to the acids which shall represent those masses of them
which severally neutralise one and the same mass of any
specified alkali.

In determining the *equivalent weights* of the alkalis it is
customary to take one reacting weight (or we may say one
molecule) of hydrochloric acid as the unit mass of standard
acid. The reacting weight of hydrochloric acid (HCl) is
36·5: one reacting weight of this acid is neutralised by (in
round numbers) 56 parts by weight of potash, 40 of soda, and
24 of lithia, respectively. The mass of sulphuric acid which
is neutralised by each of these masses of potash, soda, or lithia
is 49; the mass of chloric acid is 84·5; the mass of ortho-
phosphoric acid is 32·6; the mass of metaphosphoric acid is
80; the mass of pyrophosphoric acid is 44·5; &c. &c. The
numbers 36·5, 49, 84·5, 32·6, 80, 44·5 represent masses of the
acids mentioned which are equivalent as regards power of
neutralising 56 parts by weight of potash, or 40 parts of soda,
or 24 of lithia.

346   The notion of equivalency may be extended to the elements. If we determine the masses of a series of metals
which severally combine with 16 parts by weight of oxygen,
we shall have determined the *equivalent weights* of these
metals as regards this particular reaction.

Or we might cause a number of metals to interact with
hydrochloric acid, and determine the mass of each metal which
thus produced 1 gram of hydrogen; these masses would
represent *equivalent weights* of the metals as regards this
particular reaction.

347   When therefore we speak of the equivalent weight of an
element or compound there is always implied a comparison of
the specified substance with some other substance as regards
power of performing a definite chemical operation. Equivalent weights represent quantities of elements or compounds
which can be exchanged in some specified chemical process.

348   The expression *equivalent weight of an element* is frequently
used somewhat loosely. In order to determine equivalent
weights, elements are generally compared as regards their
combination with oxygen, and 8 parts by weight is usually
chosen as the standard mass of oxygen. Hence the equivalent
weight of an element generally means the mass of it which
combines with 8 parts by weight of oxygen. In the cases of
elements which do not combine with oxygen, hydrogen is

generally chosen as the standard element, and the equivalent weight is taken to be that mass of the element which combines with 1 part by weight of hydrogen, or sometimes (especially in the case of metals) that mass of the element which interacts with a dilute acid to produce 1 part by weight of hydrogen.

Let us apply these various definitions of equivalent weight to the metal tin. Experiment proves; (1) that 59 parts by weight of tin combine with 8 parts by weight of oxygen to produce stannous oxide; (2) that 29·5 parts by weight of the same element combine with 8 parts by weight of oxygen to produce stannic oxide; (3) that 59 parts by weight of tin interact with hydrochloric acid to produce 1 part by weight of hydrogen, stannous chloride being formed at the same time.

We therefore get two values for the equivalent weight of **349** tin. But had we accurately defined the meaning to be given to the term *equivalent weight* we should have got over this difficulty. Let tin and lead be compared as regards the formation of oxides having similar properties and similar compositions. Each metal forms a protoxide MO, and a dioxide MO_2; the oxides MO are fairly similar chemically, and so are the oxides MO_2. Experiment shews that 59 parts by weight of tin are equivalent to 103·5 parts by weight of lead as regards power of combining with 8 parts by weight of oxygen to produce oxides belonging to the type MO; and that 29·5 parts by weight of tin are equivalent to 51·75 parts by weight of lead as regards power of combining with 8 parts by weight of oxygen to produce oxides of the type MO_2.

The notion of the equivalency of elements is fairly simple; **350** but it is often very difficult to apply it accurately. The difficulty consists in finding a standard chemical change wherein a specified mass of one element may be exchanged for a specified mass of another without altering the essential character of the reaction.

The conception of equivalency has been extended to the **351** elementary atoms.

Let the standard action be ability to combine with one, and only one, atom of hydrogen to produce a gaseous molecule; let all atoms which do this be classed together as equivalent. Then the atoms of *hydrogen, chlorine, bromine, iodine,* and probably *fluorine**, are equivalent; the evidence is the existence of the gaseous molecules H_2, HCl, HBr, HI, [and HF]*; and

------

* There is still some doubt whether the molecule of gaseous hydrogen fluoride is HF or H_2F_2.

the non-existence of gaseous molecules composed of one atom
of chlorine, bromine, &c. and more than a single atom of
hydrogen.

352   The atoms of hydrogen, chlorine, bromine, iodine and
fluorine are placed together in one class, and are called *monovalent atoms.*

353   It is evident that in asserting these atoms to be equivalent,
we have relaxed the strict meaning of the term equivalency.
An atom of hydrogen, or an atom of chlorine, or an atom of
bromine, &c. combines with only one atom of hydrogen to
produce a gaseous molecule; in this respect the atoms are of
equal value in exchange. In defining the meaning of the
terms divalent, trivalent, &c. atoms, we must a little further
relax the definition of equivalency. We assume that an atom
which combines with 2 atoms, and not more than 2 atoms, of
chlorine, &c. is equivalent to an atom which combines with
not more than 2 atoms of hydrogen, &c. By doing this we
arrive at the definition of *divalent atoms* as atoms which
combine with not more than two monovalent atoms to form
gaseous molecules.

354   Applying these definitions of monovalent, divalent, &c.
atoms to all the elements compounds of which with hydrogen,
chlorine, bromine, iodine, or fluorine, have been gasified, we
arrive at the following classification of atoms.

Standard Monovalent atoms; H, F, Cl, Br, I.

I.   *Monovalent atoms;* i.e. atoms which combine with
one standard monovalent atom to form gaseous molecules
......K, Tl, **Hg**.

II.   *Divalent atoms;* i.e. atoms which combine with **two**
standard monovalent atoms to form gaseous molecules......O, S,
Se, Te, Be, Cd, Zn, **Hg, Sn**, Pb.

III.   *Trivalent atoms;* i.e. atoms which combine with
three standard monovalent atoms to form gaseous molecules...
...B, N, **P**, As, Sb, Bi, In.

IV.   *Tetravalent atoms;* i.e. atoms which combine with
four standard monovalent atoms to form gaseous molecules
......C, Si, Ti, Ge, Zr, V, **Sn**, Th, U.

V.   *Pentavalent atoms;* i.e. atoms which combine with
five standard monovalent atoms to form gaseous molecules
......P, Nb, Ta, Mo, **W**.

VI.   *Hexavalent atoms;* i.e. atoms which combine with **six**
standard monovalent atoms to form gaseous molecules......**W**.

This table includes about half the elements; the valencies of the atoms of the other elements cannot yet be determined for want of data.

The data on which this classification of atoms is based are presented in the following list of gaseous molecules :—

KI, $TlCl$, $HgCl$; OH_2, OCl_2, SH_2, SeH_2, TeH_2, $BeCl_2$, $BeBr_2$, $CdBr_2$, $ZnCl_2$, $HgCl_2$, $HgBr_2$, HgI_2, $SnCl_2$, $PbCl_2$; BF_3, BCl_3, BBr_3, NH_3, PH_3, PCl_3, AsH_3, $AsCl_3$, AsI_3, $SbCl_3$, SbI_3, $BiCl_3$, $InCl_3$; CH_4, CCl_4, SiF_4, $SiCl_4$, SiI_4, $GeCl_4$, GeI_4, $TiCl_4$, $ZrCl_4$, VCl_4, $SnCl_4$, $SnBr_4$, $ThCl_4$, UBr_4, UCl_4; PF_5, $NbCl_5$, $TaCl_5$, $MoCl_5$, WCl_5; WCl_6.

Of the 35 elements classified in the table, four viz. P, Sn, **355** W, and Hg, are found each in two classes. The atom of P is trivalent and pentavalent; the atom of Sn is di- and tetravalent; that of W is penta- and hexa-valent; and that of Hg is mono- and di-valent. As we found that some elements have · more than one equivalent weight, so now we find that the atoms of certain elements are sometimes equivalent to one number, and sometimes to another number, of monovalent atoms. In determinations both of equivalent weights and of the equivalency of atoms, the conception of equivalency is rather vaguely used.

We may *for the present* define the *maximum valency of an* **356** *atom* to be, *the maximum number of atoms of hydrogen, fluorine, chlorine, bromine, or iodine, with which the specified atom combines to form a gaseous molecule.*

This definition indicates the data which must be obtained before the maximum valency of an atom can be determined.

To say that a specified atom is divalent has generally been **357** regarded as synonymous with saying that the atom in question is *equivalent to* 2 atoms of hydrogen, fluorine, chlorine, bromine, or iodine ; and that, therefore, any atom which combines with one divalent atom is thereby proved to be itself a divalent atom. Thus, the existence of the gaseous molecules OH_2 and OCl_2 proves the atom of oxygen to be divalent : one atom of carbon combines with one atom of oxygen to form the gaseous molecule CO, and with 2 atoms of oxygen to form the gaseous molecule CO_2 ; hence, it is argued, the atom of carbon is divalent and tetravalent. Of late years many chemists have abandoned such arguments as this. They have recognised the possibility of determining the maximum valencies of the atoms of elements which form gasifiable compounds with hydrogen, fluorine, chlorine, bromine, or iodine, and of such elements only.   They have been content, for the present, to admit that a divalent atom is not necessarily strictly equivalent—i.e. of equal value in performing a chemical change—to 2 monovalent atoms, a trivalent atom to 3 monovalent atoms, a tetravalent atom to 4 monovalent or to 2 divalent atoms, &c.   *s.* also par. 372.

358   Let us now see whether the conception we have gained of valency enables us to give a definite meaning to the phrase 'arrangement of the parts of a molecule' used in par. 344.

When it is said that the atom of carbon is tetravalent because the gaseous molecules CH_4, CCl_4, $CHCl_3$, &c. exist, it is asserted that each of these molecules is formed by the union of one atom of carbon with 4 monovalent atoms.   This assertion means that in each of these molecules there is direct interaction of some kind between the atom of carbon and each of the 4 monovalent atoms.   What the nature of this interaction is, we do not know.   The molecular formulae CH_4, CCl_4, &c. do not prove the existence of this direct interaction ; there may be direct interaction between the atom of carbon and only one, two, or three, of the monovalent atoms; but the hypothesis of direct interaction between the atom of carbon and each of the 4 monovalent atoms is the simplest and most workable hypothesis that has been tried.

Adopting this hypothesis, let us indicate the existence of some kind of direct interaction between each monovalent atom and the atom of carbon by the symbols

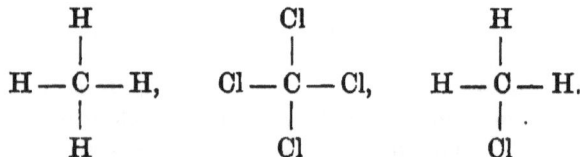

$$
\begin{array}{ccc}
H & Cl & H \\
| & | & | \\
H-C-H, & Cl-C-Cl, & H-C-H. \\
| & | & | \\
H & Cl & Cl
\end{array}
$$

It is of the greatest importance that the student should clearly understand what is meant to be conveyed by such symbols as these.   The symbols

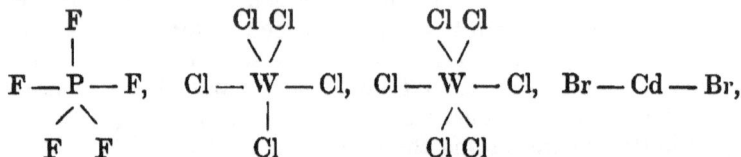

$$
\begin{array}{cccc}
F & Cl\ \ Cl & Cl\ \ Cl & \\
| & \diagdown\diagup & \diagdown\diagup & \\
F-P-F, & Cl-W-Cl, & Cl-W-Cl, & Br-Cd-Br, \\
\diagup\diagdown & | & \diagup\diagdown & \\
F\ \ \ F & Cl & Cl\ \ Cl &
\end{array}
$$

represent the atomic compositions of certain gaseous molecules; and they assert that there is direct interaction between the atom of phosphorus and each atom of fluorine, between the atom of tungsten and each atom of chlorine, and between the atom of cadmium and each atom of bromine, in the respective molecules.

Now, as no gaseous molecule has been obtained composed of one atom of carbon and more than 4 atoms of hydrogen, fluorine, chlorine, bromine, or iodine; as no gaseous molecule has been obtained composed of one atom of phosphorus and more than 5 atoms of hydrogen, fluorine, &c.; as no gaseous molecule has been obtained composed of one atom of tungsten and more than 6 atoms of chlorine, &c.; and as no gaseous molecule has been obtained composed of one atom of cadmium and more than 2 atoms of chlorine, &c.; we seem justified in asserting that an atom of carbon can directly interact with not more than 4 monovalent atoms, an atom of phosphorus with not more than 5 monovalent atoms, an atom of tungsten with not more than 6 monovalent atoms, and that an atom of cadmium can directly interact with not more than 2 monovalent atoms, in gaseous molecules.

We thus arrive at the notion of a limit to the number of **359** atoms between which direct interaction can occur in a gaseous molecule.

If we now apply this notion not only to gaseous molecules composed of one polyvalent atom united with monovalent atoms, but to all gaseous molecules, we find ourselves in possession of a good working hypothesis regarding the arrangement of atoms in molecules.

The hypothesis may be stated thus;—*each atom in a gaseous molecule can directly interact with a limited number of other atoms.*

Let us at once widen our conception of atomic valency, **360** and say that *the valency of an atom is a number expressing the maximum number of other atoms between which and the given atom there is direct interaction in any gaseous molecule.* But while thus widening the meaning of the term valency, let us agree to determine the valency of any atom by finding the maximum number of *monovalent* atoms with which it directly interacts, i.e. in this case with which it combines, in a gaseous molecule.

We have thus a perfectly definite method for finding the valencies of atoms, and we have also a wide range of application for these values.

It is customary to place small roman numerals above the **361** symbol of an element to represent the valency of an atom of that element; thus O^{II}, C^{IV}, Bi^{III}, W^{VI}, mean a divalent atom of oxygen, a tetravalent atom of carbon, a trivalent atom of bismuth, a hexavalent atom of tungsten, respectively. When a numeral is not placed over the symbol of an atom that atom is taken to be monovalent.

362 The table in par. 354 gives the results of the determinations of the valencies of atoms; the applications of these values will now be shewn by one or two examples.

Two compounds exist each having the composition expressed by the formula C_2H_6O; as both compounds have been gasified, this formula represents the atomic composition of the molecule of either compound. From the data given in the table in par. 354 it is evident that the atom of carbon is tetravalent, and the atom of oxygen is divalent; the atom of hydrogen is one of the standard monovalent atoms. In other words an atom of carbon can directly interact with not more than 4 other atoms, an atom of oxygen can directly interact with not more than 2 other atoms, and an atom of hydrogen can directly interact with not more than one other atom, in a gaseous molecule. We have agreed to represent direct interaction between two atoms in a gaseous molecule by the use of lines proceeding from the symbols of these atoms. How then can we represent the arrangement of one divalent 2 tetra- and 6 monovalent atoms? Let us assume (1) that each carbon atom directly interacts with the other carbon atom; (2) that neither carbon atom directly interacts with the other carbon atom. On assumption (1) the only possible representation of the atomic arrangement of the molecule $C_2^{IV}H_6O^{II}$ is

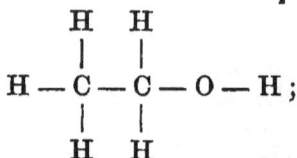

$$
\begin{array}{ccc}
\text{H} & \text{H} \\
| & | \\
\text{H}-\text{C}-\text{C}-\text{O}-\text{H}\,; \\
| & | \\
\text{H} & \text{H}
\end{array}
$$

we shall call this compound I. On assumption (2) the only possible representation is

$$
\begin{array}{ccc}
\text{H} & & \text{H} \\
| & & | \\
\text{H}-\text{C}-\text{O}-\text{C}-\text{H}\,; \\
| & & | \\
\text{H} & & \text{H}
\end{array}
$$

we shall call this compound II.

Now two compounds, and only two, exist having each the molecular composition C_2H_6O. So far the facts are in keeping with the deduction from the hypothesis of valency. But how can we tell which of the two compounds is I. and which is II. ? If the chemical properties of a molecule depend, partly, on the arrangement of the atoms which constitute the molecule, the chemical properties of compound I. must differ from those of compound II. To determine which compound is represented by formula I. and which by formula II., we must study the chemical properties of each isomeride C_2H_6O. The isomerides in question are called *ethylic alcohol*, and *methylic ether*. Each interacts with phosphorus pentachloride, but the products of the interactions are very different : ethylic alcohol interacts thus,

$$C_2H_6O + PCl_5 = POCl_3 + HCl + C_2H_5Cl ;$$

methylic ether interacts thus,

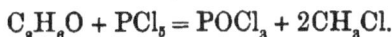

$$C_2H_6O + PCl_5 = POCl_3 + 2CH_3Cl.$$

The first interaction consists in the withdrawal of an atom of oxygen and an atom of hydrogen from the molecule C_2H_6O, and the putting in the place·of these atoms of an atom of chlorine ; the second interaction consists in the withdrawal of an atom of oxygen from the molecule C_2H_6O, the putting in the place of this atom of 2 atoms of chlorine, and the simultaneous separation of the group of atoms $C_2H_6Cl_2$ into two parts, each of which is composed of one atom of carbon united with 3 atoms of hydrogen and one atom of chlorine. Now symbol I. represents an atom of oxygen as directly interacting with an atom of carbon and also with an atom of hydrogen ; as the atom of chlorine can directly interact with a single other atom only, the withdrawal of the group of atoms OH, and the substitution for it of the atom Cl, seems a very probable change. If this change proceeds the resulting molecule will be represented

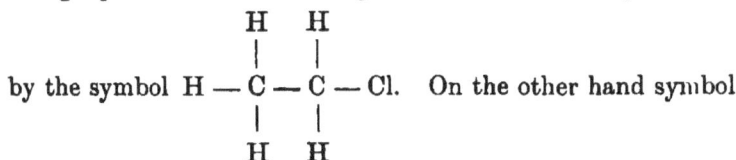

by the symbol

$$\begin{array}{ccc} & \text{H} & \text{H} \\ & | & | \\ \text{H} - & \text{C} - \text{C} & - \text{Cl.} \\ & | & | \\ & \text{H} & \text{H} \end{array}$$

On the other hand symbol

II. represents an atom of oxygen as directly interacting with 2 atoms of carbon, and with these atoms only ; if this atom of oxygen were withdrawn and 2 atoms of chlorine put in its

place we should get the molecule

$$H - C - Cl - Cl - C - H;$$

with H above and below each C atom;

but this molecule cannot exist (by hypothesis) because the chlorine atom is monovalent; hence the substitution of two atoms of chlorine for a single atom of oxygen in the molecule represented by symbol II. must result in the production of two

molecules each represented by the symbol

$$H - C - Cl.$$

with H above and below the C.

We therefore conclude that the molecule of ethylic alcohol

is represented by the symbol

$$H - C - C - O - H,$$ and the

with H above and below each C atom,

molecule of methylic ether by the symbol

$$H - C - O - C - H.$$

with H above and below each C atom.

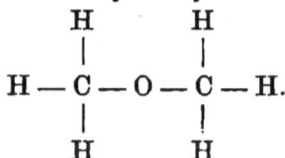

This conclusion is borne out by a further study of the properties of the two isomerides. Thus, sodium rapidly interacts with ethylic alcohol to produce $C_2H_5NaO + H$; but sodium and methylic ether do not interact. The formula given for ethylic alcohol represents one, and only one, of the 6 hydrogen atoms as directly interacting with an atom of oxygen; we should therefore expect that sodium would replace either 5 atoms, or one atom, of hydrogen from the molecule of ethylic alcohol. As sodium replaces only a single atom of hydrogen we conclude that the atom replaced is that which is represented in the formula as directly interacting with an atom of oxygen. But if this is so, we should conclude that none of the hydrogen atoms in the molecule of methylic oxide would be replaceable by sodium. Another reaction which favours this conclusion regarding the interaction of sodium with ethylic alcohol is that which occurs between sodium and water. This change is formulated thus $H_2O + Na = NaOH + H$. Now the only possible way of representing the molecule H_2O^{II} is $H - O - H$. The interaction between this molecule and an atom of sodium must be represented thus,

$$H - O - H + Na = Na - O - H + H;$$

the atom of hydrogen which is replaced by an atom of sodium must be represented as directly interacting with an atom of oxygen. Therefore, as only one hydrogen atom in the molecule of ethylic alcohol is represented as interacting directly with an oxygen atom, it is fairly probable that this is the atom of hydrogen which is replaced by sodium.

Let us take another example of the application of the **363** conception of atomic valency, that is that each atom in a gaseous molecule can directly interact with a limited number of other atoms. A certain hydrocarbon has the molecular composition C_2H_6. Can more than one compound exist having this composition? In other words, can we represent the arrangement of the atoms $C_2^{IV}H_6$ in more than one way? As the atom of carbon is tetravalent, and that of hydrogen is monovalent, we must represent the two atoms of carbon in the gaseous molecule C_2H_6 as directly interacting; the only possible way of doing this is to write the formula thus,

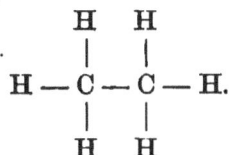

$$
\begin{array}{ccc}
\text{H} & \text{H} \\
| & | \\
\text{H} - \text{C} - \text{C} - \text{H.} \\
| & | \\
\text{H} & \text{H}
\end{array}
$$

Therefore the hypothesis of valency, when applied to the compound C_2H_6, asserts that one and only one compound having this molecular composition can exist. As a matter of fact only one compound C_2H_6 does exist.

There is another hydrocarbon C_2H_4; what are the ways **364** in which 2 tetra- and 4 mono-valent atoms can be arranged? Each carbon atom in the molecule C_2H_4 must interact with another carbon atom; we may write the formula as

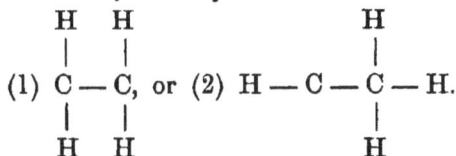

$$
\begin{array}{cccc}
\text{H} & \text{H} & & \text{H} \\
| & | & & | \\
(1)\ \text{C} - \text{C,} & \text{or } (2)\ \text{H} - \text{C} - \text{C} - \text{H.} \\
| & | & & | \\
\text{H} & \text{H} & & \text{H}
\end{array}
$$

Formula (1) represents each carbon atom in the molecule C_2H_4 as directly interacting with another carbon atom and with 2 hydrogen atoms, i.e. as directly interacting with 3 other atoms. Formula (2) represents one of the carbon atoms as directly interacting with 2 other atoms, and the other carbon atom as directly interacting with 4 other atoms. In par. 360 we defined the *valency* of an atom to be *the number expressing the maximum number of other atoms between which and the given atom there is direct interaction in any gaseous molecule.* In accordance with this definition, we may say that *formula (1) represents each atom of carbon as trivalent in the molecule C_2H_4, and formula (2) represents one atom of carbon as divalent, and one atom of carbon as tetravalent, in the molecule C_2H_4.* As it is impossible to represent both atoms of carbon as tetravalent, i.e. as directly interacting with 4 other atoms, in the molecule C_2H_4, it is evident that although the maximum valency of a carbon atom is 4, yet the actual valency of an atom of this element in a specified molecule may be less than 4.

· The student should particularly notice that the statement, an atom of carbon may act in a specified molecule as a trivalent or divalent atom, only holds good when we attach to the term valency of an atom the meaning given in par. 360.

The hypothesis of valency then points to the possible existence of two ' isomerides C_2H_4. But only one compound C_2H_4 is known to exist. Which of the two formulae given above shall we assign to this compound? The compound in question is called *ethylene.* Ethylene readily combines with chlorine to form the dichloride $C_2H_4Cl_2$. If the formula of

$$\begin{matrix} H & H \\ | & | \\ \end{matrix}$$
ethylene is C —— C, the formula of the dichloride is almost
$$\begin{matrix} | & | \\ H & H \end{matrix}$$

$$\begin{matrix} H & H \\ | & | \\ \end{matrix}$$
certainly Cl —— C —— C —— Cl; if the formula of ethylene is
$$\begin{matrix} | & | \\ H & H \end{matrix}$$

$$\begin{matrix} H \\ | \\ \end{matrix}$$
H —— C —— C —— H, the formula of the dichloride is almost
$$\begin{matrix} | \\ H \end{matrix}$$

$$\text{certainly } H - \overset{\overset{\displaystyle Cl}{|}}{\underset{\underset{\displaystyle Cl}{|}}{C}} - \overset{\overset{\displaystyle H}{|}}{\underset{\underset{\displaystyle H}{|}}{C}} - H. \quad \text{Only two formulae are possible}$$

for the molecule $C_2H_4Cl_2$; the formation of $Cl - \overset{\overset{\displaystyle H}{|}}{\underset{\underset{\displaystyle H}{|}}{C}} - \overset{\overset{\displaystyle H}{|}}{\underset{\underset{\displaystyle H}{|}}{C}} - Cl$

is much more likely than the formation of $H - \overset{\overset{\displaystyle Cl}{|}}{\underset{\underset{\displaystyle Cl}{|}}{C}} - \overset{\overset{\displaystyle H}{|}}{\underset{\underset{\displaystyle H}{|}}{C}} - H$

from $\overset{\overset{\displaystyle H \quad H}{| \quad |}}{\underset{\underset{\displaystyle H \quad H}{| \quad |}}{C - C}}$; because, as the maximum valency of a

carbon atom is 4, and as each carbon atom in the molecule $\overset{\overset{\displaystyle H \quad H}{| \quad |}}{\underset{\underset{\displaystyle H \quad H}{| \quad |}}{C - C}}$ is represented as trivalent, it is only necessary

for each carbon atom to interact directly with one of the chlorine atoms brought into contact with the molecule C_2H_4

in order to produce the molecule $Cl - \overset{\overset{\displaystyle H \quad H}{| \quad |}}{\underset{\underset{\displaystyle H \quad H}{| \quad |}}{C - C}} - Cl$; but if

the molecule $H - \overset{\overset{\displaystyle Cl}{|}}{\underset{\underset{\displaystyle Cl}{|}}{C}} - \overset{\overset{\displaystyle H}{|}}{\underset{\underset{\displaystyle H}{|}}{C}} - H$ is produced a considerable

rearrangement of the interactions of the atoms of carbon and hydrogen must occur. The only safe rule to adopt in studying the applications of valency to isomerism is, that no rearrange-ment of the interactions of atoms must be assumed to take place unless the facts absolutely require it.

Two compounds having the molecular composition $C_2H_4Cl_2$ exist : one is formed by the direct addition of chlorine to ethylene, it is called *ethylene chloride*; the other is produced by the interaction of chlorine with the hydrocarbon *ethane* C_2H_6, [thus, $C_2H_6 + 2Cl_2 = C_2H_4Cl_2 + 2HCl$], it is called *ethy-lidene chloride*. To which of these compounds must we assign the formula

$$\begin{array}{ccc} Cl & & H \\ | & & | \\ H - C & - & C - H? \\ | & & | \\ Cl & & H \end{array}$$

Ethylidene chloride is also produced by the interaction of phosphorus pentachloride with ethylic aldehyde ; thus $C_2H_4O + PCl_5 = C_2H_4Cl_2 + POCl_3$. In this reaction one atom of oxygen has been removed from the molecule C_2H_4O and 2 atoms of chlorine have been put in its place ; therefore, unless distinct reasons can be shewn to the contrary, it is likely that the 2 atoms of chlorine in the molecule $C_2H_4Cl_2$ are related to the rest of the molecule in a way similar to that in which the atom of oxygen in the molecule C_2H_4O is related to the rest of the molecule. We shall now assume that the formula of ethylic aldehyde is

$$\begin{array}{cc} H & O \\ | & | \\ H - C & - C. \\ | & | \\ H & H \end{array}$$

This formula rests on a large number of reactions; there is very little doubt as to its correctness. Now the replacement of the atom of oxygen in this mole-cule by 2 atoms of chlorine will produce the molecule*

$$\begin{array}{cc} H & Cl \\ | & | \\ H - C & - C - Cl. \\ | & | \\ H & H \end{array}$$

But the compound produced is ethy-lidene chloride ; hence the formula of ethylidene chloride is

---

* Compare this interaction of PCl_5 and C_2H_4O with that of PCl_5 and C_2H_6O (methylic ether) given in par. 362.

$$\begin{array}{cc} H & Cl \\ | & | \\ H - C - C - Cl\, ; \\ | & | \\ H & H \end{array}$$ and hence the formula of the isomeric

ethylene chloride is $\begin{array}{cc} & H \quad H \\ & | \quad | \\ Cl\!\!-\!\!C - C\!\!-\!\!Cl, \\ & | \quad | \\ & H \quad H \end{array}$ because these are the

only possible formulae for the molecule $C_2H_4Cl_2$. But ethylene chloride is produced by the direct addition of chlorine to

ethylene; hence the formula of ethylene is $\begin{array}{cc} H & H \\ | & | \\ C - C. \\ | & | \\ H & H \end{array}$

Two compounds having the molecular composition C_2H_4 **365** may exist according to the hypothesis of valency : only one actually exists; but derivatives of both, e.g. chlorides, are known. It is possible that the compound the molecule of

which would have the composition $\begin{array}{c} H \\ | \\ H - C - C - H \\ | \\ H \end{array}$ may be

produced; but it is not likely, because the many attempts made to form it have all resulted in the production of the

isomeric compound $\begin{array}{cc} H & H \\ | & | \\ C - C. \\ | & | \\ H & H \end{array}$

This is an illustration of the proposition, that it is not always possible to obtain every one of the isomerides of a given composition the existence of which is indicated by the hypothesis of valency; or, in other words, that the existence of a compound of specified composition is not conditioned solely by the valencies of the atoms which form the molecule of this compound.

366   The hypothesis of valency leads to the conception of the molecule as a structure, the parts of which are related to each other in a definite manner.

Formulae such as those given in the preceding paragraphs for ethylic alcohol, methylic ether, ethylene chloride, and ethylidene chloride, are called *rational* or *structural formulae*; they are contrasted with *empirical formulae* (C_2H_6O and $C_2H_4Cl_2$) which express the percentage and atomic composition of molecules.

Structural formulae are attempts to summarise the chief reactions of formation and decomposition of compounds in the highly symbolical language of a special hypothesis resulting from the application of the molecular and atomic theory to the chemical phenomena of isomerism.

A structural formula may be found for any gasifiable compound the molecule of which is composed of atoms of known valencies; but the structural formula to be of any value must be the outcome of many experiments on the interactions of the compound to which it is given. The value of the formula consists in its suggestiveness of reactions, and in the extent to which it exhibits the analogies between the compound formulated and other compounds. The structural formulae of carbon compounds have been greatly developed. There can be no doubt that the chemistry of these compounds would not have advanced as it has done without the aid of structural formulae ; indeed the remarkable predictions which have been made, and verified, regarding classes of chemical changes among carbon compounds afford satisfactory evidence that the conceptions on which structural formulae are based are accurate and well founded.

367   It is generally possible to shew that the characteristic properties of a group of similar carbon compounds are connected with a certain arrangement of some of the atoms in the molecules of these compounds, which arrangement is common to all the members of the group, and can be expressed in a structural formula. Thus, a great many alcohols behave similarly when oxidised ; the molecule of each loses 2 atoms of hydrogen thereby producing an aldehyde, and this aldehyde is then oxidised to an acid the molecule of which is composed of the same number of carbon atoms as the molecule of the alcohol. Such alcohols are called *primary alcohols*. The following formulae give examples of the oxidation of primary alcohols.

<table>
<tr><td></td><td>Primary alcohol.</td><td colspan="2">Oxidation-products.</td></tr>
<tr><td></td><td></td><td>(1) Aldehyde</td><td>(2) Acid.</td></tr>
</table>

	Primary alcohol.	Oxidation-products. (1) *Aldehyde*	(2) *Acid.*
Ethylic	C_2H_6O	C_2H_4O	$C_2H_4O_2$
Propylic	C_3H_8O	C_3H_6O	$C_3H_6O_2$
Butylic	$C_4H_{10}O$	C_4H_8O	$C_4H_8O_2$
Allylic	C_3H_6O	C_3H_4O	$C_3H_4O_2$.

The study of the reactions of these, and of other, primary alcohols has led to structural formulae in all of which the group

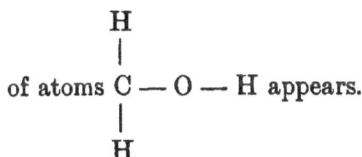

$$\begin{array}{c} H \\ | \\ \text{of atoms} \quad C - O - H \quad \text{appears.} \\ | \\ H \end{array}$$

Thus the structural formulae of the 4 alcohols in the above table are these ;—

$CH_3.CH_2.OH$; $CH_3.CH_2.CH_2.OH$; $CH_3.CH_2.CH_2.CH_2.OH$; and $CH_2.CH.CH_2.OH$; respectively.

[These formulae are shorter than, but have exactly the same meanings as, the more developed formulae;

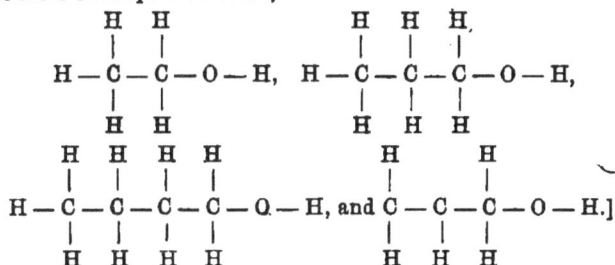

$$\begin{array}{ccc} & H \quad H & \\ & | \quad | & \\ H - & C - C - O - H, \\ & | \quad | & \\ & H \quad H & \end{array} \qquad \begin{array}{ccc} & H \quad H \quad H & \\ & | \quad | \quad | & \\ H - & C - C - C - O - H, \\ & | \quad | \quad | & \\ & H \quad H \quad H & \end{array}$$

$$\begin{array}{c} H \quad H \quad H \quad H \\ | \quad | \quad | \quad | \\ H - C - C - C - C - O - H, \text{ and } \end{array} \qquad \begin{array}{c} H \quad \quad H \\ | \quad \quad | \\ C - C - C - O - H.] \\ | \quad | \quad | \\ H \quad H \quad H \end{array}$$

A primary alcohol is sometimes defined as an alcohol the molecule of which is composed of atoms of carbon and hydrogen in union with the atomic group $CH_2.OH$; or it is sometimes said that the molecules of all primary alcohols contain the group of atoms $CH_2.OH$. By these statements we understand that the study of the chemical changes undergone by those alcohols which are classed together as primary, because of their behaviour on oxidation, has led to structural formulae which represent the molecules of these alcohols as always containing at least one atom of carbon directly interacting with 2 atoms of hydrogen and one of oxygen, which atom of oxygen also directly interacts with an atom of hydrogen.

The examination of the aldehydes obtained by oxidising

the primary alcohols has shewn that the best structural formulae which can be assigned to these aldehydes always 'contain the group' $C \underset{O}{\overset{H}{<}}$ [CHO]; and the structural formulae given to the acids obtained by oxidising these aldehydes always 'contain the group' $C \underset{O-H}{\overset{O}{<}}$ [CO.OH].
For instance, the oxidation of ethylic alcohol to aldehyde and then to acetic acid, is represented thus in structural formulae;—

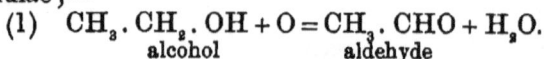

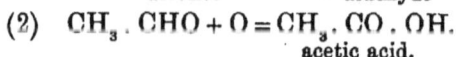

(1) $CH_3 . CH_2 . OH + O = CH_3 . CHO + H_2O.$
　　　alcohol　　　　　　　aldehyde

(2) $CH_3 . CHO + O = CH_3 . CO . OH.$
　　　　　　　　　acetic acid.

368　　Now suppose a new alcohol is discovered; analyses, and determinations of the spec. gravity of the gaseous alcohol, enable an empirical formula to be given to it. The behaviour of the alcohol on oxidation is now examined; it is found to lose 2 atoms of hydrogen per molecule and to form an aldehyde, one molecule of which then combines with an atom of oxygen and forms an acid. The new alcohol therefore belongs to the class of primary alcohols. The structural formula of the alcohol will therefore, *very probably*, 'contain the group' $CH_2 . OH$. The reactions of the alcohol are studied, and if possible a structural formula is found for it which represents the molecule as containing the atomic group $CH_2 . OH$. This formula tells a great deal about the alcohol; for instance it suggests the formulae, and therefore many of the reactions, of the aldehyde and acid which are produced by oxidising the alcohol.

In applying structural formulae in this way it is always to be remembered that a compound may be produced which exhibits most of the class-marks of a certain group but which nevertheless does not belong to that group. Thus an alcohol might be formed which oxidised to an aldehyde, and then to an acid containing the same number of carbon atoms per molecule as the alcohol, but which was not a true primary alcohol, and could not be justly represented by a molecular formula containing the group $CH_2 . OH$ characteristic of the primary alcohols.

369　　We have spoken of the molecules of certain classes of compounds as all *containing the same group of atoms*. This conception of a group of atoms forming part of a molecule, and exerting a definite influence on the properties of the molecule, is of much importance.

The hydrocarbon *ethane*, C_2H_6, interacts with chlorine to form chlorethane and hydrogen chloride; thus $C_2H_6 + Cl_2 = C_2H_5Cl + HCl$: chlorethane and caustic potash interact to produce ethylic alcohol and potassium chloride; thus $C_2H_5Cl + KOH = C_2H_6O + KCl$: when ethylic alcohol is oxidised ethylic aldehyde is formed, and when this is oxidised acetic acid is produced; thus $C_2H_6O + O = C_2H_4O + H_2O$, and $C_2H_4O + O = C_2H_4O_2$. We already know the structural formulae of most of the carbon compounds taking part in these changes; let us write the equations expressing the changes in structural formulae;—

(1) $CH_3 . CH_3 + Cl_2 = CH_3 . CH_2Cl + HCl$.
(2) $CH_3 . CH_2Cl + KOH = CH_3 . CH_2 . OH + KCl$.
(3) $CH_3 . CH_2 . OH + O = CH_3 . CHO + H_2O$.
(4) $CH_3 . CHO + O = CH_3 . CO . OH$.

All the molecules of these carbon compounds are represented as containing the atomic group CH_3. If the formulae are correct, then the group of atoms CH_3 has remained intact during this series of changes. If the sodium salt of acetic acid is prepared, mixed with solid caustic soda, and heated, we get methane (CH_4) and sodium carbonate formed; this change is represented in structural formulae thus

$$CH_3 . CO . ONa + NaOH = CH_3 . H + Na_2CO_3.$$

Here again the atomic group CH_3 has remained undecomposed.

A group of atoms which forms a part of several molecules, **370** and which remains undecomposed through a series of reactions undergone by these molecules, is called a *compound radicle*.

The following compounds have been obtained, the passage from one to the other has been effected, and the structural formula given to each is the outcome of quantitative experiments on the methods of preparation and the interactions of each compound :—$C_2H_5 . Cl$, $C_2H_5 . Br$, $C_2H_5 . CH_2 . OH$, $C_2H_5 . CN$, $C_2H_5 . NH_2$, $C_2H_5 . C_2H_3O_2$. The group of atoms C_2H_5 is therefore an example of a compound radicle.

The study of the reactions of acetic acid affords a good example of the meaning of the term compound radicle, and of the use of structural formulae. The empirical formula of acetic acid is $C_2H_4O_2$. The acid is monobasic; from this we conclude that one of the 4 hydrogen atoms is related to the rest of the molecule differently from the other 3 hydrogen atoms; we therefore adopt the formula $C_2H_3O_2$. H. Phosphorus pentachloride interacts with acetic acid; the interaction consists in the removal of one oxygen and one hydrogen atom from the molecule $C_2H_4O_2$ and the putting of one chlorine atom in their place; this interaction is thus expressed in an equation, $C_2H_3O_2 + PCl_5 = C_2H_3OCl + POCl_3 + HCl$. From this we conclude that one of the oxygen atoms in the molecule $C_2H_4O_2$ directly interacts with an atom of hydrogen, and that the relation of this oxygen atom to the rest of the molecule is different from that of the other oxygen atom to the rest of the molecule; we therefore adopt the formula $C_2H_3O . OH$, and we express the interaction with phosphorus pentachloride thus,

$$C_2H_3O . OH + PCl_5 = C_2H_3O . Cl + POCl_3 + HCl.$$

If this interpretation of the mechanism of these changes is correct, then the group of atoms C_2H_3O is a compound radicle; this group remains unchanged when acetic acid interacts with phosphorus pentachloride, and it is common to the 2 molecules $C_2H_3O . OH$ and $C_2H_3O . Cl$. If acetic acid has the formula $C_2H_3O . OH$ the formula of sodium acetate is $C_2H_3O . ONa$: when this salt is mixed with solid caustic soda and the mixture is heated, sodium carbonate (Na_2CO_3) and methane (CH_4) are produced. Therefore in this change the atomic group C_2H_3O is decomposed; one of the carbon atoms and the oxygen atom are removed, and the remaining CH_3 combines with an atom of hydrogen to form CH_4. The change is most simply represented thus,

$$CH_3 . CO . ONa + NaOH = Na_2CO_3 + CH_3H.$$

Because of this reaction we adopt for acetic acid the structural formula $CH_3 . CO . OH$; and we say that the molecule of this acid is composed of the compound radicle CH_3 combined with the other compound radicles CO and OH. The

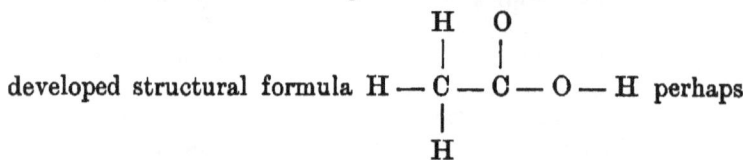

developed structural formula

$$\begin{array}{ccc} H & O & \\ | & | & \\ H - C - C - O - H \\ | & & \\ H & & \end{array}$$ perhaps

makes this plainer.

Had we stopped this investigation after examining the interaction between PCl_5 and $C_2H_4O_2$, we should have given to acetic acid the structural formula $C_2H_3O.OH$, and we should have said that the molecule of this acid is composed of the compound radicles C_2H_3O and OH. Further investigation however obliges us to modify this conclusion, inasmuch as it shews that the atomic group C_2H_3O is itself composed of the simpler groups CH_3 and CO. The formula $CH_3.CO.OH$ expresses all that is expressed by the formula $C_2H_3O.OH$, and it also suggests the interaction in which methane is produced from sodium acetate.

It appears then that a compound may have more than **371** one structural formula; that formula is the best which tells most about the characteristic reactions of the compound.

In Chap. XI. pars. 210 and 211 we glanced at the reactions of compounds of ammonia, NH_3. We found that these reactions were analogous to those of the alkali potash, KOH; to bring out these analogies we wrote the formulae of the compounds produced by the interaction of an aqueous solution of ammonia with acids as compounds of the hypothetical compound radicle *ammonium*, NH_4. The interpretation of these reactions given by the molecular and atomic theory is that in the molecule of an ammonium compound e.g. NH_4Cl, $NH_4.NO_3$, $(NH_4)_2SO_4$, $(NH_4)_2CO_3$, &c. we have always direct interaction between an atom of nitrogen and 4 atoms of hydrogen; in other words, we have the *compound radicle* or atomic group, NH_4.

As we speak of the valency of an atom in this or that molecule, meaning thereby the number of other atoms with which the specified atom directly interacts in the molecule, so we speak of the valency of an atomic group or compound radicle in a molecule. The atomic groups CH_3, C_2H_5, C_3H_7, &c. are monovalent; the group CH_2OH is also monovalent; the group CO is divalent; and so on.

In par. 357 it was mentioned that an atom which combines **372** with one divalent atom to form a molecule is often regarded as thereby proved to be itself divalent.– Thus the atom of oxygen is divalent because of the existence of the gaseous molecules OCl_2 and OH_2; one atom of carbon combines with one atom of oxygen to form the gaseous molecule CO; therefore, it has been urged, the atom of carbon is divalent in the molecule CO. Again, one atom of carbon combines with 2 atoms of oxygen to form the gaseous molecule CO_2; therefore, it is said, the atom of carbon is tetravalent in

M. E. C.   17

the molecule CO_2.   The formulae $C = O$ and $O = C = O$ are generally used to express these statements.

These formulae are evidently based on a meaning of atomic valency different from that we have been giving to this expression in preceding paragraphs.   It is rather difficult to grasp the exact meaning of the statement, 'the atom of carbon is divalent in the molecule CO and tetravalent in the molecule CO_2.'   The statement seems to imply that an atom of carbon is capable of directly combining with either 2 or 4 hydrogen, fluorine, chlorine, bromine, or iodine, atoms, or with that number of other atoms which is equivalent to 2 or 4 atoms of chlorine, &c.   The statement seems to assert that one atom of oxygen is truly equivalent sometimes to 2 atoms of hydrogen, or 2 atoms of chlorine, &c. and sometimes to 4 atoms of hydrogen, &c. But this assertion is scarcely capable of proof, because it seems impossible to define the exact meaning to be given to the expression *equivalent to,* as used with regard to atoms.

It has even been asserted that the atom of carbon is tetravalent in the molecule CO.   If this is so, then one atom of oxygen is *equivalent to* 4 atoms of chlorine, &c. when oxygen and carbon combine to form the compound CO: but 2 atoms of oxygen are *equivalent to* 4 atoms of chlorine, &c. when oxygen and carbon combine to form the compound CO_2.   We see here the extreme difficulty, if not impossibility, of giving an exact and invariable meaning to the expression *equivalent to,* as applied to atoms.

373   It is generally the custom in writing structural formulae to represent each atom whose maximum valency is greater than one with as many lines proceeding from the symbol as correspond to the maximum valency of the atom.   The atom of carbon, for instance, is generally represented with 4 lines proceeding from it, the atom of oxygen with 2 lines, and so on.   Thus the structural formulae for ethylic aldehyde and acetic acid are generally written thus ;—

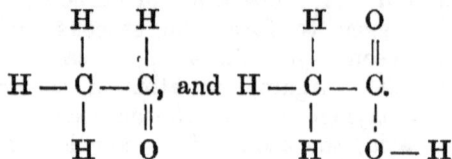

$$\begin{array}{ccccccc}
\text{H} & \text{H} & & & \text{H} & \text{O} \\
| & | & & & | & \| \\
\text{H} - \text{C} - \text{C}, & \text{and} & \text{H} - \text{C} - \text{C}. \\
| & \| & & & | & | \\
\text{H} & \text{O} & & & \text{H} & \text{O} - \text{H}
\end{array}$$

Similarly the structural formulae of ethylene and the hypothetical isomeride ethylidene are generally written thus;—

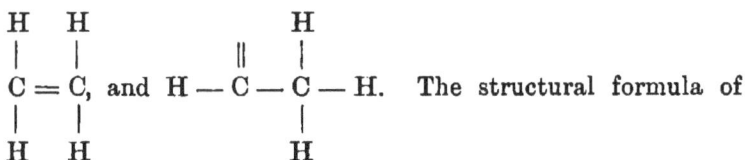

$$
\begin{array}{cc}
\text{H} \quad \text{H} & \text{H} \\
| \quad | & \| \quad | \\
\text{C} = \text{C, and } & \text{H} - \text{C} - \text{C} - \text{H. The structural formula of} \\
| \quad | & | \\
\text{H} \quad \text{H} & \text{H}
\end{array}
$$

the hydrocarbon acetylene, C_2H_2, is put into this form

$$\text{H} - \text{C} \equiv \text{C} - \text{H};$$

whereas the formula $\text{H} - \text{C} - \text{C} - \text{H}$ would be used when the meaning given to valency is that explained in par. 360.

It would be out of place in an elementary book to discuss the possible meanings of these so called '*double bonds*' and '*treble bonds.*' In the opinion of several chemists they have done much to hinder the advance of chemistry, by leading chemists to trust in names, and in far-fetched analogies, instead of in realities, and in well established and accurately defined points of resemblance and difference. On the other hand the employment of '*double and treble linkings*' or '*bonds*' has some points in its favour. It continually reminds the chemist of the maximum valency of each atom, and by doing this it suggests the possibility of reactions.

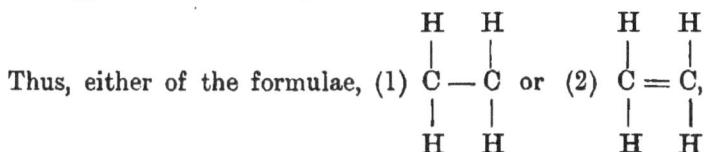

$$
\text{Thus, either of the formulae, (1)} \quad
\begin{array}{cc}
\text{H} & \text{H} \\
| & | \\
\text{C} & - & \text{C} \\
| & | \\
\text{H} & \text{H}
\end{array}
\quad \text{or (2)} \quad
\begin{array}{cc}
\text{H} & \text{H} \\
| & | \\
\text{C} & = & \text{C,} \\
| & | \\
\text{H} & \text{H}
\end{array}
$$

represents each carbon atom as directly interacting with only 3 other atoms, but formula (2) tells us that a carbon atom can directly interact with 4 other atoms, and hence suggests the possibility of adding 2 monovalent atoms to the molecule

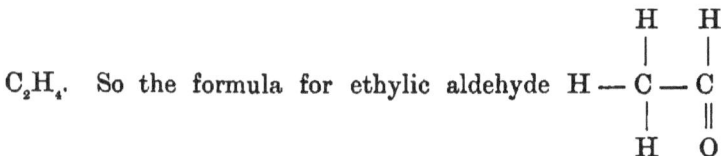

C_2H_4. So the formula for ethylic aldehyde

$$
\text{H} -
\begin{array}{cc}
\text{H} & \text{H} \\
| & | \\
\text{C} & - & \text{C} \\
| & \| \\
\text{H} & \text{O}
\end{array}
$$

visibly suggests the possibility of putting 2 monovalent atoms, *e.g.* Cl_2, in place of the atom of oxygen.

$$\text{The formulae}\quad \begin{array}{cc} \text{H} & \text{H} \\ | & | \\ \text{C} - \text{C} \\ | & | \\ \text{H} & \text{H} \end{array} \text{ and } \begin{array}{cc} \text{H} & \text{H} \\ | & | \\ \text{H} - \text{C} - \text{C} \\ | & | \\ \text{H} & \text{O} \end{array} \quad\text{of course suggest}$$

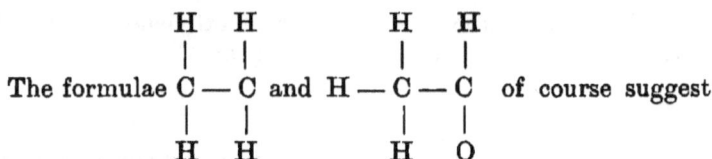

the same reactions as the formulae with *double bonds*, if we remember that the maximum valency of the carbon atom is 4, and that of the oxygen atom is 2. These formulae have the great advantage over those with *double* or *treble bonds* that they are based on a definite hypothesis regarding atomic valency.

374   The essential part of the hypothesis of valency is the conception of direct action and reaction between each atom in a molecule and a limited number of other atoms. As the whole molecule is held together by the mutual interactions of the atoms, there probably is what we may call indirect action and reaction between all the atoms which constitute the molecule. The hypothesis gives us a definition of the maximum valency of an atom, as the maximum number of monovalent atoms (*i.e.* atoms of H, F, Cl, Br, or I) with which the given atom directly interacts (*i.e.* in these cases combines) in any molecule; and it teaches that the specified atom never directly interacts with a greater number of other atoms, whatever be their valencies, than is expressed by the maximum valency as thus defined.

375   The hypothesis of valency is meaningless apart from the theory of atoms and molecules; it is based on this theory, and all the results gained by using it are expressed in the language of the theory.

The theory of atoms and molecules is strictly applicable, at present, only to gases; therefore the hypothesis of valency, and all the terms to which it has given birth, are strictly applicable, at present, only to gases. But just as we made use of the molecular and atomic theory as a general guide in studying the chemical changes which occur among liquid and solid substances, so may we make use of the hypothesis of valency, provided we exercise sufficient caution, as a general guide in attempts to learn something regarding the structure of those aggregations of atoms which form the reacting weights of solid and liquid substances. But it would be going too far afield to attempt to indicate here even the lines on which the hypothesis of valency may probably be usefully employed in discussions regarding the structure of the reacting weights of solid compounds.

Our conception of chemical composition has been widened **376** by the examination of the phenomena of isomerism.

A statement of the composition of a compound should tell the percentage composition of the compound; it should also tell the composition of a reacting weight stated in numbers of combining weights of each element, and the composition of a gaseous molecule stated in numbers of atomic weights of each element; if the compound is gasifiable, it should also give such an indication of the arrangement of the parts of the molecule relatively to each other as can be gained by studying the interactions of the compound and expressing these in a structural formula based on the hypothesis of atomic valency.

The formula which best expresses the composition of a compound also tells a great deal about the properties of the compound. A satisfactory structural formula suggests many of the characteristic reactions of the compound the composition of which it expresses.

The structural formulae of many compounds of carbon **377** which are acids have been determined; we are therefore able to trace some of the connexions between the properties of this class of compounds and their composition, using *composition* in the widest sense we have given to the term.

In the molecules of the greater number of the carbon acids an atom of carbon probably directly interacts with an atom of oxygen and with the atomic group OH; this statement is usually expressed by saying that these molecules contain the

group CO.OH $\left(\mathrm{C} {<}^{\mathrm{O}}_{\mathrm{O - H}} \right)$. The following structural formulae illustrate the statement concerning the composition of many carbon acids which has just been expressed in the symbolic language of the hypothesis of valency.

Acid.	Structural formula.	Acid.	Structural formula.
Acetic	$CH_3.CO.OH$	Benzoic	$C_6H_5.CO.OH$
Acrylic	$C_2H_3.CO.OH$	Phthalic	$C_6H_4.(CO.OH)_2$
Succinic	$C_2H_4.(CO.OH)_2$	Mellitic	$C_6.(CO.OH)_6$

The basicity of these acids is connected with, and is measured by, the number of CO.OH groups in the molecule; thus acetic, acrylic, and benzoic, acids are monobasic, succinic and phthalic acids are dibasic, and mellitic acid is hexabasic.

We can thus, in a great many cases, connect the group of properties connoted by the term *acid* with a certain arrangement of the parts of the molecules of the compounds which exhibit these properties.

In Chap. XIII. we found that a number can be assigned to each acid, called the relative affinity of the acid, which tells how much of a definite chemical change can be accomplished by that acid under defined conditions. An acid with a large affinity-constant is called a *strong* acid ; an acid with a small affinity-constant is called a *weak* acid. If definite relations can be established between the values of the affinity-constants, and the compositions, of acids, a great advance will be made in solving the essential problem of chemistry, which is to connect changes of composition with changes of properties. Investigations have been made of late years in this direction, and many results have been obtained. Thus the relative affinities of the acids HCl, HBr, HI, HF, H_2S, HCN, in aqueous solutions, are approximately in the ratio 89 : 89 : 89 : 30; 0·1 : 0·2. The change of composition from HCl to HBr or HI is attended with practically no change in the *strength* of the acid ; but when an atom of fluorine is put in the place of an atom of chlorine, bromine, or iodine, in the molecule HX (X = Cl, Br, or I) this change is attended with a great decrease in the strength of the acid. Change of composition from HCl, &c. to H_2S or HCN is accompanied by a very great decrease in the strength of the acid.

The ratio of the affinities of the three acids, of similar composition, H_3PO_2, H_3PO_3, and H_3PO_4, is approximately 77·8 : 74·5 : 61·8. The ratio of the affinities of the four similar acids H_2SO_3, H_2SO_4, $H_2S_2O_6$, and $H_2S_4O_6$, is approximately 66·5 : 150·5 : 178 : 181·5 ; and the ratio for the pair of acids H_2SeO_3 and H_2SeO_4 is 45 : 158. The basicities of the acids H_3PO_2, H_3PO_3, and H_3PO_4, are 1, 2, and 3, respectively ; but all the acids of sulphur or selenion enumerated above are dibasic. In one series, H_3PO_2 to H_3PO_4, a decrease in the value of the affinity-constant is accompanied by an increase of the basicity, and also by an increase in the number of oxygen atoms in the reacting weights, of the acids. In the other series, the basicity remains constant, and an increase in the number of oxygen atoms is accompanied by an increase in the value of the affinity-constant.   .

Relations have also been traced between the structural formulae of isomeric acids and the values of their affinity-constants. But enough has been said to shew the lines on which investigation is proceeding, and the importance of the results which are likely to be obtained.

The phenomena of *isomerism* exhibited by compounds are **378** more or less similar to those of *allotropy* exhibited by elements. The only instance of allotropy of elements to which the molecular and atomic theory can at present be applied in detail is that of oxygen and ozone (*s.* Chap. xvi., par. 323); in the other cases, allotropy is exhibited by elements in the solid state.

The application of the theory of atoms and molecules to the case of oxygen and ozone led to the conclusion, that the properties of gaseous molecules composed of 3 atoms of oxygen are different from the properties of gaseous molecules composed of 2 atoms of oxygen : the theory led us to associate change of properties with variations in the numbers, unaccompanied by variations in the nature, of the atoms constituting a gaseous molecule. The application of the theory to the cases of isomerism examined led to the conclusion that two gaseous molecules may be composed of the same number of the same atoms and yet differ in properties : the theory led us to associate change of properties with variations in the arrangement, unaccompanied by variations in the nature or number, of the atoms constituting a gaseous molecule.

The term *polymerism* is used to indicate the existence of **379** two or more different compounds, having the same composition but different molecular weights, and each capable of being formed from the other by some simple reaction, usually by raising or lowering the temperature. Thus, the molecular weight and composition of ethylic aldehyde are expressed by the formula C_2H_4O; when a little sulphuric acid is added to this compound, much heat is produced and the *polymeride* (para-ethylic aldehyde), $C_6H_{12}O_3$, is produced; when this compound is distilled with a little acid, ethylic aldehyde, C_2H_4O, is obtained. Again, when amylene C_5H_{10} is heated a portion of it is *polymerised* to diamylene $C_{10}H_{20}$.

Polymerism and allotropy are evidently more nearly allied than polymerism and isomerism.

The term allotropy is sometimes applied to compounds; it then signifies the existence of two or more varieties, or forms, of the same compound in the solid or liquid state. The differences between the various forms are physical rather than chemical; *e.g.* differences in specific gravity, melting point, crystalline form, &c. Thus calcium carbonate ($CaCO_3$) crystallises in rhombic, and also in rhombohedral, forms; the rhombic crystals separate from hot solutions, the rhombohedral from cold solutions; when the rhombic crystals are heated they change to the rhombohedral forms.

The molecular and atomic theory cannot as yet be applied to give a detailed explanation of the allotropy of solid elements or compounds.

380   The molecular and atomic theory gives then a partial explanation of the phenomena of isomerism; it represents the change from one isomeride to another as a change in the configuration of a system of atoms; and in the hypothesis of valency it supplies a means whereby these changes may be pictured to the mind with some degree of clearness, and may be expressed in a consistent language. Structural formulae are the most developed forms of chemical language; they express much but not all; of them it may emphatically be said "they are wise men's counters but the money of fools."

CHAPTER XVIII.

THE PERIODIC LAW.

WE have now to some extent studied the more important **381** relations which exist between changes of composition and of properties of compounds, and we have seen that there are definite points of connexion between the properties of a compound and the properties of the elements which compose it. The meaning of the term composition widened as we proceeded. At first the composition of a compound was a statement of the quantity of each element combined in a definite quantity of the compound; then it was a statement of the number of combining weights of each element combined in one reacting weight of the compound; then it was a statement of the number of atoms of each element combined in a molecule of the compound; then it was not only a statement of the number, but also an attempted representation of the arrangement, of the atoms which formed the molecule of the compound.

As the meaning of composition has become wider, so have the terms in which composition is expressed become more symbolic.

The properties of elements and compounds studied have been the properties exhibited in those actions and reactions which result in the production of new elements and compounds. Chemistry considers elements and compounds when they form members of systems undergoing change, rather than when they are isolated from other kinds of matter.

Attempts have been made to indicate the methods by which chemists will probably be able to find, and state in quantitative terms, the connexions which certainly exist between the compositions and the chemical functions of compounds.

We have also glanced at the changes of energy which accompany changes of composition, and we have seen that when equilibrium is attained in a chemical system there is a certain distribution of the members of the system, and also of the energy of the system.

We have arranged compounds in classes, such as *acids*, *alkalis, salts, acidic oxides*, &c. &c. We have found that all the members of each class shew certain similarities of composition. We have traced some of the connexions between the composition of each class and the typical property or properties of the class.

It has been found possible to assign to each member of certain classes of compounds a number, called the affinity-constant of the compound, which tells how much of a specified chemical change will be accomplished under definite conditions by that compound; and it has been found possible to establish the fact that there is a definite relation between this value and the composition of the compound, and, in some cases, to state the nature of this relation.

382 We have traced certain elements through the changes which they undergo by combining with other elements to form compounds belonging to different classes; we have seen the properties of the elements modified by the nature, and relative quantities, of the other elements with which they are combined; but we have not altogether lost sight·of the original elements in these changes. Each element to some extent impresses its own properties on all the compounds of which it forms a part.

Is it possible then to connect the properties of the compounds of an element, using the term properties in its widest sense, with some one definite and measurable property of the element itself? Can it be shewn that the properties of the compounds vary with variations in the chosen property of the element?

If this can be done, we shall have the basis of a satisfactory method of chemical classification both of elements and compounds.

383 We have had instances of a regular variation of properties of similar compounds of a group of allied elements accompanying variations in the atomic weights of these elements (*s.* especially Chap. xi. par. 182).

Various chemists, among whom Newlands must be especially mentioned, from time to time have drawn attention

to points of connexion between the properties and the atomic
weights of the elements.

In 1864 Newlands arranged a number of elements in order
of their atomic weights, and shewed that these elements might
be divided into groups of seven, each of which groups to some
extent repeated the properties of the next group. "The eighth
element" said Newlands "starting from a given element is
a kind of repetition of the first, like the eighth note of an
octave in music." In subsequent papers Newlands insisted
on the general applicability of what he called the 'law of
octaves.'

It is however chiefly to Mendelejeff that we owe the
systematic correlation of the atomic weights with the chemical
and physical properties of the elements, and the properties of
their compounds.

The properties of the elements and their compounds vary **384**
periodically with variations in the atomic weights of the elements.
This statement is the outcome of the work of Mendelejeff,
Lothar Meyer, Newlands, and many other chemists.

This statement, or a statement equivalent to this, is known
as *the periodic law.* The rest of this book will be devoted
to an attempt to amplify and explain the periodic law.

The properties of the chemical elements and their com- **385**
pounds is the phenomenon to be examined; the variable is
the atomic weight of the elements : the *law* asserts that with a
continuous change of the variable the phenomenon repeats itself
at definite intervals.

A quantitative value cannot be given to the phenomenon
'properties of the elements and their compounds.' To illustrate the statement of the periodic law, it is necessary to
choose a definite measurable property of the elements, or
of a series of similar compounds, and to trace the relation
between the variation in the values of this property and
the variation in the values of the atomic weights of the
elements.

We shall choose two properties, one physical, the other
partly chemical and partly physical. The properties are (1)
the *melting points,* (2) the *atomic volumes,* of the elements.
The atomic volume of an element is defined to be

$$\frac{\text{spec. gravity of solid element}}{\text{atomic weight}};$$

this quotient represents the volume occupied by a mass of the solid element proportional to the atomic weight of the element.

The most striking way of exhibiting the connexion between the variable, atomic weight, and each of the variants, melting point and atomic volume, is to represent the values of the former as lengths marked off on a horizontal line, and the values of the latter as lengths on a vertical line; then, from each pair of points so marked to produce lines until they meet, and to draw a curve cutting the points of intersection of these lines. The curves thus obtained are drawn in the plate on page 269. The melting points are calculated from the initial temperature -273°; the numbers so obtained are divided by 7 in order to bring the curve within manageable limits. The values of the atomic volumes are multiplied by 4 to make the scale of the curve comparable with that of the curve of melting points. Lack of data is indicated by a broken line, or by a gap in the curve. Thus if the elements are arranged in order of increasing atomic weights, nitrogen, oxygen, and fluorine come after carbon and before sodium; but the atomic volumes of these elements are unknown, hence the dotted line in the curve of atomic volumes. Similarly a number of elements come between didymium and tantalum; the gap in the curve indicates that the atomic volumes of these elements have not been determined. The melting points of only about two-thirds of the elements have been determined; hence many parts of the curve of melting points are shewn as dotted lines.

386   The curves shew that the melting points, and the atomic volumes, of the elements vary periodically with variations in the atomic weights of the elements. The value of either variant does not exactly repeat itself at definite intervals; but the elements fall into periods, in each of which the values of the melting point and the atomic volume decrease from a maximum to a minimum, and then again increase to a maximum. The nature of this periodical variation is best shewn by the curve of atomic volumes, as the data are here more abundant. The first period comprises the elements from lithium to sodium; the second, the elements from sodium to potassium; the third, the elements from potassium to rubidium; the fourth, the elements from rubidium to caesium; after caesium there is a great want of data. Elements, the values of whose atomic weights place them about midway between the first and last element of a period, have atomic volumes

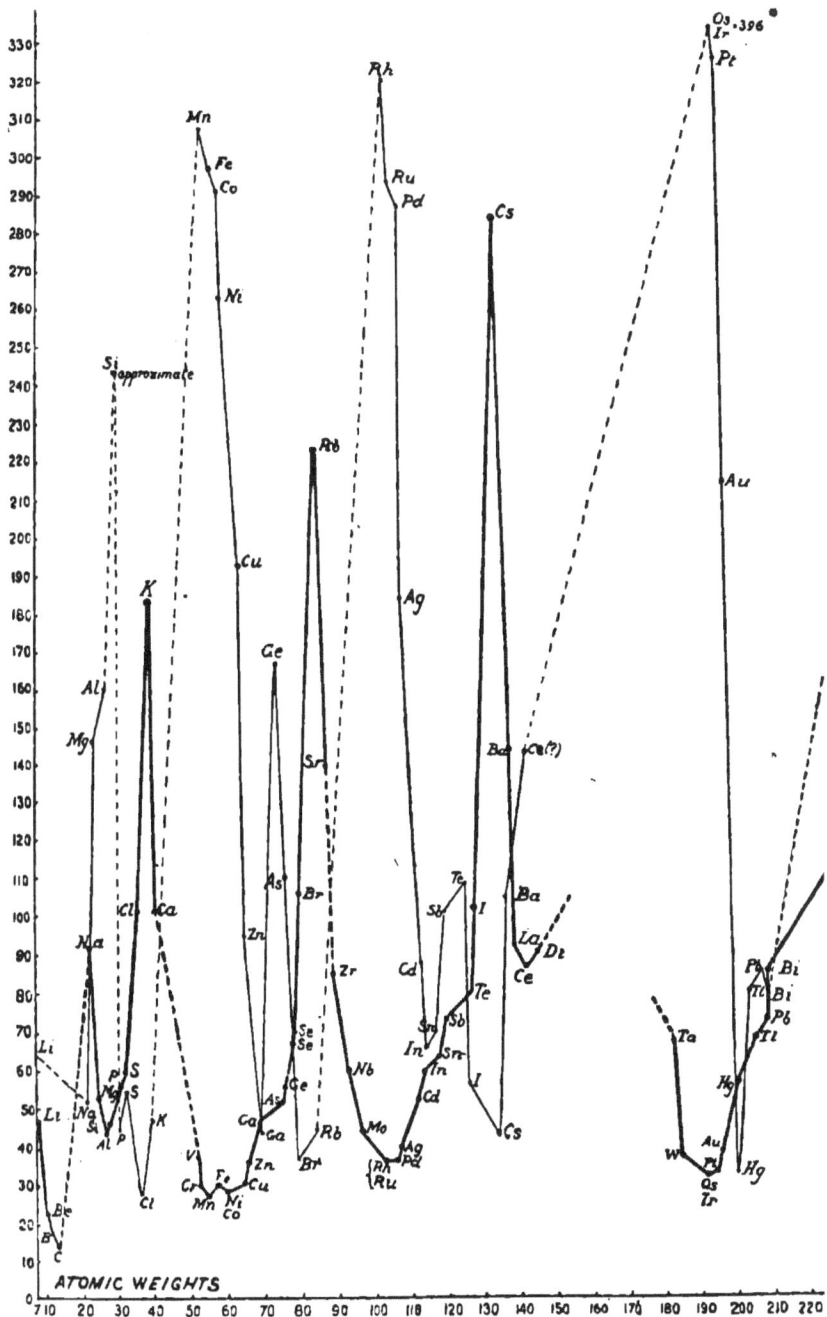

Thick line curve shews atomic volumes.
Thin   "   " melting points.
Dotted lines indicate that data are wanting

* THIS POINT SHOULD BE PLACED 66 DIVISIONS HIGHER

approximately equal to those of elements which occupy a similar position in another period ; compare, for instance, the positions and atomic volumes of chromium, manganese, iron, nickel, and cobalt, with the positions and atomic volumes of rhodium, ruthenium, palladium, and silver ; or compare the positions and atomic volumes of sulphur, selenion, and tellurium.

387   If the only properties of elements which it was necessary to study were their melting points and atomic volumes, it is evident that the connexion between the values of these and the values of the atomic weights of the elements is so marked and definite that a system of classification might well be based on this connexion. But the periodic law asserts that the properties of the elements and their compounds in general, and not only one or two properties in particular, vary periodically with variations in the atomic weights of the elements.

Let the 14 elements from lithium $(Li = 7)$ to chlorine $(Cl = 35 \cdot 5)$ be arranged in two series or periods of seven in each ; thus—

$$Li = 7 \quad Be = 9 \quad B = 11 \quad C = 12 \quad N = 14 \quad O = 16 \quad F = 19.$$
$$Na = 23 \quad Mg = 24 \quad Al = 27 \quad Si = 28 \quad P = 31 \quad S = 32 \quad Cl = 35 \cdot 5.$$

The difference in the values of the atomic weights of two consecutive elements varies from 1 to 3·5, the mean difference is about 2. The difference between the values of the atomic weights of two elements placed one under the other varies from 15 to 17 ; the mean difference is about 16.

The following statement gives a general indication of the chemical properties of lithium and sodium, beryllium and magnesium, boron and aluminium, &c.

Lithium and sodium : very light, soft, easily melted, metals ; rapidly decompose cold water, thus

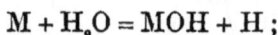

$$M + H_2O = MOH + H ;$$

oxides, M_2O, alkali-forming and strongly basic ; form salts M_2SO_4, M_2CO_3, MCl, &c. ; do not combine with H.

Beryllium and magnesium : fairly hard metals, of low spec. gravity but high melting points ; oxides basic but not alkali-forming ; form salts MSO_4, MCO_3, MCl_2, &c. ; do not combine with H.

Boron and aluminium : Al metallic, B non-metallic ; B forms BH_3, Al does not combine with H ; Al forms salts

$Al_2 3SO_4$, Al_2Cl_6, &c. B forms BCl_3 and perhaps $B_2 3SO_4$; oxides are M_2O_3, B_2O_3 is acidic, Al_2O_3 basic but feebly acidic towards strong alkalis; both dissolve in KOHAq forming borate or aluminate of K and evolving H.

Carbon and silicon: non-metals; both exhibit allotropy; neither forms salts by replacing H of acids; compounds are MH_4, MCl_4, MO_2, &c.; oxides are acidic; acids H_2MO_3 are very weak.

Nitrogen and phosphorus: non-metals; P exhibits allotropy; neither forms salts by replacing H of acids; both form strong oxyacids HMO_3; N also forms HNO_2, and P forms H_3PO_2, H_3PO_3, &c.; oxides M_2O_3 and M_2O_5 are acidic; both combine with H forming MH_3.

Oxygen and sulphur: strongly negative non-metals; both exhibit allotropy; compounds with H are MH_2, one neutral, the other a feeble acid; form many analogous compounds, e.g. P_2M_5, As_2M_3, CuM, &c. &c.

Fluorine and chlorine: non-metals; very negative; neither exhibits allotropy; compounds with H, MH, are strong acids; oxides Cl_2O and ClO_2 are acidic, no oxide of F is known.

This statement shews that there is a very similar gradation of properties in each of these series or periods of seven elements; the first member of each period is a strongly positive metal, the last is a markedly negative non-metal; there is a regular decrease in the metallic, and an increase in the non-metallic, character of the members as each period is ascended, i.e. as the atomic weights increase. The relations between the chemical properties of a pair of consecutive members of one series are on the whole very similar to those of the corresponding pair of consecutive members of the other series. Thus if the symbol of an element is used to represent the *general chemical character* of that element, then we may say that Li : Be = Na : Mg; or C : N = Si : P; or O : F = S : Cl.

If the elements are arranged in order of atomic weights **388** from hydrogen (H = 1) to uranium (U = 240), and if they are then marked off into series or periods each of which contains 7 elements, it is found that in some cases the properties of one period are to a great extent a repetition of the properties of the preceding period, but in other cases no such repetition of properties is to be noticed; in other words, it is found that series of 7 elements sometimes form periods in which the properties vary periodically with the variation of the atomic weights of the elements, but that other series do not shew any clear connexion of a periodic kind between the variation of atomic weights and the variation of properties. But if it is assumed that there are several gaps in the list of elements, to be filled up by the discovery of elements at present unknown, and if a peculiar, and at first sight abnormal, position is assigned to about a dozen elements, (s. Chap. XXVI.), then it is possible to arrange the elements in order of increasing atomic weights in periods, so as to shew a distinct connexion of a periodic kind between variation of atomic weights and variation of properties.

389   The following table (p. 273) shews the arrangement of the elements in accordance with the periodic law. The values of atomic weights are given in round numbers. The elements in a vertical column form a *Group*; the elements in a horizontal column form a *Series*; the connotation of these terms will be discussed later. Hydrogen is placed in a series by itself. The elements in Group VIII. must be considered to some extent apart from the other elements.

390   All the elements in the same *group* resemble each other in their chemical properties; there is a gradation of properties from the first member (that with smallest atomic weight) to the last. The elements in a *series* differ from each other; the difference becomes more marked as the series is ascended, that is as atomic weight increases, so that the greatest difference is that between the first and last members of the series.

Thus, taking *Group V.* we find that all the elements in this group form oxides of the composition M_2O_5 (where $M=N$, P, V, &c.); that these oxides as a class are acidic, but that when M is one of the higher members of the group (Di to Bi) M_2O_5 are very feebly acidic, and at the same time are also basic; that the lower members of the group are non-metals, the intermediate members are both metallic and non-metallic, and the element with highest atomic weight (Bi) is a metal; and that these elements taken as a whole are more like each other than they are like any other element or class of elements.

Then, taking say *Series* 2, we have already seen how very different fluorine, the last member of the series, is from lithium, the first member of the series; and we have learned that no two of these elements could be placed in the same

THE PERIODIC LAW. TABLE I.

Groups

Series	I	II	III	IV	V	VI	VII	VIII
1	H 1							—
2	Li 7	Be 9	B 11	C 12	N 14	O 16	F 19	
3	Na 23	Mg 24	Al 27	Si 28	P 31	S 32	Cl 35·5	
4	K 39	Ca 40	Sc 44	Ti 48	V 51	Cr 52	Mn 55	Fe 56, Ni 58·6, Co 59.
5	Cu 63	Zn 65	Ga 69	Ge 72	As 75	Se 79	Br 80	
6	Rb 85	Sr 87	Y 89	Zr 90	Nb 94	Mo 96	(? 100)	Rh 104, Ru 104·5, Pd 106.
7	Ag 108	Cd 112	In 114	Sn 118	Sb 120	Te 125	I 127	
8	Cs 133	Ba 137	La 139	Ce 140	Di 144	? 149	? 150	? 152—156, 3 Elements?
9	? 170	? 172	? 4 Elements 156 to 162? Yb 173	? 178	Er 166	? 167	? 169	Ir 192·5, Os 193, Pt 194.
10					Ta 182	W 184	? 190	
11	Au 197	Hg 200	Tl 204	Pb 207	Bi 208	? 2 Elements 212 to 220?		
12	? 3 Elements 220 to 230?			Th 232	? 237	U 240	? 245	

M. E. C.

class if similarity of properties is to be the distinguishing
mark of the members of a class.

391   Omitting hydrogen, it may be said that the properties of the
members of a *series* vary much from the first to the last member, and that each series is to a great extent a repetition
of that which precedes it; and that the properties of the
members of a *group* vary from the first to the last member,
but that all the members are more like each other than they
are like any other elements.

392   The properties of any element may be determined by considering (1) the properties of the group to which it belongs, (2)
the properties of the series in which it occupies a place, (3) the
position of the element in the group and in the series, (4) the
relations between the properties of elements situated similarly to
the given element and those of the other members of the group
and of the series in which these elements occur, (5) the
relations of the group and the series in which the specified
element occurs with other groups and series.

We cannot thoroughly grasp these generalisations until we
have examined in some detail the properties of several groups
and series, but their general bearing will be made more intelligible by glancing at the position assigned to one specified
element in the scheme of classification based on the periodic law.
Let us choose the element *antimony*.

393   The chemical properties of an element may be fairly summarised by stating (*a*) the compositions, and (*b*) the properties,
of its more important compounds, such as oxides, hydrides,
haloid compounds, salts, &c.

We must look in the first place at the properties of the
group of elements to which antimony belongs. The compositions of the typical compounds of the elements of this
group are represented by the general formulae M_2O_3, M_2O_5,
MH_3, MCl_3, MBr_3, MI_3, $MOCl$, and, in the cases where the
elements form salts, $M_2(SO_4)_3$, $M_2 6NO_3$, &c. The oxides are
acidic, but become less acidic and more basic as the atomic
weight of M increases. The hydrides, when the elements
combine with hydrogen, are gases, solutions of which in
water are more or less alkaline; these hydrides are oxidised
by mixing with oxygen and heating, giving M_2O_3 (or M_2O_5)
and H_2O. ·

We must look in the second place at the properties of the
series in which antimony finds a place. Arranging the elements of the series in order of increasing atomic weights we

find them forming oxides, haloid compounds, hydrides, salts, 'thus ;—

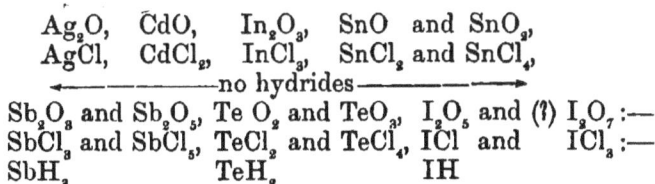

$$Ag_2O, \quad CdO, \quad In_2O_3, \quad SnO \text{ and } SnO_2,$$
$$AgCl, \quad CdCl_2, \quad InCl_3, \quad SnCl_2 \text{ and } SnCl_4,$$
$$\longleftarrow \text{————no hydrides————} \longrightarrow$$
$$Sb_2O_3 \text{ and } Sb_2O_5, \text{ Te } O_2 \text{ and } TeO_3, \quad I_2O_5 \text{ and } (?) \text{ } I_2O_7 :—$$
$$SbCl_3 \text{ and } SbCl_5, \quad TeCl_2 \text{ and } TeCl_4, \text{ ICl} \quad \text{and} \quad ICl_3 :—$$
$$SbH_3 \qquad\qquad TeH_2 \qquad\qquad IH$$

Ag_2SO_4&c., $CdSO_4$ &c., $In_2 3SO_4$&c., $Sn(SO_4)$and $Sn(SO_4)_2$&c., $(?)Sb_2 3SO_4$, salts of Te very unstable, no salts of I.

A glance at the third determining condition, viz. the position of antimony in the group and series, shews ; (1) that this element comes in Group V. following after elements whose oxides are acidic, most of which form hydrides, and which as a class do not form stable salts by replacing the hydrogen of acids, and most of which combine with hydrogen and oxygen to form well marked acids; (2) that the element is succeeded in Group V. by elements whose highest oxides are feebly acidic but at the same time interact with acids to form salts, which do not form hydrides, and the compounds of which with hydrogen and oxygen are easily separated into oxides and water and on the whole are basic rather than acidic.

If we now choose *tin* as an element similarly situated to antimony, we find that this element forms the oxides SnO and SnO_2; that the elements preceding and succeeding it in Group IV. form oxides of similar composition; that the oxides CO_2 and SiO_2 are decidedly acidic, the oxides TiO_2 GeO_2 and ZrO_2 are less acidic and are also basic, and that CeO_2 PbO_2 and ThO_2 are basic with perhaps very feebly marked acidic properties; that the only elements of the group which form hydrides are carbon and silicon; and that all the elements of the group except carbon and silicon replace the hydrogen of acids and thereby produce salts.

Looking back on these facts we see that antimony is well placed in Group V., Series 7; that it is more metallic than the elements, taken as a whole, which precede it, and less metallic than those which succeed it, in the group; that it is more negative or nonmetallic than the elements which precede it in the series, and less negative than those which come after it; that its highest typical oxide (Sb_2O_5) is composed of a greater number of oxygen atoms relatively to the atoms of antimony than the highest oxide of the elements which precede it in

18—2

Series 7, (Ag_2O, Cd_2O_2, In_2O_3, Sn_2O_4), but that it is succeeded in the series by elements whose highest typical oxides (Te_2O_6, I_2O_7) are composed of a greater number of atoms of oxygen relatively to the atoms of the other element than Sb_2O_5.

Finally, we ought to examine the relations of Group V. and Series 7 to other groups and series. It is impossible to do this at present except in the merest outline. Looking at this subject broadly, it may be said that the members of Group I. shew greater differences among themselves, and the members of Group VII. are more closely related among themselves, than the members of the intermediate groups; and that the variation of properties from the first to the last member of a series is very marked in Series 2, but becomes, on the whole, less marked as we pass through Series 3, 4, 5, to Series 11. If we may apply so vague a generalisation as this, we should conclude that antimony ought to exhibit very well marked analogies with the other members of the group in which it occurs; and that although it must widely differ from the other elements in its series, yet it will probably not differ to so very marked an extent as, say, nitrogen differs from the other members of Series 2, or phosphorus from the other members of Series 3. These tentative and somewhat vaguely worded conclusions, are fairly borne out by the actual relations between antimony, the members of Group V. on the one hand, and the members of Series 7 on the other hand.

394   In the sketch which has now been given of the periodic law, each group of elements has been treated as a whole. But the more detailed study of these groups will shew us that each, except Group VIII., is more or less sharply divided into two sub-groups; one sub-group contains the elements belonging to even series, the other sub-group contains the elements which are placed in odd series.

This division of the groups into sub-groups is sometimes marked, e.g. in Group VI.; sometimes it is almost hidden by the distinct way in which all the members of the group are stamped with the characteristics of the groups, e.g. in Group V. Some groups, for instance Group II., exhibit very clearly both the general characteristics of the group, and the division into two sub-groups the members of either of which are more like each other than they are like the members of the other sub-group.

This division into sub-groups is rendered very clear in the annexed table.

THE PERIODIC LAW. TABLE II.

Series

Groups	1	2	3	4	5	6	7	8	9	10	11	12
I	H 1	Li 7	Na 23	K 39	Cu 63	Rb 85	Ag 108	Cs 133	—	—	Au 196	—
II		Be 9	Mg 24	Ca 40	Zn 65	Sr 87	Cd 112	Ba 137	—	—	Hg 200	—
III		B 11	Al 27	Sc 44	Ga 69	Y 89	In 114	La 139	—	Yb 173	Tl 204	—
IV		C 12	Si 28	Ti 48	Ge 72	Zr 90	Sn 118	Ce 140	—	—	Pb 207	Th 232
V		N 14	P 31	V 51	As 75	Nb 94	Sb 120	Di 144	—	Ta 182	Bi 208	—
VI		O 16	S 32	Cr 52	Se 79	Mo 96	Te 125	—	Er 166	W 184	—	U 240
VII		F 19	Cl 35·5	Mn 55	Br 80	—	I 127	—	—	—	—	—
VIII				Fe 56 Ni 58·6 Co 59		Ru 104 Rh 104 Pd 106				Ir 192·5 Os 193 Pt 104		

Typical Elements

Series

The name *family* is often applied to a sub-group ; thus it may be said that in Group V. the group-character preponderates over the family-character, but that the family-character is more marked than the group-character in Group VI.

Those groups in which neither character is much in the ascendancy are best suited for exhibiting the general applications of the periodic law. For this reason we shall begin our detailed study of this law by considering Group II.

395 The position of an element in the scheme of classification arising out of the periodic law is indicated by the use of Roman numerals to express the group, and Arabic numbers to express the series ; thus the positions of antimony, nitrogen, and iodine, respectively, are defined by the notation V. - 7, V, - 2, and VII. - 7.

CHAPTER XIX.

THE ELEMENTS OF GROUP II.

		2	4	6	8	10	12
Group II.	Even series.	Be = 9·08	Ca = 39·9	Sr = 87·3	Ba = 136·8	· -	—
		3	5	7	9	11	
	Odd series.	Mg = 24	Zn = 64·9	Cd = 112	—	Hg = 199·8	

Even-series elements	BERYLLIUM.	CALCIUM.	STRONTIUM.	BARIUM.
Atomic weights	9·08	39·9	87·3	136·8
		The *molecular weights* of these elements are unknown.		
Sp. grs. (approx.)	1·7	1·58	2·5	3·75
Sp. heats	·6 (at abt. 500°) Increases rapidly as temp. increases.	·17	not determined.	not determined.
Melting points (approx.)	not determined.	full red-heat; above Sr.	above Ba; moderate red-heat.	below red-heat.
Atom. weights spec. gravs.	5·4	25·3	34·9	36·5
Colour, appearance, &c.	White, lustrous, hard.	Whitish-yellow; abt. as hard as lead, very ductile, but becomes brittle when hammered.	Clear whitish-yellow; harder than lead, ductile and malleable.	Gold-yellow; fairly ductile.
Occurrence and preparation	Not widely distributed. Oxide occurs in a few rocks. Prepared by reducing fused $BeCl_2$ by Na, *not* by electrolysis of $BeCl_2$.	Carbonate, phosphate, sulphate, silicate, &c. very widely diffused in rocks, water, plants, and animals. Prepared by electrolysis of mixture of $CaCl_2$ with $SrCl_2$ and NH_4Cl, or by reducing $CaCl_2$ by Zn − Na amalgam.	Carbonate and sulphate occur in some rocks and water, but not very widely diffused. Prepared by electrolysis of fused $SrCl_2$, or by reducing $SrCl_2$ by Zn − Na amalgam.	Carbonate, sulphate, and silicate, occur in some rocks, water, and plants, but not very widely diffused. Prepared by electrolysis of $BaCl_2$ mixed with NH_4Cl, or by reducing $BaCl_2$ by vapour of K.
General chemical properties	Not oxidised in ord. air; even when heated in O is only superficially oxidised. Does not decompose H_2O even at red heat. Combines with Cl, Br, and I, at high temps.; does not combine directly with S. Dissolves in KOHAq forming BeO and H. Oxide (BeO) basic but not alkaline. Distinctly metallic.	Quickly oxidises in moist air; decomposes cold H_2O rapidly; burns in air at red heat. Combines with Cl, Br, I, P, and S, at high temperatures. Oxide (CaO) strongly basic and alkaline. Strongly positive metal.	Closely resembles Ca; decomposes cold H_2O more rapidly. Oxide (SrO) strongly basic and alkaline.	Closely resembles Ca. Oxide (BaO) very strongly basic and alkaline.

397   General formulae and chemical characters of compounds.
(M = Be, Ca, Sr, or Ba). MO, MO_2H_2, MO_2 (no BeO_2 known),
MS, MS_2H_2 (no BeS_2H_2 known), MX_2 ($X = F$, Cl, Br, I),
MSO_4, $M2NO_3$, MCO_3, &c.   The only compounds which have
been gasified are $BeCl_2$ and $BeBr_2$.

398   The oxides MO may be prepared by direct combination
of metal with oxygen, or by decomposing the hydroxides
(MO_2H_2) by heat (BaO_2H_2 is not decomposed by heat alone).
The hydroxides MO_2H_2 where $M = Ca$, Sr, or Ba, are
obtained by combining water with the oxides MO, or by
precipitating solutions of salts of M by potash or soda.
Beryllium hydroxide, BeO_2H_2, is prepared by precipitating
an aqueous solution of a salt of Be by NH_3Aq, and drying at
about 100°.   The peroxides MO_2 ($M = Ca$, Sr, or Ba) are
produced by interactions between H_2O_2Aq and solutions of
salts of M; the compounds $MO_2 . xH_2O$ thus obtained lose
water when dried, when $M = Ba$ the drying is conducted
over sulphuric acid *in vacuo*, when $M = Sr$ the hydrated
peroxide is dried at 100°, and when $M = Ca$ the temperature
is raised to 130°. BaO_2 is also obtained by heating BaO in
oxygen at about 200°; the other oxides MO do not directly
combine with oxygen.

The oxides CaO, SrO, and BaO are somewhat soluble
in water; the solubility increases from CaO to BaO.   The
solutions are alkaline towards litmus paper; they interact
with acids to produce salts and water; they precipitate
hydrates of iron, copper, manganese, and many other heavy
metals, from solutions of salts of these metals; they absorb
and combine with carbon dioxide.   These oxides combine with
water forming hydroxides which are very stable compounds.
Beryllium oxide, BeO, is insoluble in water; it does not
directly combine with water.   This oxide has no alkaline
properties; it interacts with acids to form salts and water.
None of these oxides, except BeO, is easily reduced, e.g. by
heating with C, or in H or CO.

The hydroxides MO_2H_2, where $M = Ca$, Sr, or Ba, are
fairly soluble in water; the solubility increases as the atomic
weight of M increases; CaO_2H_2 is decomposed by heat (to
$CaO + H_2O$) at 300°—400°; SrO_2H_2 at a higher temperature;
BaO_2H_2 is not decomposed even at a full red heat.   These
hydroxides do not interact with solutions of the alkalis (potash,
soda, ammonia).   They form compounds with water (hydrates);
the most marked of these hydrates have the composition

$MO_2H_2 . 8H_2O$. Beryllium hydroxide, BeO_2H_2, is easily changed by heat to $BeO + H_2O$; it is insoluble in water and has not an alkaline reaction towards litmus. When freshly precipitated this hydroxide dissolves in KOHAq or NaOHAq, but is reprecipitated on heating. The freshly precipitated hydroxide also dissolves in $(NH_4)_2CO_3Aq$ and K_2CO_3Aq; by boiling the solution in $(NH_4)_2CO_3Aq$ a basic carbonate of beryllium is precipitated. BeO_2H_2 forms a number of hydrates the compositions of which vary with small variations in the conditions under which they are prepared; they are all readily decomposed by heat.

The peroxides MO_2 ($M = Ca$, Sr, Ba) shew no alkaline reaction towards litmus; they are insoluble in water, but all form compounds (hydrates) with water. They are all decomposed by heat to $MO + O$; BaO_2 is the most stable towards heat. No peroxide of beryllium has been obtained.

The oxides and hydroxides are all white solids; BeO_2H_2 is much more gelatinous than the other hydroxides. The specific gravities of the oxides MO are, approximately, 3·08 for BeO, 3·15 for CaO, 4·5 for SrO, and 5·4 for BaO; of the hydroxides MO_2H_2, 2·08 for CaO_2H_2, 3·62 for SrO_2H_2, and 4·49 for BaO_2H_2. The heats of formation of these oxides and hydroxides, in aqueous solutions, from calcium to barium, increase as the atomic weights of the metals increase.

The sulphides CaS, SrS, and BaS are prepared (1) by **399** heating the oxides MO in a stream of carbon disulphide mixed with carbon dioxide, (2) by heating a mixture of the sulphates MSO_4 with charcoal; $MSO_4 + 4C = MS + 4CO$.

The hydrosulphides MS_2H_2 are usually obtained by reactions between the hydroxides, generally in aqueous solution, and sulphuretted hydrogen; thus

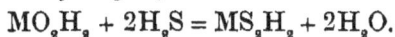

$$MO_2H_2 + 2H_2S = MS_2H_2 + 2H_2O.$$

These compounds are also formed when the sulphides MS interact with a little water;

$$2MS + 2H_2O = MS_2H_2 + MO_2H_2.$$

The sulphides and hydrosulphides are white, or yellowish white, solids; they are more easily decomposed by heat, and by interaction with water, than the corresponding oxygen compounds. Besides the compounds MS and MS_2H_2 the following sulphides are known :—CaS_4, CaS_5, SrS_4, BaS_3, $BaS_4 . H_2O$. '

Beryllium does not combine directly with sulphur, nor is a sulphide produced by heating BeO in carbon disulphide vapour, or by heating $BeSO_4$ in hydrogen or sulphuretted hydrogen.

400   The haloid compounds MX_2,—where $X = F$, Cl, Br, or I, and $M = Ca$, Sr, or Ba—are obtained by interactions between the oxides, hydroxides, or carbonates, of the metals and aqueous solutions of the haloid acids HX. The corresponding beryllium compounds, $BeCl_2$, $BeBr_2$, and BeI_2, are prepared by strongly heating an intimate mixture of beryllium oxide and finely divided carbon in chlorine, bromine, or iodine vapour, respectively; $BeO + C + X_2 = BeX_2 + CO$. Solutions of beryllium oxide, or carbonate, in haloid acids yield oxyhaloid compounds on evaporation, e.g. $BeO.BeCl_2$.

The compounds MX_2 where M is Ca or Sr usually crystallise with $6H_2O$, i.e. the crystals have the composition $MX_2.6H_2O$; when $M = Ba$ the compounds MX_2 usually crystallise with $2H_2O$. The crystallised fluorides however of all the metals Ca to Ba seem to be anhydrous. Hydrates of BeX_2 crystallise from solutions of the three compounds in the haloid acids HX. The compounds MX_2 are white solids, which melt at high temperatures; e.g. M.P. of $CaCl_2 = 719^0$, $CaBr_2 = 676^0$, $CaI_2 = 631^0$; M.P. of $SrCl_2 = 825^0$, $SrBr_2 = 630^0$, $SrI_2 = 507^0$; M.P. of $BeCl_2$ = abt. 600^0. $BeBr_2$ and $BeCl_2$ have been gasified without decomposition. The haloid compounds of Ca, Sr, and Ba, with the exception of the fluorides, are very soluble in water, and are also soluble in alcohol; the Be compounds are not soluble in water. The Be compounds easily form oxyhaloid compounds of more or less complex composition expressible by the general formula $xBeX_2.yBeO$; the haloid compounds of Ca, Sr, Ba, form only a few similar compounds, which are obtained by boiling solutions of MX_2 with the oxides MO. Most of the haloid compounds of Ca, Sr, and Ba, absorb and combine with ammonia, generally producing compounds of the form $MX_2.6NH_3$; these are easily decomposed by heat. These haloid compounds, except the fluorides, do not shew any great tendency to combine with other haloid compounds and so produce double salts.

401   The salts of the metals Be, Ca, Sr, Ba, i.e. the compounds obtained by replacing the hydrogen of acids by these metals, are very numerous. The salts of Ca, Sr, and Ba, as a class are very definite and stable compounds; the oxides and hydroxides of these metals interact with most acids in aqueous solutions to form normal salts. With the weak acids, especially with boric acid, these metals also form *basic salts*: e.g. $3CaO.5B_2O_3$, and $3SrO.B_2O_3$. A few *double salts* are known of these three metals; e.g. $CaSO_4.K_2SO_4.H_2O$; and $SrSO_4.K_2SO_4$. The greater number of these double salts are derivatives of the weaker acids; e.g.

$$Ba(NH_4)_2AsO_4, \text{ and } 2CaO.B_2O_3.Na_2B_4O_7.15H_2O.$$

The salts of beryllium are less definite compounds than those of the other three metals we are considering: many are *basic salts*, e.g.

$$3BeO.CO_2; \quad 3BeO.SO_3; \quad 2BeO.SO_3; \quad 7BeO.3SeO_2.14H_2O.$$

The normal Be salts are more easily decomposed by heat, or by heating in presence of water, than the salts of Ca, Sr, and Ba. Beryllium also forms many *double salts*, e.g.

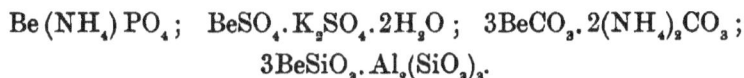

$$Be(NH_4)PO_4; \quad BeSO_4.K_2SO_4.2H_2O; \quad 3BeCO_3.2(NH_4)_2CO_3;$$
$$3BeSiO_3.Al_2(SiO_3)_3.$$

The carbonates MCO_3 are all decomposed by heat alone to $MO + CO_2$; their stabilities towards heat increase as the atomic weight of M increases.

The existence of the gaseous molecules $BeCl_2$ and $BeBr_2$ **402** indicates that the atom of beryllium is divalent; as the vapour densities of no compounds of the metals Ca, Sr, Ba, have as yet been determined, we cannot be certain as to the valency of the atoms of these elements, but judging from the analogies between these three, and the other, members of Group II., it is probable that the atoms of these elements are divalent.

- The three elements, calcium, strontium, and barium, are evidently very closely related; they are much more like each other than any of them is like beryllium.

403   We shall now consider the *odd-series* members of Group II.

Odd-series elements	MAGNESIUM.	ZINC.	CADMIUM.	MERCURY.
Atomic weights	24	64·9	112	199·8
Molecular weights	—	64·9	112	199·8
Sp. grs. (approx.)	1·75	7·2	8·6	13·6
Sp. heats	·25	·095	·056	·032 (for solid Hg)
Melting points (approx.)	500°—700°	420°	320°	−39°·5
$\frac{Atom.\ weights}{Sp.\ gravs.}$	13·7	9	13	14·7
Vapour-densities (H=1)	—	64·9	112	199·8
Colour, appearance, &c.	White, lustrous, fairly hard; ductile; crystalline; may be distilled at very high temperature.	White with slight blue tinge, lustrous; rather brittle, but malleable at 100°—150°; crystalline; readily distilled; vapourised at 900°—1000°.	White, lustrous; malleable and ductile; easily crystallised; vapourised at 750°—800°.	Liquid above −39°·5. White, lustrous; crystalline. Boils at 350°.
Occurrence and preparation.	Carbonate, sulphate, chloride, and oxide of Mg are widely distributed in rocks and water. Prepared by electrolysis of molten MgCl₂, or by reducing MgCl₂ by Na.	Sulphide, carbonate, and oxide are fairly widely distributed in rocks. Prepared by deoxidising ZnO by charcoal.	Usually accompanies Zn in its ores. Prepared by reducing CdO by charcoal.	The metal occurs in small quantities. HgS is found in considerable quantities, but not widely distributed. Prepared by heating HgS in air, or with Fe or CaO, and condensing Hg.
General chemical properties. †	Oxidised by heating in air or O. Decomposes water at nbt. 100°. Combines directly with Cl, Br, I. Oxide is basic and slightly alkaline. Metal interacts with acids to form salts and H.	Oxidised by heating in air or O to redness. Decomposes steam at red heat. Combines directly with Cl, Br, I, but not with S. Reacts with acids to form salts and H. Oxide is basic but not alkaline; it is soluble in KOHAq. Metal has been gasified, and molecule found to be monatomic.	Closely resembles Zn in properties; but oxide is not soluble in KOHAq. Combines directly with S. Metal has been gasified, and molecule found to be monatomic.	Oxidised by heating in air or O. Oxide is decomposed to Hg and O at full red heat. Does not decompose water or steam. Readily combines with Cl, Br, I, and S. Forms two series of salts. Oxide HgO is soluble in molten KOH. Metal has been gasified, and molecule found to be monatomic.

404   General formulae and chemical characters of compounds.

Oxides and hydroxides. MO, MO_2H_2, when $M = Mg, Zn, Cd$. M_2O and MO, when $M = Hg$; no hydroxides of Hg known.

Sulphides. MS; and also M_2S when $M = Hg$.

Haloid compounds. MX_2; and also MX when $M = Hg$.

Salts. MSO_4, $M2CO_3$, $M2NO_3$, &c.; and also M_2SO_4, MNO_3, &c. when $M = Hg$.

No compound of Mg has been gasified; several haloid compounds, MX_2, of the other metals have been gasified.

The oxides MO are obtained by heating the metals in **405** oxygen, or by precipitating solutions of their salts by an alkali, and heating the hydroxides MO_2H_2 thus formed; the oxide HgO is obtained when an alkali is added to the solution of a mercuric salt, e.g. to $HgCl_2Aq$ or $Hg(NO_3)_2Aq$. Mercuric oxide (HgO) is a red or yellow solid; CdO is brown; ZnO is pale yellowish-white; and MgO is white; they can all be obtained in crystals. The specific gravities of these oxides are approximately 3·1 for MgO, 5·6 for ZnO, 6·96 for CdO, and 11·1 for HgO. MgO is slightly soluble in water, 1 part of MgO dissolving in about 60,000 parts of cold water; the other oxides are almost insoluble in water. MgO combines directly with water to form MgO_2H_2; this hydroxide is decomposed to $MgO + H_2O$ at a red heat. The other oxides MO do not directly combine with water; HgO appears not to combine with water under any conditions.

The oxide HgO dissolves in molten potash forming the compound $K_2O.2HgO$. The oxide exists in two forms; red HgO, obtained by heating Hg in O, and yellow HgO, obtained by precipitating the solution of a mercuric salt by KOHAq: these oxides shew considerable differences in their reactions with acids &c., e.g. red HgO scarcely interacts with chlorine, whereas yellow HgO readily interacts to produce an oxychloride of mercury and Cl_2O_2

The oxides of Zn and Cd are reduced to metal by heating with charcoal, or in H or CO; HgO is reduced to metal by heat alone; MgO is not deoxidised by ordinary reducers.

The hydroxides MgO_2H_2, ZnO_2H_2, and CdO_2H_2, are obtained by precipitating solutions of salts of the metals by an alkali, and drying; they are all decomposed by heat alone to $MO + H_2O$; their stabilities towards heat are inversely as the atomic weights of the metals. These hydroxides are all basic, MgO_2H_2 has a slightly alkaline reaction towards litmus; ZnO_2H_2 is soluble in KOHAq, but is precipitated again when a concentrated solution is placed over sulphuric acid *in vacuo*, the other hydroxides are insoluble in alkali solutions.

Mercurous oxide Hg_2O is a black solid, specific gravity 10·7, obtained by adding KOHAq to the solution of a mercurous salt, e.g. to $HgNO_3Aq$, or to solid HgCl. This oxide is decomposed by heat to mercury and mercuric oxide, and at a higher temperature to mercury and oxygen.

406   The sulphides, MS, are solids. MgS resembles the sulphides of Ca, Sr, and Ba, it is an unstable compound which interacts with water to form MgS_2H_2 and MgO_2H_2; the MgS_2H_2 is quickly changed to MgO_2H_2 and H_2S. The compounds MgS_3 and MgS_4 have also been prepared. The other sulphides, MS, where $M = Zn$, Cd, or Hg, are more definite and stable compounds. They are produced by passing H_2S into solutions of the salts of the metals; ZnS is soluble in and decomposed by most acids, the two others are insoluble in dilute acids. CdS and HgS are also obtained by directly combining the metals with sulphur, the reaction proceeds slowly and to a limited extent with Cd.

Pentasulphides of zinc and cadmium, MS_5, seem to exist; they are unstable compounds. Mercuric sulphide exists in two forms; black and amorphous, by precipitating solutions of mercuric salts by H_2S; red and crystalline, by directly combining Hg and S, or by subliming the amorphous form.

When HgS is precipitated from solutions of mercuric salts by excess of ammonium sulphide, and potash is added, the HgS dissolves and the solutions probably contain a compound of K_2S and HgS; the compound $K_2S . 2HgS$ is said to have been obtained. None of the other sulphides MS exhibit any tendency to form compounds with the sulphides of the strongly positive metals. HgS also forms many double compounds, chiefly with mercuric salts; e.g. $2HgS.HgCl_2$; $HgS.HgI_2$; $HgS.2HgSO_4$. The sulphides of Zn and Cd form a few oxysulphides, e.g. $ZnO.ZnS$.

Mercurous sulphide, Hg_2S, if it exists is extremely easily decomposed to HgS and Hg.

407   The haloid compounds, MX_2, are obtained by the direct combination of metal with halogen, or by dissolving the metals, oxides, or carbonates in aqueous solutions of the haloid acids HX. They are white solids, generally soluble in water and alcohol; the solubility decreases as the atomic weights of the metals increase. They all form oxyhaloid compounds of the general form $xMCl_2 : yMO$; a great many mercury oxychlorides have been obtained. $MgCl_2$ cannot be obtained by evaporating a solution in HClAq as when most of the water has been removed this solution is decomposed to MgO and HCl.

The haloid compounds all melt at temperatures below 1000°; e.g. $MgCl_2$ melts at about 700°, $ZnCl_2$ at approximately $500^\circ—600^\circ$, $CdCl_2$ at about 550°, and $HgCl_2$ at about

300°. They all combine with ammonia, and also with various other haloid compounds, to form double compounds; e.g. $ZnCl_2.2NH_4Cl$; $2CdCl_2.SrCl_2$; $3HgCl_2.MgCl_2$. The haloid mercury compounds form a very great number of such double compounds; they also combine with various salts, e.g. with $K_2Cr_2O_7$, $Cu(C_2H_3O_2)_2$, &c.

The *mercurous haloid compounds*, HgCl, HgI, and HgBr, are obtained by heating the corresponding mercuric compounds with mercury; they are nearly insoluble in water, and are partially decomposed by heat into mercury and the corresponding mercuric compounds, e.g. $2HgCl = HgCl_2 + Hg$.

The following haloid compounds have been gasified : $ZnCl_2$, $CdBr_2$, $HgCl_2$, HgI_2; these formulae represent the compositions of molecules of the compounds, hence the atoms of Zn, Cd, and Hg are divalent in these molecules. It is probable, but not quite certain, that the formula HgCl represents the gaseous molecule of mercurous chloride; if this formula is molecular the atom of Hg is monovalent as well as divalent.

The salts of the metals we are considering are very **408** numerous; each metal forms salts of the form $M2NO_3$, MSO_4, MCO_3, &c. and besides these Hg forms a series of mercurous salts, M_2SO_4, MNO_3, &c. Many of the salts are isomorphous; some salts of mercury are isomorphous with corresponding salts of copper. The metals all form *basic salts*, e.g. $4MgCO_3.MgO_2H_2$; $4ZnO.SO_3$; $Cd2NO_3.CdO_2H_2$; $3HgO.SO_3$; $3HgO.N_2O_5$; a very great many basic mercury salts are known; Mg seems to form fewer of these salts than any other of the four metals. *Double salts* of all these metals are numerous, especially in the case of mercury. The salts of the four metals, as a class, are stable and well defined; those of Hg, on the whole, are less soluble .in water than the others. The *mercurous salts* are considerably less stable than the mercuric salts; they are easily changed into the latter. Mercury salts form a very large number of compounds with ammonia; the composition of many of these is complex; in this respect mercury resembles copper, gold, platinum, and chromium.

The elements placed in Group II. are evidently closely **409** related in their chemical properties. The four elements Ca, Sr, Ba, and Mg, more nearly resemble one another than they resemble any of the other members of the group. This resemblance is well shewn in the alkalinity ·of their

oxides and hydroxides, in the formation of hydroxides by
the direct union of water with the oxides, in the great
stability of their oxides towards reducing agents, in the
stability of their salts and the existence of but few basic
salts, in the preparation and properties of their sulphides
and hydrosulphides, &c.   In its physical character however
Mg differs from Ca, Sr, and Ba, and resembles Be, and to
some extent Zn and Cd.   Beryllium, although occurring in
Series 2, is decidedly more like the odd-series, than the even-
series, members of the group ; the properties of the oxide and
hydroxide of Be are very similar to those of the corresponding
compounds of Zn and Cd ; in the comparatively large number
of basic salts which it forms, and in the readiness with which
oxyhaloid compounds of Be are produced, the metal also
resembles the odd-series members, especially Hg.   Some of
the haloid compounds of Be, . Zn, Cd, and Hg, are gasifiable
without decomposition at workable temperatures ; no compounds of the other elements of the group have been
gasified.

Mercury is distinguished from the other members of the
group by the fact that it forms two series of compounds,
mercurous HgX, and mercuric HgX_2,

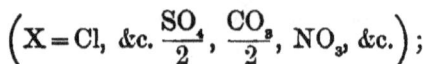

$$\left(X = Cl, \ \&c. \ \frac{SO_4}{2}, \ \frac{CO_3}{2}, \ NO_3, \ \&c. \right) ;$$

also by its physical properties, &c.

Looked at broadly, Ca, Sr, and Ba, are the most positive
members of the group, next to them comes Mg, and Hg is the
least positive.   It is to be noted that the arrangement of
Group II. in accordance with the periodic law shews a gap in
Series 9 ; if an element is discovered to fill this gap it will
probably resemble mercury on the one hand and zinc and
cadmium on the other.

Beryllium to some extent summarises in itself the properties of the other members of the group.

It cannot be said that the even-series members of Group II.
form a sub-group or family distinctly marked off from the
odd-series members of the group ; nor can it be asserted that
there is a gradual change of properties from the first to the
last member of the group.   All the members shew distinct
relations to each other ; neither the family-character nor the
group-character preponderates.

If we now turn to Group VI. we shall find that the **410**
first member of the group, oxygen, to some extent summarises in itself, or is typical of, the properties of all the
other members ; and that the other elements placed in this
group fall into two well marked families, one of which contains
the even-series members and the other contains the odd-series
members of the group.

CHAPTER XX.

THE ELEMENTS OF GROUP VI.

411

	2	4	6	8	10	12
Group VI. { Even series.	O=15·96	Cr=52·4	Mo= 95·8	—	W=183·6	U=239·?
Odd series.	3	5	7	9	11	
	S=31·98	Se=78·8	Te=125	—	—	

	CHROMIUM.	MOLYBDENUM.	TUNGSTEN.	URANIUM.
Even-series elements (omitting oxygen)				
Atomic weights	52·4	95·8	183·6	239·9
	The *molecular weights* of these elements are unknown.			
Sp. grs. (approx.)	6·7	8·5	18·5	18·7
Melting points	Above 2000°.	Infusible at full white-heat.	Softens and agglomerates at white-heat.	A full red-heat.
Sp. heats	·10 (?)	·072 (?)	·0334	·028
Atom. weights / spec. grav.	7·7	11·3	9·7	12·9
Occurrence and preparation.	Chrome-ironstone, (FeOCr₂O₃) and lead chromate, &c. occur in a few rocks; not widely distributed. Obtained by deoxidising Cr₂O₃ by C, or by action of K on Cr₂Cl₆, or by electrolysis of Cr₂Cl₄ containing Cr₂Cl₆.	Occurs in small quantities as oxide and sulphide, also as lead or cobalt molybdate. Obtained by reducing oxide or chloride by H, or the oxide by C or KCN.	Occurs very sparingly, as tungstate of Ca, of Fe and Mn, and of Pb; also as oxide. Obtained by reducing the oxide or chloride in H.	Sparingly distributed, as oxide in pitchblende, as uran ite of Ca and of Cu as carbonate of U and Ca, &c. Obtained by reduc ing the chloride by Na.
Colour, appearance, &c.	Very hard, brittle, powder composed of minute brilliant tin-white crystals. Descriptions differ much; probably the metal has not yet been obtained approximately pure.	Ashen-grey powder; when compressed, is a silver-white, lustrous, hard, brittle, infusible, metal.	Resembles iron in colour and lustre; hard and brittle; also obtained as a brown amorphous powder.	White, lustrous, metal; softer than steel; malleable, but cannot be beaten into thin plates; also obtained as a grey black powder.
General chemical properties.	Burns in stream of O; heated in air is superficially oxidised; oxidised by molten KNO₃ or KClO₃. Easily dissolved by dilute HClAq or H₂SO₄Aq, but does not react with hot conc. HNO₃. Combines readily with Cl and I when heated. Decomposes steam slowly at a red-heat. Replaces H of acids forming two series of salts. CrO₃ is an anhydride; Cr₂O₃ seems to form a few salts by heating with basic oxides. Atom of Cr is perhaps hexavalent in gaseous molecules.	Not oxidised in air at ordinary temperature but burns at low red-heat. No reaction with HClAq, HFAq, or H₂SO₄Aq; oxidised by conc. HNO₃. MoO₃; oxidised by molten KOH, but no reaction with KOHAq. Combines with Cl and Br, but not directly with I. Salts formed by replacing H of acids by Mo are scarcely known. MoO₃ is an anhydride, and also combines with many more negative anhydrides, e.g. P₂O₅. Atom of Mo is pentavalent in gaseous molecules.	Burns in air at red-heat; unchanged at ordinary temperatures. Oxidised to WO₃ by hot HNO₃Aq, HClAq or H₂SO₄. Dissolves in hot KOHAq to form K tungstate and H. Combines with Cl at high temperatures. Does not seem to form salts by replacing H of acids. WO₃ is an anhydride, and it also combines with other more negative anhydrides, e.g. SO₃. Atom of W is penta and hexavalent in gaseous molecules.	Slowly tarnishes in air; oxidised rapidly in air at 150°—200°. Combines with C and Br when heated and very slowly with hot l vapour. Dissolves in most F aqueous acids with evolution of H and formation of salts. Forms two series of salts, members of one of which always contain O in addition to U and the acid radicle. UO₃ is an anhydride. Atom of U is tetra valent in gaseous molecules.

General formulae and chemical characters of compounds. **412**
The compositions of the more important compounds of these
four metals are expressed by the following formulae; but
representatives of each formula are not known for all the
elements, thus sesquioxides, M_2O_3, of tungsten or uranium
have not been obtained, and the formula MX_6 is represented
by WCl_6 only.

Oxides. MO, M_2O_3, MO_2, MO_3: hydrates of some of
these are known.

Sulphides. MS, M_2S_3, MS_2, MS_3, MS_4.

Haloid compounds; chiefly chlorides. MX_2, MX_3,
MX_4, MX_5, MX_6.

Acids. H_2MO_4, $H_2M_2O_7$, &c., H_2MS_4.

Salts. MSO_4, $M2NO_3$, &c.; $M_2(SO_4)_3$, $M_2(NO_3)_6$, &c. when
$M = Cr$. $M(SO_4)_2$, $M(NO_3)_4$, $MO_2(SO_4)$, &c. when $M = U$.
Salts of Mo and W are scarcely known.

The oxides MO are scarcely known; hydrates of CrO and **413**
MoO are obtained by adding a solution of potash in air-free
water to solutions of Cr_2Cl_4 and Mo_2Cl_4, respectively. These
hydrates are rapidly oxidised in the air; neither yields
corresponding salts by its reactions with acids.

The sesquioxides M_2O_3 are stable compounds when $M = Cr$
or Mo (chromic and molybdic oxides); no sesquioxide of W
or U is known. Cr_2O_3 is prepared by precipitating a solution
of a chromic salt by ammonia, washing, drying, and heating;
in the case of Mo_2O_3, the hydrated oxide precipitated by potash
is heated in hydrogen. Both oxides form dark coloured
solids, insoluble or nearly insoluble in acids. Hydrated
$Cr_2O_3(Cr_2O_3.3H_2O)$ dissolves readily in acids forming chromic
salts, e.g. Cr_23SO_4. This oxide also seems to combine with a
few basic oxides, e.g. with CaO. It is therefore basic but
also slightly acidic.

The dioxides MO_2 are dark coloured solids, obtained by
reducing the oxides MO_3 directly or indirectly. CrO_2 is more
easily formed by passing nitric oxide into an aqueous solution
of potassium dichromate ($K_2Cr_2O_7$); MoO_2 and WO_2 by heating
MoO_3 and WO_3 in hydrogen to low redness, or by digesting
a solution of MoO_3, or WO_3, in hydrochloric acid with copper
or zinc and then precipitating by ammonia. UO_2 may be
prepared by digesting UCl_4 with water.

CrO_2 is decomposed by heat at $300°$ with evolution of
oxygen and production of Cr_2O_3; the other oxides MO_2 are
oxidised to MO_3, MoO_3 by heating with nitric acid, WO_2 by

heating in air, and UO_2 by the action of air at ordinary temperatures. These oxides MO_2 are slightly soluble in acids ; UO_2 gives salts, e.g. $U(SO_4)_2$, but no definite salts have yet been certainly obtained corresponding to any of the other oxides MO_2, although such salts seem to exist.

The trioxides MO_3 are anhydrides. CrO_3 is prepared by adding a sufficient quantity of concentrated sulphuric acid to a solution of potassium dichromate, MoO_3 and WO_3 are obtained by oxidising the lower oxides, or better from ammonium molybdate and tungstate, respectively, by heating with nitric acid and then washing out the ammonium nitrate formed. UO_3 is obtained by heating uranyl nitrate

$$(UO_2)(NO_3)_2.$$

CrO_3 is very soluble in water forming a markedly acid liquid; under special conditions the hydrate $CrO_3 . H_2O$—i.e. H_2CrO_4, *chromic acid*—can be obtained from this liquid.

CrO_3 interacts with acids to form oxygen and salts corresponding with the oxide Cr_2O_3; thus

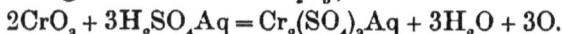

$$2CrO_3 + 3H_2SO_4Aq = Cr_2(SO_4)_3Aq + 3H_2O + 3O.$$

This oxide readily parts with part of its oxygen and therefore acts as an oxidiser, e.g. when it is heated, Cr_2O_3 and oxygen are produced.

MoO_3 is much less soluble in water than CrO_3; WO_3 is only very slightly soluble in water; and UO_3 is insoluble. Hydrates of these oxides exist and exhibit acidic properties (*s.* Acids, par. 416), but none of them is obtained by the direct addition of water to the oxide. The oxides MoO_3 and WO_3 form various complex compounds with several anhydrides such as SO_3, P_2O_5, B_2O_3, &c. The oxide UO_3 interacts with a few acids to form salts (*s.* Salts, par. 417).

The oxides M_2O_3 and MO_2 are on the whole basic ; the oxides MO_3 are acidic, but their acidic character is less marked as the atomic weight of M increases. The change from MO_2 to MO_3 is effected the more easily and directly the greater the atomic weight of M. Of the oxides MO_2, UO_2 shews the most clearly marked basic character. Of the oxides MO_3, UO_3 is the most stable towards heat and reducing agents, and CrO_3 is the least stable.

414   The most important sulphides are Cr_2S_3; MoS_2, MoS_3, MoS_4; WS_2, WS_3; and US_2. *Chromic sulphide*, Cr_2S_3, is prepared by passing sulphuretted hydrogen over hot chromic oxide (Cr_2O_3); it is not obtainable by reactions between compounds in solution.   This sulphide is feebly acidic ; it combines with certain more basic sulphides, e.g. with ZnS, CaS, &c.

When excess of sulphuretted hydrogen is passed into the solution of an alkaline molybdate, e.g. K_2MoO_4, and an acid is then added, *molybdenum trisulphide* MoS_3, is precipitated. This sulphide is distinctly acidic; it interacts with K_2S, &c. to form thio- (or sulpho-) salts, e.g. K_2MoS_4.

When potassium thiomolybdate, K_2MoS_4, is heated with MoS_3, a salt having the composition K_2MoS_5 is formed ; this salt interacts with acids to produce *molybdenum tetrasulphide* MoS_4. *Molybdenum disulphide*, MoS_2, is obtained by heating together MoO_3 and sulphur.

Tungsten trisulphide WS_3 is obtained similarly to MoS_3. The *disulphide* WS_2 is formed by heating together tungsten and sulphur. WS_3 is distinctly acidic, forming thiotungstates, e.g. K_2WS_4, $BaWS_4$, &c.

Uranium disulphide, US_2, is formed similarly to WS_2; it exhibits no acidic functions.

As CrS_3 has not not yet been prepared it is difficult to compare corresponding sulphides of the four elements ; but on the whole it appears that the sulphides become more acidic as the atomic weight of the metals increases.

The haloid compounds of the four elements we are **415** considering are important.   Their compositions are shewn by the following formulae ; $CrCl_2$, $CrCl_3$, $CrBr_2$, $CrBr_3$, CrI_2, CrI_3, CrF_3 ; $MoCl_2$, $MoCl_3$, $MoCl_4$, $MoCl_5$, $MoBr_2$, $MoBr_3$, $MoBr_4$; WCl_2, WCl_4, WCl_5, WCl_6, WBr_2, WBr_5, WI_2 ; UCl_3, UCl_4, UCl_5, UBr_3, UF_4.   The following have been gasified ; $MoCl_5$, WCl_4, WCl_6, UCl_4, UBr_4; these formulae are therefore molecular.   The formulae of the other compounds are the simplest that can be given, but they are not necessarily molecular.

Chromic chloride, $CrCl_3$, is obtained by heating an intimate mixture of chromic oxide and carbon in chlorine ; *chromous chloride*, $CrCl_2$, is formed by heating $CrCl_3$ in hydrogen. The higher chloride is stable in the air, but when strongly heated it gives Cr_2O_3; solutions of this chloride when heated give precipitates of various oxychlorides $Cr_xO_yCl_z$; the most important *oxychloride of chromium* is CrO_2Cl_2.   *Chromic chloride* forms either violet crystals by subliming in chlorine or hydrochloric acid gas, or a greenish solid by dissolving chromic hydrate in hydrochloric acid, evaporating nearly to dryness, and heating in chlorine.   The violet form is almost insoluble in water; the green form readily dissolves in water. *Chromous chloride*, $CrCl_2$, is very unstable; it removes chlorine readily from various chlorides, e.g. $HgCl_2$, and absorbs oxygen rapidly from the air.

When *molybdenum* is strongly heated in chlorine the *pentachloride*, $MoCl_5$, is formed. By heating this chloride in carbon dioxide the *tetra-* and *di-chlorides*, $MoCl_4$ and $MoCl_2$, are obtained: the *trichloride* is also got from the penta-chloride, by heating in hydrogen. Various *oxychlorides of molybdenum* are known; the more important are MoO_2Cl_2 and $MoOCl_4$.

Tungsten hexachloride, WCl_6, is produced by strongly heating tungsten in a stream of dry chlorine; hot water decomposes it to WO_3 and hydrochloric acid; heated in air it yields $WOCl_4$. The *penta-* and *di-chlorides*, WCl_5 and WCl_2, are obtained from WCl_6 by heating in hydrogen; the *tetra-chloride* WCl_4 is produced by heating a mixture of WCl_6 and WCl_5 in hydrogen or carbon dioxide. The *oxychlorides* WO_2Cl_2 and $WOCl_4$ are known.

When *uranium* dioxide, UO_2, is mixed with carbon and heated in chlorine the *tetrachloride* UCl_4 is formed; this chloride is decomposed by hot water to UO_3 and hydrochloric acid; it is an energetic reducing agent, e.g. it reduces ferric chloride to ferrous chloride. The *pentachloride* UCl_5 is formed by the direct addition of chlorine to UCl_4; but when UCl_5 is heated to 230° in hydrogen or carbon dioxide it is again separated into the tetrachloride and chlorine. Only a few *oxyhaloid compounds of uranium* have been prepared; the chief are UO_2X_2 where $X = Cl$, Br, or F.

416 Acids and salts derived therefrom. Many of the hydroxides of chromium, molybdenum, tungsten, and uranium, are acidic.

The precipitate obtained by adding ammonia to a solution of a chromic salt varies in composition according to the conditions, but it is always a hydrate of the oxide $Cr_2O_3(Cr_2O_3.xH_2O)$. This compound is basic, as it interacts with acids to form salts. By dissolving CrO_3 in a little cold water, warming, and again cooling, crystals of the hydrate $CrO_3.H_2O(=H_2CrO_4)$ are said to be formed. This compound is distinctly acidic, from it is derived a well marked series of salts, the chromates, $MCrO_4$, $M = K_2$, Ba, &c.

Ammonia ppts. $UO_2.xH_2O$ from solutions of the tetrachloride UCl_4; this hydrate dissolves in acids to form salts, e.g. $U(SO_4)_2$;

it is therefore basic. The hydrate $UO_3.H_2O(=H_2UO_4)$ is obtained indirectly from $UO_2(NO_3)_2$; this hydrate interacts with some acids to form salts, e.g. $UO_2(SO_4)$ &c. and it also interacts with alkalis to form salts of the form MUO_4 ($M=Na_2$, Ba, &c.), it is therefore both basic and acidic.

Hydrates of the dioxide and sesquioxide of molybdenum $MoO_2.xH_2O$ and $Mo_2O_3.xH_2O$ are known, but their interactions with acids have been little examined; they appear however to possess only basic properties. Various hydrates of MoO_3 are prepared indirectly (that is not by addition of water to the oxide); the chief are $MoO_3.H_2O$ and $MoO_3.2H_2O = H_2MoO_4$ and H_4MoO_5; these compounds are acidic; they may however also shew basic functions; their interactions with acids have not been much investigated.

Two hydrates of tungsten trioxide are known, $WO_3.H_2O$ and $WO_3.2H_2O = H_2WO_4$ and H_4WO_5; these compounds are acidic, and possibly also basic. Another compound $H_2W_4O_{13}.7H_2O$ has been obtained; it is acidic.

The chief salts derived from the acidic hydroxides of the elements under consideration, by replacing hydrogen by metals, are the *chromates* and *dichromates*, the *molybdates*, *di- tri- tetra-* &c. *molybdates*, the *di- tri-* &c. *tungstates*, and the *uranates*.

The *chromates* $MCrO_4$ where $M=$ Ba, Ag_2, K_2, &c. &c. are prepared by double decomposition from potassium chromate which may be obtained by neutralising a solution of chromium trioxide with potash. The alkaline chromates are also formed by fusing chromic oxide, Cr_2O_3, or a chromic salt, with potash and a little potassium nitrate or chlorate. When potassium chromate (K_2CrO_4) is treated with dilute sulphuric acid potassium dichromate, $K_2Cr_2O_7$, is formed; from this salt a series of *dichromates* MCr_2O_7 is obtained. *Trichromates* MCr_3O_{10}, and *tetrachromates* MCr_4O_{13}, are also known; but the best marked salts are the chromates and dichromates.

The *molybdates* $MMoO_4$ are obtained by double decomposition from the alkali salts; most of these salts form non-crystalline masses. *Dimolybdates* MMo_2O_7, *trimolybdates* MMo_3O_{10}, and *tetramolybdates* MMo_4O_{13}, are obtained by boiling various metallic carbonates with molybdenum trioxide under various conditions: these salts crystallise well and are more stable and definite than the molybdates.

A few *tungstates* MWO_4 are obtained by heating WO_3 with alkali or alkaline carbonates; several series of *condensed tungstates* or *polytungstates* exist belonging to the forms MW_2O_7, $M_2W_3O_{11}$, MW_2O_{13}, $M_2W_5O_{17}$, $M_3W_7O_{24}$, and $M_5W_{12}O_{41}$, ($M = K_2$, Na_2, Ba, Ca, &c.). Many of these are obtained by boiling tungstates with tungsten trioxide.

A few *uranates* MUO_4 are known; they are produced by adding alkalis or alkaline earths to solutions of uranyl salts. The *diuranates* MU_2O_7 are more definite and marked salts than the uranates. A few *polyuranates* are also known.

417   Salts. Chromium and uranium replace the hydrogen of acids forming well marked salts; salts of molybdenum and tungsten almost certainly exist, but they have been very little investigated.

The salts of chromium form two series, the *chromous salts* CrX, and the *chromic salts* $Cr_2 3X$, where $X = SO_4$, $2NO_3$, CO_3, $2ClO_3$, $\frac{2}{3}PO_4$, $\frac{2}{3}AsO_4$, &c. A few so called *basic salts of chromium* also exist, e.g. $Cr_2O_2SO_4$, $Cr_2O(SO_4)_2$, $Cr_2O(NO_3)_4$, &c.

The salts of uranium also form two series, the *uranic salts* UX_2, and the *uranyl salts* UO_2X, where $X = SO_4$, $2NO_3$, $\frac{2}{3}PO_4$, &c. The uranyl salts belong to the class of *basic salts*.

The chromous salts are very easily oxidised to chromic salts; *chromous acetate* $Cr(C_2H_3O_2)_2.H_2O$ is obtained by adding a concentrated solution of sodium acetate to chromous chloride solution produced by reducing chromic chloride by zinc and hydrochloric acid. *Chromous sulphate* $CrSO_4.7H_2O$, *chromous oxalate* CrC_2O_4, and a few other salts may be obtained from the acetate.

The chromic salts are well marked and stable compounds; among the more important are the *sulphate* $Cr_2(SO_4)_3.18H_2O$, and the *phosphates* $Cr_2(PO_4)_2.12H_2O$, $Cr_4(P_2O_7)_3.7H_2O$, $Cr_2(PO_3)_6$. These salts are generally prepared by dissolving $Cr_2O_3.xH_2O$ in the various acids. Many chromic salts exist in two forms, one violet to red, the other green. In some cases, e.g. $Cr_2(SO_4)_3$, both varieties are known in the solid form and have the same composition; in other cases, e.g. $Cr_2(NO_3)_6$, only a violet salt is known in crystals but a green solution is obtainable from these crystals. Aqueous solutions of most of the violet salts become green when boiled; many of these solutions become violet again on cooling or on standing for some time. Only the violet solutions yield crystalline salts; the green solutions give amorphous gummy solids on evaporation. Various hypotheses have been suggested to account for these changes of colour; that which seems to rest on the best experimental evidence asserts that the violet salts are the normal salts and that they are partially decomposed on boiling in aqueous solution into green basic salts and a little free acid.

Chromic sulphate forms double salts with the alkali sulphates of the composition $Cr_2(SO_4)_3.M_2SO_4.24H_2O$ where $M = Na$, K, &c.; these salts are *alums*.

The *uranic salts* have not been fully investigated; the sulphate $U(SO_4)_2$, obtained by dissolving $UO_2.xH_2O$ in sulphuric acid and evaporating in contact with excess of acid, is one of the most important. These salts are generally unstable and easily changed to basic salts.

A fair number of *uranyl salts* have been prepared. The nitrate $UO_2(NO_3)_2.6H_2O$ is obtained by dissolving $UO_3.H_2O$ in nitric acid and evaporating; various salts are obtained by treating the nitrate with different acids, e.g.

$$UO_2SO_4,\ UO_2SeO_3.2H_2O\ ;$$

others are obtained by dissolving UO_3H_2O in acids, or by double decomposition from the nitrate, e.g.

$$UO_2HPO_4,\ (UO_2)_3(AsO_4)_2.12H_2O.$$

Many of these compounds form double salts especially with salts of the alkali and alkaline earth metals.

A very large number of compounds of chromium with **418** chlorine, bromine, SO_4, NO_3, &c. and ammonia are known. These compounds are of complex compositions; most of them belong to one or other of the six following general forms;—

$M = X = Cl,\ Br,\ I,\ \dfrac{SO_4}{2},\ \dfrac{CrO_4}{2},\ NO_3$, &c. X can be easily replaced by other negative radicles, M can be replaced only with difficulty.

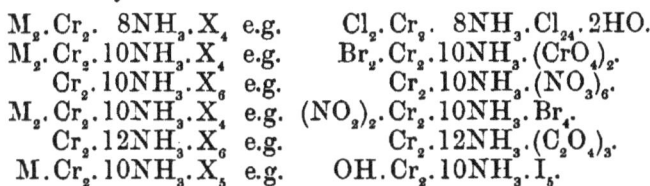

$M_2.Cr_2.\ 8NH_3.X_4$ e.g. $Cl_2.Cr_2.\ 8NH_3.Cl_{24}.2HO.$
$M_2.Cr_2.10NH_3.X_4$ e.g. $Br_2.Cr_2.10NH_3.(CrO_4)_2.$
$Cr_2.10NH_3.X_6$ e.g. $Cr_2.10NH_3.(NO_3)_6.$
$M_2.Cr_2.10NH_3.X_4$ e.g. $(NO_3)_2.Cr_2.10NH_3.Br_4.$
$Cr_2.12NH_3.X_6$ e.g. $Cr_2.12NH_3.(C_2O_4)_3.$
$M.Cr_2.10NH_3.X_5$ e.g. $OH.Cr_2.10NH_3.I_5.$

The four even-series elements of Group VI. are, then, **419** evidently very similar in their chemical properties. The compositions of the compounds of chromium differ to some extent from the compositions of the corresponding compounds of the other elements; thus the most marked oxide of chromium is Cr_2O_3, but tungsten and uranium are characterised by the oxides MO_2 and MO_3; similarly Cr_2S_3 is the highest sulphide of chromium known with certainty, but the important sulphides of the other elements are MS_2 and MS_3. Again the chlorides of chromium are $CrCl_2$ and $CrCl_3$, but the chlorides MCl_4, MCl_5, and even MCl_6, are characteristic of the other elements of the series.

The oxides MO_3 are all acidic; but the acidic character becomes less marked as the atomic weight of M increases. This decrease in acidic character is shewn by the production of such a salt as $UO_2(SO_4)$ from the oxide UO_3, and by the fact that the most definite and stable molybdates, tungstates, and (probably) uranates, belong to the form $XMO_4 . xMO_3 (X = K_2,$ Ba, &c.; $M = Mo,$ W, U), whereas the most marked chromates are the normal salts $XCrO_4$. In other words, the combination of a relatively large quantity of the acidic oxides MO_3 with basic oxides seems to be necessary for the production of stable salts when MO_3 is MoO_3, WO_3, or UO_3.

Thio-salts, usually of the composition $MXS_4 (M = K_2,$ Ba, &c. $X = Cr,$ Mo, W), of all the elements except uranium are known.

The salts of molybdenum and tungsten have been so little examined that no generalisations regarding them can be made; salts of chromium and uranium are numerous, many of them are basic, and several form double salts chiefly by combining with salts of the alkali and alkaline earth metals.

420 The odd-series members of Group VI. are SULPHUR, SELENION, and TELLURIUM. The properties of these elements have been already considered (Chap. XI. pars. 170 to 179); it will suffice to summarise these properties here.

Sulphur and *selenion* are distinctly non-metallic in their chemical properties; *tellurium* inclines towards the metals but it is decidedly less metallic than chromium, molybdenum, tungsten, or uranium.

The existence of stable gaseous hydrides MH_2; the distinctly acidic functions of the oxides MO_2 and MO_3; the non-existence of salts produced by replacing the hydrogen of acids by M; the existence of strong acids H_2MO_3 and H_2MO_4, each giving a series of definite salts; these among other properties, mark the non-metallic character of the elements sulphur, selenion, and tellurium.

The negative character of *sulphur* and *selenion* is further marked by the fact that these elements exhibit allotropy; by their physical properties; by the possibility of forming oxychlorides directly from the acids H_2MO_3 and H_2MO_4; by the slightly acidic functions of the hydrides MH_2.

Tellurium does not exist in more than one form; the acids H_2TeO_3 and H_2TeO_4 are not produced by the direct inter-actions of the oxides TeO_2 and TeO_3 with water; the hydride TeH_2 shews no acidic properties; the anhydride TeO_2 combines with some acids (e.g. $TeO_2 \cdot 2HCl$ is known); some of the physical properties of tellurium approximate to those of the metals.

The existence of the stable gasifiable tetrachloride $TeCl_4$; the formation not only of ditellurates MTe_2O_7, but also of tetratellurates MTe_4O_{13}, and of salts of the form MTe_2O_5 ($M = K_2$, Ba, &c.); the unreadiness to enter into chemical reaction with alkalis or alkaline carbonates of the oxide TeO_2: these are some of the properties in which tellurium approaches the higher members (W and U) of the even series of Group VI.

OXYGEN is the first member of the even-series of the group **421** now under consideration. Oxygen is a typical non-metallic or negative element both in its chemical and physical properties (*s.* Chap. VIII.). Nevertheless the properties of some of the compounds of this element suggest the properties of the other elements of the group in which oxygen occurs. Thus oxygen forms two compounds with hydrogen, water H_2O and hydrogen peroxide H_2O_2, but neither is acidic; oxygen combines with the positive elements to form oxides the composition of which is frequently similar to that of the sulphides and selenides of the same elements, compare for instance the formulae MO and MS where $M = Cu$, Fe, Mg, Ni, Co, Mn, Ca, Ba, Sr, K_2, Na_2, &c.; most of the oxides of positive elements are basic, some however as we have seen are acidic; most of the sulphides of positive elements interact with acids to form salts and hydrogen sulphide, some however interact with alkaline sulphides to form thio-salts.

The compounds of oxygen with chlorine ClO_2 and (?) Cl_2O_2 do not resemble the chlorides of the other members of Group VI. in composition, but the oxide OCl_2 is analogous in composition to MCl_2 when $M = S$, Se, Te, Cr, Mo, or W. The compounds of oxygen with chlorine are very easily decomposed by heat and reagents generally; in this they resemble the compounds of sulphur and selenion with chlorine. The existence of the gaseous molecules O_2 and O_3, Se_2 and Se_3, S_2 and S_6, emphasises the resemblance between oxygen, sulphur, and selenion. Oxygen forms a solid, stable, compound with iodine, O_5I_2; although the composition of this compound is not similar to that of the iodides of the other elements of Group VI., the fact of the existence of this stable compound suggests the existence of the stable iodides of tellurium, chromium, and tungsten. The existence of many definite and stable compounds of oxygen with non-metallic elements (oxides of As, B, C, I, N, P, Si) shews that oxygen is to some extent positive in its chemical properties : the compositions of these oxides are very frequently similar to those of the sulphides of the same elements when sulphides of these elements exist.

422   Group VI., then, is evidently divided into two well marked sub-groups or families; one of these families consists of the elements *chromium, molybdenum, tungsten,* and *uranium*; the other is formed of *sulphur, selenion,* and *tellurium*; *oxygen*, which is the first member of the group, to some extent summarises the properties of both families, but at the same time it differs from all the other members of the group. At the same time the elements of Group VI. taken as a whole more closely resemble one another than they resemble any other elements.

423   If we now turn to Group V. we shall find a group of ten elements shewing a gradation of properties from the first to the last member ; we shall find that the group-character preponderates over the family-character, so that although the even-series members are on the whole more like each other than they are like the odd-series members, yet it is not possible to divide Group V., as we have divided Group VI., into two distinct sub-groups or families.

424

	2	4	6	8	10	12
Even series.	N=14·01	V=51·2	Nb= 94	Di=144	Ta=182	—
Group V.	3	5	7	9	11	
Odd series.	P=30·96	As=74·9	Sb=120	Er=166	Bi=208	

	Nitrogen.	Vanadium.	Niobium.	Didymium*.	Tantalum.
Even-series elements					
Atomic weights	14·01	51·2	94	· 144	182

The *molecular weight* of nitrogen is 28·02; the molecular weights of the other elements are unknown.

	Nitrogen	Vanadium	Niobium	Didymium	Tantalum
Sp. grs. (approx.)	·97 if air=1; liquid; s.o.=·885	5·5	7 (?)	6·5 (?)	11 (?)
Sp. heats	—	not determined.	not determined.	·0456	not determined.
Atom. weights / spec. gravs.	—	9·3	18·4 (?)	22·3 (?)	16·5(?)

The *melting points* of most of these elements have not been determined.

	Nitrogen	Vanadium	Niobium	Didymium	Tantalum
Occurrence and preparation.	In large quantities in air. Many compounds, especially ammonia and nitrates, also occur in large quantities and widely distributed. Prepared by removing O from air.	In a few minerals, not widely distributed, chiefly as oxides and sulphides.			
		Prepared by long continued heating VCl_2 in hydrogen.	Prepared by heating $NbCl_5$ repeatedly in hydrogen.	Prepared by heating $DiCl_3$ with potassium; or by electrolysing molten $DiCl_3$.	Prepared by heating K_2TaF_7 with potassium.
Appearance, and general physical properties.	Colourless, tasteless, odourless, gas. Liquefied at very low temperature and under great pressure; liquid boils at abt. – 195°.	Grey, crystalline, powder.	Steel-grey, lustrous, solid.	White solid with slightly yellow tinge. Ductile, hard.	Grey, lustrous, solid: not yet obtained approximately pure.
General chemical properties.	Combines directly with few if any elements at ordinary temperatures; but at very high temperatures combines directly with B, Si, Cr, Mg, V, and a few other elements. If electric discharge is passed through mixture of N with O, or H, a very little NO_2, or NH_3, is formed. Strongly negative. Oxides are generally anhydrides. Hydride, NH_3, is strongly alkaline. Atom trivalent in gaseous molecules.	Burns in air to V_2O, then V_2O_3. Burns in Cl to VCl_4. Combines directly with N forming VN. Dissolved by conc. H_2SO_4, and HNO_3Aq. Molten alkalis form vanadates. Some oxides are basic, others are basic and also acidic. V acts both as a metallic and a non-metallic element. No hydride known. Atom is tetra- (and perhaps also tri-) valent in gaseous molecules.	Burns in air to Nb_2O_5. Combines directly with Cl to form $NbCl_5$. Dissolved by conc. H_2SO_4 not by $HClAq$ or HNO_3Aq. Nb_2O_5 is an anhydride. Salts of Nb are not known but the subject has not been thoroughly investigated. No hydride known with certainty. Atom pentavalent in gaseous molecules.	Burns in air to Di_2O_3. Oxides seem to act only as bases; Di_2O_3 forms $NbCl_5$ acts as a peroxide. No compounds have been gasified. No hydride known.	Burns in air to Ta_2O_5. Combines directly with Cl to form $TaCl_5$. Dissolved only by conc. HFAq or by mixture of H_2SO_4Aq and HFAq. Ta_2O_5 is an anhydride, and has apparently no basic properties. No hydride known. Atom pentavalent in gaseous molecules.

* There is some doubt whether the body known as didymium is or is not a mixture of two or more elements; many of the properties of compounds of didymium are probably the properties of mixtures.

Odd-series elements	PHOSPHORUS.	ARSENIC.	ANTIMONY.	ERBIUM.	BISMUTH.
Atomic weights	30·96	74·9	120	166	208
Molecular weights	123·84 and 61·92	299·6 and 149·8	?360 or 240	◄—————unknown—————►	
Sp. grs. (approx.)	1·9	5·7	6·7	—	9·9
Melting points (approx.)	45°	500° (under pressure)	430°	—	270°
$\dfrac{\textit{Atomic weights}}{\textit{spec. gravs.}}$	16·3	13·2	18	—	20·5
Sp. heats	·202	·083	·053	not determined.	·0308
Occurrence and preparation.	Many phosphates occur very widely distributed in rocks and waters; also in bones and in parts of plants. Compounds of P with Cl, N, and O occur in nerve and brain matter. Prepared by heating $Ca(PO_3)_2$ with charcoal.	As_2O_3 and As_2S_3 occur; also compounds of As with Te and S, with Ni, Co, &c. occur in small quantities widely distributed. Prepared by heating As_2O_3 with charcoal.	Chief naturally occurring compound is Sb_2S_3; found in comparatively small quantities in very various parts. Prepared by heating Sb_2O_3 with charcoal.	Er_2O_3 occurs in small quantities in *ytterbite*, a Swedish mineral. Metal not yet obtained.	Bismuth is found native; also Bi_2O_3 and Bi_2S_3, &c., but not in large quantities. Prepared by heating Bi_2O_3 with charcoal.
Appearance, and general physical properties.	Soft wax-like solid. Crystalline. Non-conductor of electricity.	Grey, hard, brittle, solid. Crystalline. Fair conductor of electricity.	Grey, lustrous, brittle, very crystalline, solid. Fair conductor of electricity.	—	Grey, with faintly reddish tinge; crystallises easily; brittle. Bad conductor of electricity.
General chemical properties.	Burns in air to P_2O_3 and P_2O_5. Combines directly with Cl, Br, and I. Oxidised by HNO_3. Does not form salts by interacting with acids. Hydride, PH_3, resembles NH_3 but is less alkaline. Oxides are anhydrides. Atom trivalent and pentavalent in gaseous molecules. Phosphorus shews allotropy (*s.* Chap. XI., par. 220).	Burns in air to As_2O_3. Combines directly with Cl, Br, and I. Oxidised by HNO_3. No salts known derived from acids. Hydride, AsH_3, is scarcely if at all alkaline. Oxides are anhydrides. Atom trivalent in gaseous molecules. Arsenic exists in allotropic forms (*s.* Chap. XI., par. 222).	Burns in air to Sb_2O_3. Combines directly with Cl, Br, and I. Oxidised by HNO_3. A few salts derived from acids seem to exist. Hydride, SbH_3, not alkaline. Oxides are anhydrides, but also slightly basic. Atom trivalent in gaseous molecules.	Oxide, Er_2O_3, is basic, forming salts, e.g. $Er_2 3SO_4$. No hydride known. No oxide known to act as an anhydride. No compound yet gasified.	Burns in air to Bi_2O_3. Combines directly with Cl, Br, and I. Oxidised by HNO_3, and at same time $Bi.3NO_3$ is formed. Many salts known derived from acids. No hydride known. Oxides are basic; Bi_2O_5 shews very slight acidic functions. Atom trivalent in gaseous molecules.

425 **General formulae and chemical characters of compounds of elements of Group V.**

Hydrides. MH_3; M = N, P, As, Sb.

Oxides. M_2O; M = N, V. M_2O_2; M = N, V, Nb, Bi. M_2O_3; M = any metal of the group except Nb and Ta. M_2O_4; M = N, P, V, Nb, Sb, Ta, Bi. M_2O_5; M = any element of the group except Er. Hydrates of many of these oxides are known.

Sulphides. No sulphides of niobium or erbium have yet been prepared. $M_2S_4^{v'}$; M = any element of the group except N, Ta, (Nb or Er). M_2S_5; M = P, V, As, Sb. A few other sulphides are known, e.g. N_2S_3, Ta_2S_4.

Haloid compounds. No haloid compounds of erbium have yet been prepared; the haloid compounds of nitrogen are very unstable and explode with great violence, their composition is still doubtful.

· MX_3; M = any element of the group except Ta (? N) or Er. MX_5; M = P, Nb, Sb, Ta; X = Cl and in some cases also Br or I. A few other haloid compounds exist, e.g. P_2I_4, VCl_4.

Acids. The following are the chief compounds of hydrogen with oxygen and an element of Group V. which are acids (when Aq is added to a formula it indicates that the acid is known only in aqueous solution).

$HN\Theta Aq$, HNO_2Aq, HNO_3. H_3PO_2Aq, H_3PO_3Aq, H_3PO_4, HPO_3, $H_4P_2O_7$, HVO_3, $H_4V_2O_7$. ? H_3AsO_3Aq, $HAsO_2$, H_4AsO_4, $H_4As_2O_7$. H_3SbO_3, $HSbO_3$, H_3SbO_4, $H_4Sb_2O_7$. $H_4Ta_2O_7$.

Salts. The only elements of the group which are known to form series of definite salts by replacing the hydrogen of acids are vanadium, didymium, erbium, and bismuth. The salts of vanadium are all basic salts, they belong for the most part to the series VOX and $(VO)_2X_3$; the salts of didymium and erbium belong to the series M_23X; the normal salts of bismuth belong to the series Bi_23X, but besides these a great many basic salts exist. $(X = SO_4, 2NO_3, \&c.)$. Antimony forms a few compounds which may be regarded as basic salts, especially $xSb_2O_3.ySO_3$ and $xSb_2O_3.yN_2O_5$; tartar-emetic, $KSbC_4H_4O_7$, is probably to be classed as a double antimony-potassium salt.

The following are the principal compounds of elements of **426** Group V. which have been gasified, and the relative densities of the vapours of which have been determined; NH_3, PH_3, AsH_3, PCl_3, PI_3, PF_5, VCl_4, $AsCl_3$, AsI_3, $NbCl_5$, $SbCl_3$, SbI_3, $TaCl_5$, $BiCl_3$. The statements made in the table in par. 424 regarding the valencies of the atoms of the elements of Group V. are based on the existence and compositions of these gaseous molecules.

Arsenious and antimonious oxides have been gasified; the

compositions of the gaseous molecules of these compounds are expressed by the formulae As_4O_6 and Sb_4O_6. The formulae of the following oxides of nitrogen are molecular N_2O, NO, NO_2, N_2O_4. A few oxychlorides, e.g. $NbOCl_3$, have also been gasified.

The formulae of the other compounds are not necessarily molecular; they are the simplest formulae that can be given, consistently with the determined values of the atomic weights of the elements, and with the reactions of the compounds.

427 The hydrides MH_3 are gases under ordinary conditions.

Ammonia, NH_3, is obtained in small quantities by the direct union of nitrogen and hydrogen under the influence of the induced electric discharge; also in many reactions in which hydrogen is produced in contact with nitrogen, e.g. when steam and nitrogen are passed over hot iron, or when nitrogen is produced in contact with hydrogen, e.g. when hydrogen and nitric oxide are passed over hot finely divided platinum. Ammonia is usually prepared by heating a mixture of ammonium chloride and lime;

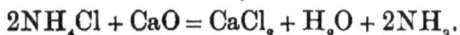

$$2NH_4Cl + CaO = CaCl_2 + H_2O + 2NH_3.$$

Phosphoretted hydrogen or *phosphine*, PH_3, may be obtained by heating a mixture of phosphonium iodide and an alkali; $PH_4I + KOHAq = KIAq + H_2O + PH_3$; it is however more usually prepared by heating a solution of an alkali with phosphorus; $3KOHAq + 3H_2O + 4P = 3KH_2PO_2Aq + PH_3$ (potassium hypophosphite remains in solution, phosphine is evolved as a gas).

Arsenuretted hydrogen or *arsine*, AsH_3, is obtained by producing hydrogen in contact with a solution of an arsenic compound. *Antimonuretted hydrogen*, or *stibine*, SbH_3, is obtained by a similar method. Thus,

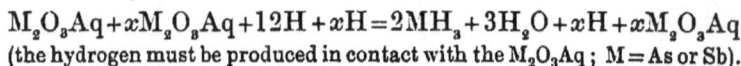

$$M_2O_3Aq + xM_2O_3Aq + 12H + xH = 2MH_3 + 3H_2O + xH + xM_2O_3Aq$$

(the hydrogen must be produced in contact with the M_2O_3Aq; $M = As$ or Sb).

Ammonia, NH_3, combines with most acids to form *ammonium salts*; e.g. $NH_3 + HCl = NH_4Cl$; $2NH_3 + H_2SO_4 = (NH_4)_2SO_4$. Phosphine, PH_3, combines with hydriodic acid, and with hydrochloric acid under increased pressure, to form *phosphonium salts*; e.g. $PH_3 + HI = PH_4I$. (Regarding ammonium salts s. Chap. XI. pars. 210, 211, and Chap. XVII. par. 371.) Arsine and stibine do not combine with acids.

The hydrides MH_3 can be oxidised to oxides and water; ammonia is oxidised by mixing it with much oxygen and bringing a flame to the mixture; water, nitrogen, and a little ammonium nitrate (NH_4NO_3) are formed; phosphine is oxidised to phosphorus pentoxide (P_2O_5) and water by mixing it with air or oxygen and raising the temperature; arsine and stibine are oxidised by mixing with air or oxygen and bringing a flame to the mixture; if much oxygen is present the products are water and arsenious or antimonious oxide; if little oxygen is present, water, arsenic or antimony, and a little arsenious or antimonious oxide, are formed.

Arsine and stibine are decomposed by heat; ammonia and phosphine are much more stable towards heat.

The most important and best studied oxides of the elements of Group V. are those whose composition is expressed by the formulae M_2O_3 and M_2O_5, respectively. **428**

The *trioxides* M_2O_3 may be formed by the direct union of their elements; Nb_2O_3 and Ta_2O_3 have not yet been prepared, and as erbium has not been isolated it is uncertain whether the oxide Er_2O_3 would or would not be formed by heating erbium in oxygen. Nitrogen and oxygen combine only when a mixture of the gases is submitted to the continued action of electric sparks, and then only a small quantity of N_2O_3 is formed; the other elements of the group are readily oxidised to M_2O_3 by heating in air or oxygen.

The *oxides* M_2O_3 may be divided into three classes: (1) acidic oxides, N_2O_3, P_2O_3; (2) basic oxides, Di_2O_3, Er_2O_3, Bi_2O_3; (3) oxides which are both acidic and basic, As_2O_3, Sb_2O_3, (? V_2O_3). The oxides N_2O_3 and P_2O_3 dissolve in water to form solutions of nitrous acid HNO_2, and phosphorous acid H_3PO_2, respectively; neither oxide shews the smallest tendency to interact with acids and form salts. The oxides Di_2O_3, Er_2O_3, and Bi_2O_3 interact with acids to produce salts; they are all insoluble in, and unchanged by contact with, water. *Arsenious oxide*, As_2O_3, dissolves in water and the solution interacts with caustic soda or potash to form salts of the composition M_3AsO_3 (M = Na or K); but no oxyacid derived from As_2O_3 has been obtained. *Antimonious oxide*, Sb_2O_3, is slightly soluble in water and the solution reacts with potash or soda to form antimonites M_3SbO_3; the acid H_3SbO_3 is known as a solid, but it is not obtained by the direct interaction of water with Sb_2O_3. *Arsenious oxide* interacts with concentrated hydrochloric acid to form arsenious chloride, $AsCl_3$; it is also said to form a salt $AsKC_4H_4O_7$ by reacting with solution of potassium-hydrogen tartrate; *antimonious oxide* forms $SbCl_3$ by reacting with concentrated hydrochloric acid, it forms $SbKC_4H_4O_7$ by dissolving in $KHC_4H_4O_6Aq$ and crystallising, and it probably interacts with concentrated sulphuric acid to form $Sb_2(SO_4)_3$. *Vanadium trioxide* V_2O_3 is said to interact with acids to produce a series of unstable, easily oxidised, salts.

The *pentoxides*, M_2O_5, can be obtained by heating the elements in oxygen when M = P, V, Nb, or Ta; by evolving oxygen in contact with M_2O_3 when M = As, Sb, Di, or Bi; and by withdrawing water from nitric acid in the case of N_2O_5 ($2HNO_3 - H_2O = N_2O_5$).

The oxides N_2O_5 and P_2O_5 dissolve in water to form acid solutions from which the acids HNO_3 (from N_2O_5), HPO_3 H_3PO_4 and $H_4P_2O_7$ (from P_2O_5) can be obtained. The oxides As_2O_5 and Sb_2O_5 are slightly soluble in water forming acid solutions from which salts can be obtained, e.g. Ag_3AsO_4, $KSbO_3$; V_2O_5 is scarcely soluble, Nb_2O_5, Ta_2O_5, Di_2O_5, and Bi_2O_5 are insoluble, in water; salts of the forms MVO_3, $MTaO_3$, and $M_4Nb_2O_7$ &c. are obtained by heating the pentoxides of V, Ta, and Nb, with molten potash or soda; no salts derived from Di_2O_5 or Bi_2O_5 are known; Bi_2O_5 however dissolves in much molten potash and possibly forms very easily decomposed salts.

Of the remaining oxides, *nitrous oxide* N_2O, *nitric oxide* NO, *nitrogen dioxide* NO_2, and *nitrogen tetroxide* N_2O_4, are gases; they are all obtainable by reactions between aqueous solutions of nitric acid and various reducing agents; the oxide NO_2 exists at moderately high temperatures, when this oxide is cooled it becomes very dark reddish black in colour and the vapour density now shews that its molecular composition is expressed by the formula N_2O_4. All the oxides of nitrogen are acidic; none shews any basic properties.

429   The haloid compounds are generally obtained by the direct union of their elements; also in many cases by dissolving the trioxides, M_2O_3, in haloid acid and evaporating. Those elements of the group which form *pentachlorides* MCl_5, are P, Nb, Sb, and Ta; the compounds MCl_5 are produced by heating these elements in excess of chlorine. *Trichlorides* MCl_3, are formed by heating arsenic or bismuth in chlorine; vanadium heated in a stream of chlorine forms the tetrachloride VCl_4. Pentahaloid compounds (MX_5) of nitrogen, vanadium, arsenic, didymium, and bismuth, have not yet been obtained; no trihaloid compound (MX_3) of tantalum is known; haloid compounds of erbium have not as yet been isolated. Vanadium is the only member of the group which forms a haloid compound of the form MX_4 stable in the state of gas. Niobium and tantalum pentachlorides can be gasified; the pentachlorides of phosphorus and antimony are dissociated to $MCl_3 + Cl_2$ when heated.

The haloid compounds are all decomposed by water, generally forming solutions of haloid acid and the hydrated oxide of the element; thus PCl_3 gives H_3PO_3Aq, $AsCl_3$ gives As_2O_3Aq, VCl_4 gives V_2O_4Aq, $NbCl_5$ gives $Nb_2O_5.xH_2O$, $TaCl_5$ gives $Ta_2O_5.xH_2O$, $SbCl_3$ gives Sb_2O_3Aq if much warm water is used but solid $SbOCl$ if less water is added,—in each case $HClAq$ is also formed; $BiCl_3$ gives $BiOCl$ and $HClAq$ whatever be the quantity, or the temperature, of the water employed.

Many *oxyhaloid compounds* of Group V. exist; the more important belong to the forms $MOCl$ and $MOCl_3$.

The sulphides, M_2S_3, of the lower members of the group **430** generally interact with alkaline sulphides to form thio- (or sulpho-) salts; thus As_2S_3 dissolves in ammonium sulphide solution to form a solution of ammonium thio-arsenite, $(NH_4)_3AsS_3$; Sb_2S_3 under similar conditions forms a solution of NH_4SbS_2. The best studied and apparently most distinctly acidic sulphides are As_2S_3 and Sb_2S_3; P_2S_3 and V_2S_3 also interact with alkaline sulphides to form thiosalts. Ta_2S_3 is not known, but Ta_2S_4 is said not to react with alkaline sulphides. Bi_2S_3 shews no acidic properties. Nitrogen forms the sulphide N_2S_2; it is prepared by passing ammonia into sulphur dichloride (SCl_2); it is very easily decomposed by heat. Phosphorus forms several sulphides; P_4S, P_2S, P_2S_3, P_2S_5, &c.; they are generally obtained by direct union of phosphorus and sulphur. The highest sulphide of vanadium V_2S_5 is formed by passing sulphuretted hydrogen over hot V_2O_5; Ta_2S_4 is produced by passing carbon disulphide (CS_2) over hot Ta_2O_5. The sulphides As_2S_3, Sb_2S_3, and Bi_2S_3 are produced as solid precipitates when sulphuretted hydrogen is passed into acidulated solutions of the oxides M_2O_3.

The acids formed by the combination of elements of **431** Group V. with hydrogen and oxygen are numerous and important. The following table shews the compositions of the best marked of these acids, and exhibits the relations of composition between them and their corresponding oxides.

Acid.	Corresponding Oxide.	Remarks.
$HNOAq$*	N_2O	Acid not formed from oxide; but aqueous solution of acid boiled gives the oxide.
HNO_2Aq	N_2O_3	Oxide dissolves in water to form a solution of the acid.
HNO_3	N_2O_5	Oxide reacts with water to form the acid; oxide also obtained from the acid by withdrawing water, but not by action of heat.
$H.H_2PO_2Aq$	—	No oxide known corresponding to acid.
$H_2.HPO_3Aq$ $HPO_3,\ H_3PO_4,\ H_4P_2O_7$	P_2O_3 P_2O_5	Oxides interact with water to form the acids; oxides not obtained by heating the acids.
$HVO_3,\ H_4V_2O_7$ (also salts of form M_3VO_4)	V_2O_5	Acids not obtained directly from oxide; oxide obtained by heating the acids.
Salts of form $MAsO_2$ and M_3AsO_3	As_2O_3	Solution of oxide in water interacts with alkalis &c. to form salts.
$HAsO_3,\ H_3AsO_4,\ H_4As_2O_7$	As_2O_5	Acids not obtained from the oxide by action of water; oxide is obtained by heating the acids.
No definite acid of Nb, only hydrates $Nb_2O_5.xH_2O$; most salts are complex, may be represented as $xNb_2O_5.yRO$ where $RO=K_2O,\ CaO,$ &c.	Nb_2O_5	Oxide obtained by heating the hydrates; hydrates not obtained directly from the oxide.
H_3SbO_3 (also salts of form $MSbO_2$)	Sb_2O_3	Acid not obtained directly from oxide; oxide formed by heating the acid.
$HSbO_3,\ H_3SbO_4,\ H_4Sb_2O_7$	Sb_2O_5	Acids not obtained directly from oxide; oxide formed by heating the acids.
$H_4Ta_2O_7$ (also salts of form $MTaO_3$ and various complex salts $xTa_2O_5.yRO$)	Ta_2O_5	Acid not obtained directly from oxide; oxide produced by heating the acid.

No acids, or salts derived from acids, of didymium, erbium, or bismuth, have been isolated.

* The symbol Aq here signifies that the acid to the formula of which it is added exists only in aqueous solution, and has not been isolated as a solid.

Hyponitrous acid (HNO) and *nitrous acid* (HNO_2) act as reducing agents; they readily combine with oxygen to produce nitric acid. *Nitric acid* on the other hand is very frequently used as an oxidiser; when heated it is decomposed to water, oxygen, and nitrogen dioxide ($2HNO_3 = H_2O + 2NO_2 + O$). All the nitrogen acids are monobasic.

When aqueous solutions of *hypophosphorous acid* (H_3PO_2) and *phosphorous acid* (H_3PO_3) are boiled, phosphine (PH_3) is evolved, and phosphoric acid (H_3PO_4) remains in solution. The *three phosphoric acids*, *ortho*- H_3PO_4, *meta*- HPO_3, and *pyro*- $H_4P_2O_7$, may be formed by adding water to phosphoric anhydride P_2O_5 (*s.* Chap. XI., par. 215). Hypophosphorous acid is monobasic, and phosphorous acid is dibasic. Of the three phosphoric acids, orthophosphoric H_3PO_4, forms the largest number of definite salts: sodium pyrophosphate, $Na_4P_2O_7$, is obtained by heating ordinary sodium phosphate ($2Na_2HPO_4 = H_2O + Na_4P_2O_7$); sodium metaphosphate, $NaPO_3$, may be obtained by heating sodium-ammonium phosphate ($Na(NH_4)HPO_4 = H_2O + NH_3 + NaPO_3$). When orthophosphoric acid is heated to $230°$ or so pyrophosphoric acid is obtained ($2H_3PO_4 = H_2O + H_4P_2O_7$), at a red heat metaphosphoric acid is produced ($H_3PO_4 = H_2O + HPO_3$). When meta- or pyro-phosphoric acid is boiled with water orthophosphoric acid is produced.

Metavanadic acid HVO_3, and *pyrovanadic acid* $H_4V_2O_7$, are prepared, indirectly, from salts of these acids. Besides the salts of these acids, numerous *polyvanadates* (or condensed vanadates) exist; the following are given as examples, $Na_2V_4O_{11}$, SrV_6O_{16}.

No *arsenious acid* has been isolated. An aqueous solution of arsenious oxide (As_2O_3) may contain arsenious acid; when this solution is neutralised with soda the salt $NaAsO_2$ is obtained; by adding silver nitrate to an aqueous solution of As_2O_3, Ag_3AsO_3 is precipitated. The *arsenites* are unstable salts, their composition seems to change with small variations in the conditions of their formation. *Arsenic acid*, H_3AsO_4, is formed by oxidising As_2O_3 in presence of water, either by nitric acid or by chlorine, and crystallising. This acid loses water at $150°$ or so with formation of *pyro-arsenic acid* ($2H_3AsO_4 = H_2O + H_4As_2O_7$), and at about $210°$ water is again evolved and *meta-arsenic acid* remains

$$(H_4As_2O_7 = H_2O + 2HAsO_3).$$

Both meta- and pyro-arsenic acids are at once changed to the ortho-acid by solution in water.

Antimonious oxide dissolves in hot caustic soda solution; from this solution the salt $NaSbO_2$ is obtained on cooling. *Ortho-antimonic acid* H_3SbO_3 is produced, indirectly, from tartar emetic, $KSbC_4H_4O_7$. The *antimonites* are easily oxidised; they are unstable and easily undergo change.

Antimonic acid, H_3SbO_4, is obtained by adding a little water to $SbCl_5$ and drying the solid thus obtained over sulphuric acid; at 100° water is evolved and *pyro-antimonic acid*, $H_4Sb_2O_7$, remains, and at 200° this acid again loses water with production of *meta-antimonic acid*, $HSbO_3$. These acids seem all to exist in aqueous solution; salts derived from $HSbO_3$ and $H_4Sb_2O_7$, but not from H_3SbO_4, are known.

432 Salts are obtained by replacing the hydrogen of various acids by the elements *vanadium*, *didymium*, *erbium*, or *bismuth*. The *salts of didymium* and *erbium* have not been much studied; they seem to belong to the form M_23X, where $X = SO_4$, $2NO_3$, &c. *Vanadium* pentoxide, V_2O_5, interacts with alkalis to produce *salts* of the form M_3VO_4; but it also interacts with sulphuric acid, and with a few other acids, to produce *basic salts*, e.g. $(VO)_2(SO_4)_3$. Vanadium tetroxide, V_2O_4, also interacts with sulphuric acid to form the salt $VO.SO_4$.

Bismuthous oxide, Bi_2O_3, forms a series of salts by interacting with acids; these *salts* belong to the form Bi_23X ($X = SO_4$, $2NO_3$, $\frac{2}{3}PO_4$, &c.), e.g. Bi_23SO_4, Bi_26NO_3, $BiPO_4$, $BiAsO_4$. Most of these salts are decomposed by water with formation of *basic salts;* the composition of some of these basic salts is represented by the general formula $BiOX$ where $X = NO_3$, $\frac{SO_4}{2}$, &c.; but many others are more complex, and their composition can be expressed only by such a formula as $xBi_2O_3.yR.zH_2O$ where $R =$ an acidic oxide, e.g. N_2O_5, SO_3, &c.

433 The elements of Group V. shew a gradual change of properties from the decidedly negative nitrogen to the metallic bismuth. Various small sub-classes appear in the group; thus arsenic and antimony are very closely related; so are niobium and tantalum; nitrogen and phosphorus also are very similar in many chemical properties. Vanadium appears to be more distinctly metallic in its chemical properties than the elements which succeed it in the even-series, but it is to be remembered that the properties of these elements have been very imperfectly investigated. The group cannot be divided into two families comprising the even-series and odd-series elements, respectively. Although it cannot be said that all the elements shew marked similarities, yet the group-character is impressed on them all. All the elements of this group are distinctly more like each other than they are like the elements of any other group.

Group I. presents a number of elements some of which are **434** very similar in their chemical properties, while others are so different from these and from one another that it seems at first sight quite a mistake to place them in the same group.

CHAPTER XXII.

THE ELEMENTS OF GROUP I.

	2	4	6	8	10	12
Group I. Even series.	Li=7·01	K=39·04	Rb=85·2	Cs=132·7	—	—
	1	**3**	**5**	**7**	**9**	**11**
Odd series.	H=1	Na=23	Cu=63·2	Ag=107·66	—	Au=197

	LITHIUM.	SODIUM.	POTASSIUM.	RUBIDIUM.	CAESIUM.
Even-series elements, and second member of odd-series					
Atomic weights	7·01	23	39·04	85·2	132·7
	The *molecular weights* of sodium and potassium are the same as their atomic weights; the other elements of the family have not yet been gasified, and therefore their molecular weights are unknown.				
Sp. grs. (approx.)	·59	·98	·87	1·52	1·88
Melting points (approx.)	180°	95°·5	58°—62°	38°	26°—27°
$\dfrac{\text{Atom. weights}}{\text{spec. gravs.}}$	11·9	23·5	44·9	56·1	90·6
Sp. heats	·941	·293	·166	not determined.	not determined.
Appearance and general physical properties.	Silver-white; very soft; fairly ductile; not very tenacious; not volatile at red-heat.	Silver-white; soft; very ductile at 0°; can be distilled at red-heat.	White; soft; brittle at 0°; malleable at abt. 5°; pasty at 15°; can be distilled at 700°—800°.	Silver-white; soft as wax at −10°	Silver-white; soft.
Occurrence and preparation.	Silicate and phosphate occur with same salts of other metals of the family. Compounds are widely distributed, but in small quantities, in rocks, water, plants, and some animal secretions. Prepared by electrolysing fused mixture of LiCl and NH₄Cl.	Chloride, silicate, fluoride, nitrate, &c., occur in large quantities widely distributed. Prepared by deoxidising Na₂CO₃ by hot carbon.	Nitrate, sulphate, silicate, &c. occur in large quantities widely distributed. Prepared by deoxidising K₂CO₃ by hot carbon.	Compounds occur very widely distributed, but in very small quantities, in most minerals containing salts of K and Na. Prepared by deoxidising Rb₂CO₃ by hot carbon.	Silicate occurs as a rare mineral. Minute quantities of some compounds occur in many rocks and waters. Prepared by electrolysing fused caesium-barium cyanide.
General chemical properties.	Combines directly with oxygen, but not so rapidly as other elements of the family. Decomposes cold water giving LiOHAq and H.	Oxidises rapidly in air. Decomposes cold water rapidly with evolution of H and production of NaOHAq.	Oxidises very rapidly in air. Decomposes cold water very rapidly with production of KOHAq and H; H usually takes fire.	Oxidises so rapidly in air that usually takes fire. Very rapidly decomposes cold water, giving RbOHAq and H.	Exceedingly rapidly and completely oxidised in air. Properties not yet much investigated.

These metals all combine directly and rapidly with the halogens, and with sulphur. The following compounds of these metals have been gasified and their molecular weights determined, KI, RbI, RbCl, RbI, CsCl, CsI; in these molecules the atoms of potassium, rubidium, and caesium are monovalent.

The five elements we are now considering form the family **436** of *alkali metals ;* the prominent chemical characteristics of these metals have been already examined in Chap. XI. pars. 163—168. It will suffice to summarise these characteristics here.

Oxides and hydroxides, M_2O and MOH, are strongly basic and alkaline; very soluble in water, M_2O forming $MOHAq$. The hydroxides are formed at ordinary temperatures by direct interaction of oxides M_2O with water; they are not decomposed by heat alone. Oxides of rubidium and caesium have not yet been isolated. The oxides Na_2O_2, K_2O_4, and a few others, are known.

Sulphides and hydrosulphides, M_2S and MSH, are strongly basic; they interact with many more negative sulphides to form thio-salts. No sulphide of rubidium or caesium has yet been isolated. M_2S_2, M_2S_3, M_2S_4, M_2S_5, are known, where $M = Na$ or K.

Haloid compounds, MX, are very stable solids, soluble in, and not decomposed by, water. The chlorides, except LiCl, form many double compounds with chlorides of less positive elements, e.g.

$$PtCl_4 . 2MCl; \quad SbCl_3 . 6MCl.$$

Salts, M_2X where $X = SO_4$, $2NO_3$, CO_3, $2ClO_3$, $\frac{2}{3}PO_4$, &c. are very definite, stable, bodies; very few basic salts exist. Many of the salts combine with similar salts of less positive elements forming double salts; the *alums* $M_2SO_4 . X_2 . 3SO_4 . 24H_2O$ are important (X = Al, Cr, Fe, Ga, In). Lithium does not form an alum. Most of the salts are easily soluble in water.

Lithium is less like the other members of the family than they are like each other. LiOH is much less soluble in water than the other hydroxides; Li_2CO_3 and Li_3PO_4 are also much less soluble than the other carbonates and phosphates; Li_2SO_4 does not combine with $Al_2 . 3SO_4$, a double salt $3Li_2SO_4 . Cr_2(SO_4)_3$ is known, but it is not an alum. All the elements of the family except lithium form sulphates of the form $MHSO_4$; $Li_2H_2(SO_4)_2$ is known.

The odd-series elements of Group I. shew great differences **437** in their physical and chemical properties; sodium is evidently closely allied to the even-series members; hydrogen must be considered apart from the other elements; copper, silver, and gold exhibit resemblances but also marked differences.

Odd-series elements (hydrogen and sodium omitted)	COPPER.	SILVER.	GOLD.
Atomic weights	63·2	107·66	197
	The *molecular weights* of these elements are unknown.		
Sp. grs. (approx.)	8·8	10·5	19·5
Melting points (approx.)	1050°	1000°	1200°
Sp. heats	·095	·057	·0324
$\dfrac{Atomic\ weights}{spec.\ grvs.}$	7·2	10·3	10·1
Appearance and general physical properties.	Heavy, lustrous, reddish, solid; very tenacious and ductile; malleable; very good conductor of heat and electricity. Crystallises in regular octahedra.	White, very lustrous, hard, solid; very malleable, tenacious, and ductile; good conductor of heat and electricity. Crystallises in regular octahedra.	Yellow-red, lustrous, rather soft, solid; extremely malleable and ductile; very tenacious; good conductor of heat and electricity.
Occurrence and preparation.	Metal occurs native; also as sulphide &c., frequently with similar compounds of silver, iron, &c. Fairly widely distributed and in considerable quantities. Prepared by roasting Cu_2S in air till mixture of CuO and Cu_2S is produced, then shutting off air and raising temperature, when $Cu_2S + 2CuO = 4Cu + SO_2$.	Metal occurs native; also as sulphide frequently with Cu_2S, Sb_2S_3, &c.; AgCl also occurs. Most lead ores contain small quantities of silver. Widely distributed, generally in comparatively small quantities. Prepared by heating Ag_2S with ·salt in air $(Ag_2S + 2NaCl + 4O = Na_2SO_4 + 2AgCl)$, and then agitating with iron $(2AgCl + Fe = FeCl_2 + 2Ag)$.	Metal occurs native; generally alloyed with silver, and frequently also with small quantities of copper and iron. Compounds of gold do not occur. Widely distributed, generally in small quantities. Preparing by washing away gangue &c., sometimes by mixing crushed auriferous quartz with mercury whereby a gold-amalgam is formed, and then removing Hg by heating.
General chemical properties.	Oxidised to CuO by strongly heating in air. Combines directly with the halogens and with sulphur, at moderately high temperatures. Interacts with many acids to form salts. Decomposes steam slowly at full red-heat giving CuO and H.	Slowly oxidised by direct union with oxygen at very high temperatures. Molten silver absorbs oxygen but the gas is evolved again as the metal cools. Combines directly at moderate temperatures with the halogens and with sulphur. Interacts with many acids to form salts. Does not react with steam.	Not oxidised by direct union with oxygen. Combines directly with the halogens at moderately high temperatures, but not with sulphur. Does not interact directly with many acids, insoluble in most acids; but a few salts are obtained from the oxide. No reaction with steam.

438   **General formulae and chemical characters of compounds.** The only compounds which have been gasified and the vapour-densities of which have been determined are Cu_2Cl_2 and $AgCl$. The following formulae are not necessarily molecular.

Copper forms two series of compounds, represented by the following; (1) Cu_2O, Cu_2Cl_2, Cu_2SO_3; (2) CuO, $CuCl_2$, $CuSO_4$. Silver forms one series of compounds represented by Ag_2O, Ag_2Cl_2, Ag_2SO_4.

Gold forms two series, represented by, (1) Au_2O, Au_2Cl_2, (2) Au_2O_3, Au_2Cl_6 or $AuCl_3$.

A hydride of copper, Cu_2H_2, is known.

Oxides. M_2O where $M = Cu$, Ag, Au; MO where $M = Cu$, Ag, Au; M_2O_3 where $M = Au$ only.

Sulphides. M_2S where $M = Cu$ or Ag; MS where $M = Cu$; M_2S_3 where $M = Au$.

Haloid compounds. M_2Cl_2 where $M = Cu$, Ag, Au; MCl_2 where $M = Cu$; MCl_3 where $M = Au$.

Salts. Most of the copper salts belong to the form CuX where $X = SO_4$, $2NO_3$, CO_3, $\frac{2}{3}PO_4$, $2ClO_3$, &c.; a few are known of the form Cu_2X. The definite silver salts all belong to the form Ag_2X. Very few salts of gold are known; some seem to be similar in composition to the silver salts and to be represented by the formula Au_2X, others belong to the form Au_23X.

The compositions of the compounds of copper silver and gold may be represented by the following general formulae; $X = O$, S, Cl_2, Br_2, I_2, SO_4, $2CO_3$, $2NO_3$, $\frac{2}{3}PO_4$, &c.

$$M_2X.$$

Cu_2O, Cu_2S, Cu_2Cl_2, Cu_2I_2, and a few unstable salts e.g. Cu_2SO_3.
Ag_2O, Ag_2Cl_2, Ag_2I_2, Ag_2Br_2 &c., and all salts e.g. Ag_2SO_4.
Au_2O, Au_2Cl_2, Au_2I_2, and a very few unstable salts e.g. $Au_2S_2O_3$.

$$MX.$$

CuO, $CuCl_2$, &c., and many well-marked salts.
AgO, no salts.
AuO, $AuSO_4$.

$$M_2X_3.$$

Au_2O_3, Au_2Cl_6, $Au_2(OH)_6$, and a few salts e.g. $Au(NO_3)_3.HNO_3.3H_2O$. No Cu or Ag compounds.

The oxides M_2O are prepared; in the case of Ag_2O by **439** adding potash or soda to a solution of a salt of silver (e.g. $2AgNO_3Aq + 2KOHAq = Ag_2O + 2KNO_3Aq + H_2O$); in the cases of Cu_2O and Au_2O by reducing solutions of copper or gold salts in the presence of an alkali, $CuSO_4Aq$ may be reduced by grape sugar, and $AuCl_3Aq$ by sulphur dioxide.

Cuprous oxide Cu_2O is fairly stable, but it is easily oxidised to CuO by heating in air; with acids it forms cupric salts, e.g. $CuSO_4$; the cuprous salts, which correspond to Cu_2O, are not obtained from the oxide. *Argentous oxide*, Ag_2O, is a stable compound which interacts with acids to form salts and water; e.g. $Ag_2O + 2HNO_3Aq = 2AgNO_3Aq + H_2O$; it is decomposed at about $250°$ to silver and oxygen. This oxide is strongly basic, in some of its reactions it closely resembles the oxides of the alkali metals; an hydroxide of silver has not however been isolated. *Aurous oxide*, Au_2O, is very easily decomposed; it rapidly absorbs oxygen with formation of auric oxide Au_2O_3: one or two salts corresponding to Au_2O are known, but they are not obtained from the oxide; the oxide is said to be soluble in cold water.

The oxides MO are prepared; CuO by adding an alkali to the solution of a cupric salt, and heating the hydrated oxide $CuO.H_2O$ which is precipitated (e.g. $CuSO_4Aq + 2KOHAq = CuO.H_2O + K_2SO_4Aq$; and $CuO.H_2O$ heated, even in contact with water, $= CuO + H_2O$); AgO by passing ozonised oxygen over finely divided silver; AuO by heating hydrated auric oxide, $Au_2O_3.H_2O$, to $160°$.

Cupric oxide CuO is basic, it interacts with acids to form a large series of salts $CuX(X = SO_4$, &c.); this oxide is not decomposed by heat alone; most reducing agents remove the oxygen from heated CuO. The oxide does not combine directly with water, but the hydrate $CuO.H_2O$ is obtained, indirectly, as described above. *Argentic oxide*, or *silver peroxide*, AgO (or Ag_2O_2) is easily decomposed; heated to $110°$ it suddenly evolves oxygen and silver remains; this oxide interacts with acids to form argentous salts (e.g. Ag_2SO_4, $Ag2NO_3$, &c.) and oxygen, that is, it behaves as a peroxide.

Auro-auric oxide AuO (Au_2O_3 is usually known as auric oxide) is easily decomposed; at $173°$ it is separated into gold and oxygen; it does not appear to form any salts by reacting with acids.

Auric oxide, Au_2O_3, the only representative of the M_2O_3 class of oxides, is prepared by adding an alkali to a solution of auric chloride $AuCl_3$ (magnesia seems to be the best alkali to use), drying the precipitated $Au_2O_3.H_2O$, and then carefully heating it to $100°$. The oxide is easily decomposed by heat, or by exposure to light; it dissolves in nitric acid, and the salt $Au(NO_3)_3.HNO_3.3H_2O$ can be obtained from this solution; it interacts with hydrochloric acid to form the compound $AuCl_3.HCl$, and with hydrobromic acid to form $AuBr_3.HBr$; these compounds are monobasic acids forming salts such as $KAuCl_4$ and $KAuBr_4$. Hydrated auric oxide, $Au_2O_3.H_2O$, acts as an acid towards strong alkalis; it dissolves in concentrated KOHAq, and the salt *potassium aurate* $KAuO_2.3H_2O$ may be obtained by evaporating *in vacuo*.

The sulphides M_2S (M = Cu or Ag) are obtained by the **440** direct union of the .elements, or in the case of Ag_2S by precipitating a solution of a silver salt by hydrogen sulphide. When hydrogen sulphide is passed into a solution of a salt of copper, *cupric sulphide*, CuS, is precipitated. The *two sulphides of gold*, AuS and Au_2S_3, seem to be obtained by the interaction of hydrogen sulphide with solution of auric chloride, $AuCl_3$, under varying conditions.

The sulphides Cu_2S, CuS, and Ag_2S interact with acids to form salts and hydrogen sulphide, cupric salts being obtained from both sulphides of copper. Both sulphides of gold, AuS and Au_2S_3, dissolve in solutions of alkali sulphides to form thio-salts, the salts NaAuS and $KAuS_2$ having been obtained; these sulphides are therefore acidic.

The haloid compounds M_2X_2, M = Cu, Ag, Au, are **441** obtained; in the case of copper and gold by reducing the compounds CuX_2; in the case of silver by the interaction of haloid acids with silver oxide Ag_2O. Cu_2Cl_2 is formed by heating a solution of $CuCl_2$ in HClAq with copper, or with sulphur dioxide, and pouring into water; Au_2Cl_2 is produced by dissolving gold in *aqua regia*, evaporating to dryness, and heating cautiously. By the interaction between $AuCl_3$ and hydriodic acid, *aurous iodide*, Au_2I_2, is obtained

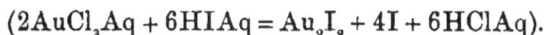

$$(2AuCl_3Aq + 6HIAq = Au_2I_2 + 4I + 6HClAq).$$

Cuprous iodide is formed by reducing $CuSO_4Aq$ by sulphur dioxide and adding potassium iodide solution.

The haloid compounds of silver, AgCl, AgBr, AgI, are very stable compounds; Cu_2Cl_2 has been gasified; Cu_2I_2 is also a stable compound; Au_2Cl_2 is separated into its elements by heat or by the action of reducing agents.

The only representatives of the class of haloid compounds MX_2 are the cupric compounds. *Cupric chloride*, $CuCl_2$, is obtained from the corresponding oxide, CuO, by dissolving in HClAq, and evaporating. This chloride combines with ammonia to form various compounds $CuCl_2.xNH_3$, it also combines with alkali chlorides to form double salts, e.g. $CuCl_2.2KCl.2H_2O$. Various *oxychlorides* are also derived from $CuCl_2$; their composition may be expressed by the general formula $xCuCl_2.yCuO$.

Gold is the only element of the family which forms haloid compounds of the form MX_3. *Auric chloride* $AuCl_3$ is best obtained by adding water to aurous chloride. Solutions of this compound are very easily decomposed by reducing agents, or by light, with separation of gold. When gold is dissolved in *aqua regia* and the solution is evaporated, crystals of the monobasic *chloro-auric acid*, $HAuCl_4$, are obtained; *bromo-auric acid*, $HAuBr_4$, has also been isolated; and from each a few definite salts have been formed, e.g. $KAuCl_4$ and $KAuBr_4$.

442   The salts of copper and silver are numerous and are very definite compounds; only a few salts of gold have been isolated, and they seem easily to undergo change.

A few *cuprous salts* are known; one of the most definite is cuprous sulphite Cu_2SO_3, prepared by adding ammonium sulphite to a solution of copper sulphate, and then passing sulphur dioxide into the solution. The *cupric salts*, CuX, e.g. $CuSO_4$, $Cu2NO_3$, $CuCO_3$, $Cu2ClO_3$, $Cu_3(PO_4)_2$, &c. are usually obtained by dissolving moist copper oxide in the various acids. Many form double compounds especially with the salts of the alkali metals, e.g. $CuSO_4.(NH_4)_2SO_4.6H_2O$. Several *basic copper salts* are known, e.g. $3CuO.N_2O_5$, $4CuO.As_2O_5$, $8CuO.2SO_3$. Many copper salts combine with ammonia, e.g. $CuSO_4.4NH_3$, $Cu(NO_3)_2.2NH_3$. The *salts of silver* all belong to the form Ag_2X, e.g. Ag_2SO_4, $AgNO_3$, $Ag_2S_2O_3$, Ag_3PO_4, Ag_2CO_3. A few *basic salts of silver* are known chiefly derived from the weaker acids, e.g. $3Ag_2O.2CO_2$; many silver salts combine with ammonia, e.g. $AgNO_3.3NH_3$, $Ag_2SO_4.4NH_3$. *Silver sulphate* forms an *alum* $Ag_2SO_4.Al_2(SO_4)_3.24H_2O$. Only a few *salts of gold* have been prepared; $Au_2S_2O_3$ is obtained combined with $Na_2S_2O_3$ by mixing concentrated solutions of sodium thiosulphate and auric chloride and adding alcohol. Hydrated auric oxide $Au_2O_3.H_2O$ dissolves in nitric acid, and from this solution the nitrates $Au(NO_3)_3.HNO_3.3H_2O$ and $AuO.NO_3$ are obtained. Sulphuric acid interacts with $AuO.NO_3$ to form the sulphate $AuO.H.SO_4$, and this on heating gives the sulphate $AuSO_4$.

Many of the compounds and salts of gold are soluble in water; e.g. $AuCl_3$, Au_2O (?), $AuSO_4$. The oxides and haloid compounds, and the greater number of the salts, of silver are insoluble in water. The oxides of copper are insoluble in water; the cupric haloid compounds are soluble, the cuprous haloid compounds are insoluble, in water; many cupric salts are soluble in water.

There are prominent differences between the three ele- **443** ments, copper, silver, and gold, placed in the odd series of Group I. Gold is marked off from the others by the great instability of its salts, by the acidic character of its hydrated oxide Au_2O_3 and its sulphide Au_2S_3, as well as by the existence of the compounds Au_2X_3 ($X = O$, Cl_2, &c.) and the acids $HAuCl_4$ and $HAuBr_4$, which find no analogues among the compounds of copper or silver. Although gold possesses the physical characters of a metal in a very marked way, yet chemically considered it rather inclines to the non-metals.

The existence and solubility in water of the oxide Au_2O, and the existence of aurous chloride $AuCl$ (or Au_2Cl_2) suggest a slight resemblance between gold and the alkali metals.

The similarities between silver and the alkali metals, although feeble, are more distinct than those between these metals and either copper or gold. The compositions of the silver compounds and of the compounds of the alkali metals are expressible by the formula M_2X, where $X = O$, Cl_2, SO_4, $2NO_3$, &c. and $M = Ag$, Li, Na, K, Rb, or Cs. In the general formula for the aluminium alums, $M_2SO_4.Al_23SO_4.24H_2O$, M may be any alkali metal except lithium, or it may represent Ag. Moist silver oxide Ag_2O may often be used, especially in organic chemistry, in place of a solution of caustic potash or soda; it reacts as an alkali. The insolubility in water of the haloid compounds, and the comparative insolubility of many salts, of silver, as well as the physical characters of the silver compounds, mark off these compounds from those of the alkali metals.

The cuprous compounds belong to the form M_2X, which includes the compounds of the alkali metals and the silver compounds, but the marked compounds of copper are the cupric compounds CuX; almost the only silver compound of this form is the unstable peroxide AgO (or Ag_2O_2); the gold compounds belonging to this form are also very few in number (AuO, $AuSO_4$).

When the compounds of gold have been more fully examined it is probable that further resemblances will be found between the chemical characters of gold and those of the other elements of Group I.

HYDROGEN is placed in Group I, series 1. The chemical **444** characters of hydrogen are peculiar. It is at once a typical metal and a typical non-metal. (For an account of some of the properties of hydrogen, *s.* Chap. VIII.) In the electrical series hydrogen stands midway between the metallic and the non-metallic elements. It forms stable compounds with all the distinctly non-metallic elements (? with boron); it also forms compounds with some of the metals and metal-like elements (e.g. Cu_2H_2, AsH_3, SbH_3). In the acids hydrogen plays the part of a metal, as it can be replaced from acids by metals with formation of compounds analogous in composition to the acids. In the hydrocarbons (CH_4, C_2H_6, C_6H_6, &c. &c.) hydrogen acts as a non-metal, as it can be replaced by non-metallic elements such as chlorine, bromine, &c. The physical properties of hydrogen are very different from those which characterise the metals as a class, yet it seems to form alloys with such metals as palladium and platinum.

To mark the peculiar chemical relationships of hydrogen this element is placed apart from the others in Series 1, of which series it is the only member.

445   In Group I, then, the family-character plainly predominates over the group-character : there is the family of the alkali metals—*lithium, sodium, potassium, rubidium,* and *caesium; silver* is more or less allied to this family; *copper, silver,* and *gold* form another family the members of which differ very considerably from each other ; for some reasons silver and copper are placed in one family, and gold is separated from them ; *hydrogen* must be placed apart from all the other elements.

It must be remembered that Group I. includes three elements which have not yet been isolated.

446   In Chap. XVIII. (par. 392) it was remarked that the properties of an element are determined by considering (1) the properties of the group to which it belongs, (2) the properties of the series in which it finds a place, (3) the position of the element in the group and in the series, (4) the relations between the properties of elements situated similarly to the given element and the properties of the other members of the groups and series to which these elements belong, and (5) the relations between the group and series in which the given element occurs and other groups and series.

Let us now apply these statements to the elements copper, silver, and gold, the positions of which in Group I. seem peculiar.

We have already considered the properties of the group to which these elements belong, and the position occupied by each element.

What are the properties of the series in which each of the three elements finds a place, and what is the position of each element in the series?

Copper is the first member of Series 5; it is succeeded by the metallic elements zinc, gallium, and germanium, which are followed by the metal-like non-metal arsenic, and the non-metallic elements selenion and bromine. Between series 4 and 5 are placed the distinctly metallic elements iron, nickel, and cobalt, forming a section of Group VIII. (s. Tables I. and II., pars. 390 and 394).

Silver is the first member of Series 7; it is succeeded by the metals cadmium, indium, and tin, which are followed by the metal-like non-metals antimony and tellurium, and the non-metallic element iodine. Silver comes immediately after the metals rhodium, ruthenium, and palladium which form a section of Group VIII.

Gold is the first member of Series 11; it is succeeded by the metals mercury, thallium, lead, and bismuth; the sixth and seventh members of this series are not yet isolated. Gold, like copper and silver, immediately succeeds a section of Group VIII., the section, namely, comprising the metals iridium, osmium, and platinum.

The positions of copper, silver, and gold are peculiar; no other known elements are similarly situated in the classificatory scheme based on the periodic law. The change from the last member of an even series to the first member of the next odd series seems to be always less sudden and abrupt than the change from the last member of an odd series to the first member of the succeeding even series. Group VIII. differs considerably from the other groups; each section of it seems to impress its character on the elements which come before and on those which succeed it. We cannot expect the relations of copper, silver, and gold, to lithium, potassium, rubidium, and caesium, to resemble the relations of zinc, cadmium, and mercury, to beryllium, calcium, strontium, and barium, except in a broad and general way. (s. also Chap. XXVI.)

If we consider the relations of group to group and series to series we shall find, speaking broadly, that in the lower groups the first member of the odd series is very like the second and succeeding members of the even series, but that this similarity becomes less and less marked as we pass to the higher groups; magnesium, for instance, closely resembles calcium, strontium, and barium, but sulphur differs considerably from chromium, molybdenum, and tungsten. We shall also find that the resemblances between the first and the succeeding even-series members of a group become on the whole less marked as we pass from lower to higher groups; beryllium, calcium, strontium, and barium, for instance, more nearly resemble each other than do nitrogen, vanadium, niobium, and didymium. Finally we shall find that the first odd-series member of a group is more like the succeeding odd-series members of the same group, when the group is one of the higher than when it is one of the lower groups; thus, the resemblances between sulphur, selenion, and tellurium, are more marked than those between magnesium, zinc, and cadmium.

If we apply these general conclusions to Group I. they would lead us to expect to find (1) marked analogies between sodium, potassium, rubidium, and caesium; (2) lithium fairly closely resembling potassium, rubidium, and caesium; (3) considerable differences between sodium on the one hand, and copper, silver, and gold, on the other hand.

The position given to copper, silver, and gold, is thus seen to be less anomalous than at first sight it appeared to be.

CHAPTER XXIII.

GROUP VII. is **unfortunately far from complete; it com- **447**
prises the four distinctly negative and non-metallic elements
fluorine, chlorine, bromine, and *iodine,* and the element *manganese* which is usually classed with the metals. We have
already considered the most important properties of chlorine,
bromine, and iodine (Chap. XI. pars. 150—159), and also of •
manganese (Chap. XI. pars. 194—203); it remains therefore
to consider fluorine, and to summarise the properties of all the
elements of the group.

FLUORINE. This element is not obtained by a process similar **448**
to that whereby chlorine, bromine, and iodine are separated from
their compounds. When liquid hydrogen fluoride is electrolysed at a low temperature, a colourless gas is evolved at the
positive pole; crystallised silicon and boron burn in this gas
to SiF_4 and BF_3, respectively; the gas interacts with water
to form ozone and a solution of hydrofluoric acid. This gas is
very probably fluorine.

The chief naturally occurring fluorine compound is *fluorspar*
which is more or less pure calcium fluoride, CaF_2. The compositions of many fluorine compounds are similar to those of
the compounds of chlorine, bromine, and iodine; thus HF,
BF_3, SbF_3, BiOF, CrO_2F_2, &c. are analogous to HCl, BBr_3, SbI_3,
BiOCl, CrO_2Cl_2, &c. In some cases a stable fluoride is known
to which there is no corresponding chloride, bromide, or
iodide; thus PF_5 exists as a gas, but the highest gasifiable
chloride of phosphorus is PCl_3. No oxide or oxyacid of
fluorine has yet been obtained; but the reactions of the
element itself have scarcely been examined as it has only
recently been isolated.   .

Hydrogen fluoride, HF, is prepared by the interaction **449** of sulphuric acid with calcium fluoride CaF_2; thus $CaF_2 + H_2SO_4Aq = CaSO_4 + 2HFAq$ (compare preparation of HCl, HBr, and HI, Chap. XI. par. 153). Hydrogen fluoride is a colourless strongly smelling and irritating gas at temps. above 20°, and a light mobile liquid at temps. under 20°. The vapour-density points to the existence of the gaseous molecule HF only at fairly high temperatures, and to the existence of the gaseous molecule H_2F_2 at temps. not very far above 20°. Liquid hydrogen fluoride chars organic matter rapidly, and dissolves many bodies which are insoluble in all other acids, e.g. strongly heated silica, titanium oxide, boron, silicon, &c. An aqueous solution of this compound interacts with metals and basic oxides similarly to aqueous solutions of hydrochloric, hydrobromic, and hydriodic, acids; fluorides, salts of the form MF ($M = K$, Na, $\dfrac{Ba}{2}$, $\dfrac{Bi}{3}$, &c.), are produced.

The metallic *fluorides* shew great readiness to combine with hydrogen fluoride and produce double compounds; e.g. $KF.HF$,
•　$BiF_3.3HF$, &c. Some of the fluorides of non-metallic elements also combine with hydrogen fluoride; the products in some cases react as acids; thus $SiF_4.2HF$ is a dibasic acid (H_2SiF_6, *fluosilicic acid*), and $BF_3.HF$ is a monobasic acid (HBF_4, *fluoboric acid*). A few similar compounds of hydrogen chloride and bromide are known, e.g. $HAuCl_4$ and $HAuBr_4$, both of which react as monobasic acids (s. par. 441.)

Hydrofluoric acid, HFAq, is an extremely weak acid; its affinity for bases is less than ·5 when that of hydrochloric acid is taken as 100 (s. Chap. XIII. pars. 251, 255).

450　　Whether fluorine does or does not interact with water and solutions of alkalis similarly to chlorine, bromine, and iodine, cannot be determined until the properties of fluorine have been more fully investigated.

The chemical properties of fluorine, so far as they have been investigated, shew that this element is very similar to the elements chlorine, bromine, and iodine; but, at the same time, there are fairly marked differences between fluorine and these three elements. No one of the four elements shews any tendencies to react as a metal.

451　　MANGANESE is the second member of the even-series of Group VII. The sketch of the chemical properties of manganese given in pars. 195—199 of Chap. XI. shews that manganese is at once metallic and non-metallic in its chemical functions. The oxides MnO, Mn_3O_4, and Mn_2O_3 are basic; MnO_2

is feebly acidic. A series of manganous salts MnX ($X = SO_4$, $2NO_3$, $\frac{2}{3}PO_4$, &c.) exists; a few manganic salts $Mn_2 3X$ are also known. Permanganic acid, $H_2Mn_2O_8$, has been isolated and a number of permanganates have been obtained as definite stable salts, generally isomorphous with perchlorates MCl_2O_8 ($M = K_2$, Ba, &c.). Many manganates, $MMnO_4$, are also known; these salts do not correspond in composition with any salts derived from acids of chlorine, bromine, iodine, or fluorine; they are similar to the sulphates, selenates, and tellurates MXO_4 ($M = K_2$, Ba, &c.; $X = S$, Se, Te).

Manganese, then, shews very feeble analogies with the other elements which are placed in the same group with it.

If the three generalisations stated in par. 446 are applied **452** to Group VII., they would lead us to expect that the unknown members of the even series of this group should resemble manganese, but should on the whole be more distinctly metallic than this element; and that the unknown members of the odd series of the group should resemble the halogen elements, but should be less decisively non-metallic than these elements; the unknown members of series 9 and 11, Group VII., might fairly be expected to form a few salts by the interactions of their oxides with acids.

THE ELEMENTS OF GROUP III.

Group III.		2	4	6	8	10	12
	Even series.	B=10·9	Sc=44	Y*=89·6	La=138·5	Yb*=173	--
		3	5	7	9	11	
	Odd series.	Al=27·02	Ga=69	In=113·4	—	Tl=203·64	

	BORON.	SCANDIUM.	YTTRIUM.	LANTHANUM.	YTTERBIUM.
Even-series elements					
Atomic weights	10·9	44	89·6	138·5	173
		The *molecular weights* of these elements are unknown.			
Sp. grs. (approx.)	2·5	—	—	6·2	—
Atom. weights / sp. grav.	4·3	—	—	22·3	—
Sp. heats	·5 (?)	not determined.	not determined.	·0449	not determined.
Appearance, and general physical properties.	Dark greenish-brown powder. Non-conductor of electricity. Has not been melted.	Not isolated.	Grey powder.	Steel-grey powder; or, when compressed, lustrous, grey-white, hard, particles. Fairly malleable and ductile. Melts at full red-heat.	Not isolated.
Occurrence and preparation	Chief compounds occurring in rocks and waters are boric oxide B_2O_3 and borax $Na_2B_4O_7$; not widely distributed, nor in large quantities. Prepared by strongly heating B_2O_3 with sodium.	Small quantities of silicate occur in a rare Swedish mineral. Element has not been isolated.	Occurs as silicate in small quantities with silicate of scandium and ytterbium. Prepared by electrolysing fused $YCl_3.xNaCl$, or by dechlorinating the same salt by sodium.	Silicate occurs, with silicates of Ce, Di, Fe, and Ca, in a few rare minerals. Prepared by reducing $LaCl_3$ by potassium.	Small quantities of silicate occur in a rare Swedish mineral. Element has not been isolated.
General chemical properties.	Burns in air or oxygen to B_2O_3. Decomposes steam at red-heat forming B_2O_3 and H. Oxidised by heating with HNO_3, H_2SO_4, molten KOH, or molten KNO_3. Combines directly with Cl, Br, I, S, and also with N. Atom of boron is tri-valent in gaseous molecules.	—	Burns when heated in air, giving Y_2O_3. Decomposes water, rapidly when warmed. Dissolves in dilute acids, also in hot KOHAq, with evolution of hydrogen.	Oxidises in ordinary air to La_2O_3. Decomposes cold water slowly, hot water rapidly, with evolution of hydrogen.	—

* There are still some doubts whether these *elements* are or are not mixtures of two or more distinct kinds of matter.

Boron is said to exist both as an amorphous powder and as metal-like, hard, lustrous, crystals; but recent investigation shews that the so-called crystalline boron is a compound of boron and aluminium.

General formulae and chemical characters of compounds. 454 The compounds BCl_3, BBr_3, and BF_3, have been gasified; these formulae are molecular. No compounds of the other even-series elements have yet been gasified; the formulae given are the simplest that express the compositions of these compounds. Gaseous aluminium and indium chlorides have the molecular composition MCl_3, and gallium chloride appears to exist in the gaseous state both as $GaCl_3$ and Ga_2Cl_6. There are indications of the existence of a hydride of boron, probably BH_3; but the compound has not been isolated. The compounds of the elements of this family, with the exception of those of boron, have not been very fully investigated : there are many points in the chemical history of boron which require elucidation.

Oxides: M_2O_3.
Haloid compounds: MX_3.
Sulphides: M_2S_3.
Acids: H_3BO_3, $H_2B_2O_4$, $H_2B_4O_7$; no acids of other elements are known.
Salts: $M_2 3X$; $X = SO_4$, $2NO_3$, $\frac{2}{3}PO_4$, &c. ; $M = Sc$, Y, La, Yb, *not* B.

The oxides M_2O_3, with the exception of B_2O_3, are obtained 455 by adding an alkali to the solution of a salt, and heating the hydrated oxide which is pptd. B_2O_3 is found in waters in various volcanic districts; it is formed by heating boron in oxygen.

Boric oxide, B_2O_3, is acidic; it dissolves in water, and from this solution crystals of *ortho-boric acid*, H_3BO_3, separate on evaporation. By heating ortho-boric acid to $100°$ *meta-boric acid*, $H_2B_2O_4$, is obtained; and at $160°$ *tetra-* or *pyro-boric acid*, $H_2B_4O_7$, is formed. Salts are known derived from each of these acids; borax, one of the most important salts, is sodium tetra-borate $Na_2B_4O_7$; the ortho-borates are very unstable salts, it is doubtful whether they exist in solution in water. A solution of boric oxide in water reacts as if it contained dibasic metaboric acid, $H_2B_2O_4$. Boric acid is an extremely weak acid. Dilute aqueous solutions of alkali borates precipitate metallic oxides, not borates, from solutions of many metallic salts; that is to say, these dilute solutions behave as if they contained boric acid and free alkali, pro-

duced by the action of the water on the borate. Borates are
ill-defined salts; many basic salts seem to exist. Boric oxide
combines with a few anhydrides of strong acids to form
compounds in which the boric oxide acts as a base; thus
$B_2O_3.P_2O_5$ (? BPO_4) is a stable compound; $B_2O_3.xSO_3$ also
exists but is decomposed by heat, or by water.

The other *oxides* M_2O_3 ($M = Sc$, Y, La, Yb) are basic;
they dissolve in acids and form salts; they are insoluble
in solutions of alkalis. Hydrates of $Y_2O_3(Y_2O_3.6H_2O)$ and
La_2O_3 ($La_2O_3.3H_2O$) have been obtained, but not by the
direct reaction of water with the oxides. Lanthanic hydrate
is said to turn red litmus blue.

456 The sulphides M_2S_3, where $M = B$ or La, are formed by
passing carbon disulphide vapour over heated B_2O_3 or La_2O_3;
they are easily decomposed by cold water to M_2O_3 and H_2S.

457 Haloid compounds of boron, yttrium, and lanthanum,
have been prepared. *Boron chloride* and *bromide*, BCl_3 and
BBr_3, are obtained by heating boron in chlorine and bromine,
respectively, or by passing chlorine, or bromine, over a strongly
heated mixture of boric oxide and carbon.

Boron fluoride, BF_3, is formed by heating boric oxide with
calcium fluoride to a full white heat.

Boron chloride and bromide are liquids, boiling without
decomposition at 17° and 90° respectively; boron fluoride is a
gas. These haloid compounds are all decomposed by water;
BCl_3 and BBr_3 to solutions of boric oxide and hydrochloric, or
hydrobromic, acid; BF_3 is partly decomposed to boric oxide
and hydrofluoric acid, but the latter combines with some
unchanged BF_3 to form *fluoboric acid* HBF_4 ($= HF.BF_3$).
Boron chloride, BCl_3, combines with many other compounds to
form stable *double compounds*; e.g. $2BCl_3.3NH_3$; $BCl_3.POCl_3$;
$BCl_3.NOCl$; $BCl_3.HCN$. *Yttrium* and *lanthanum chlorides*
are obtained as $YCl_3.6H_2O$, and $2LaCl_3.15H_2O$, respectively, by dissolving the oxides in HClAq and evaporating.
The anhydrous chlorides have also been obtained. These
chlorides form several double compounds chiefly with other
chlorides; e.g. $YCl_3.3HgCl_2.9H_2O$; $2LaCl_3.3PtCl_4.24H_2O$;
$2LaCl_3.3AuCl_3.21H_2O$. The haloid compounds YBr_3, YI_3,
$2YF_3.H_2O$, $LaBr_3.7H_2O$, and $2LaF_3.H_2O$, have been isolated.
Two *oxychlorides of boron*, $BOCl$ and $BOCl_3$, are known.

458 The salts of the elements we are considering, so far as
they have been examined, belong to the form $M_2.3X$ where
$X = SO_4$, $2NO_3$, $\frac{2}{3}PO_4$, &c. No definite salts have been
obtained by replacing the hydrogen of acids by boron, although

boric oxide combines with a few strongly acidic oxides (*s. par.* 455). Several salts of yttrium, scandium, lanthanum, and ytterbium have been prepared; few basic salts are known. The sulphates seem all to combine with sulphate of potassium to form double salts; these are not alums; they generally have the composition $M_2(SO_4)_2.3K_2SO_4$.

So far as the odd-series elements of Group III. have been ex- **459** amined, it appears that the four elements scandium, yttrium, lanthanum, and ytterbium, are closely related, and that boron is distinctly separated from the other members of the family.

Odd-series elements	ALUMINIUM.	GALLIUM.	INDIUM.	THALLIUM.
Atomic weights	27·02	69	113·4	203·64 **460**
		The *molecular weights* of these elements are unknown.		
[p. grs. (approx.)	2·6	5·9	7·5	11·8
Atom. weights spec. gravs.	10·5	11·8	15·2	17·3
Sp. heats	·225	·08	·057	·031
Melting points (approx.)	700°—800°	30°	175°	290°
Appearance, and general physical properties.	White; very malleable and ductile; hard; tough; very sonorous; good conductor of electricity.	Silver-white; crystalline; hard, melts very easily; rather brittle.	White; lustrous; soft.	Grey-white; very lustrous; soft; fairly malleable and ductile; crystalline; may be distilled in hydrogen.
Occurrence and preparation.	Oxide of aluminium occurs as *corundum*, *sapphire*, &c. Silicates occur in clay, felspar, &c. in enormous quantities. Prepared by reducing $Al_2Cl_6.2NaCl$ by sodium or potassium, or by electrolysis of the fused salt.	Compounds occur in small quantities in a few zinc blendes. Prepared by electrolysing an alkaline solution of a basic sulphate.	Compounds occur in small quantities in certain zinc blendes. Prepared by precipitation by means of pure zinc from a solution of the sulphate.	Compounds of thallium occur in small quantities in various widely distributed minerals. Prepared by electrolysis of aqueous solution of $TlNO_3$, &c.; by reduction of same solution by zinc; or by fusing TlCl with KCN.
General chemical properties.	Unchanged in air; thin pieces heated in air are burnt to Al_2O_3. Superficially oxidised by melting in oxygen. Combines at high temperatures with S, P, and As, also with N. Decomposes water at 100°, evolving H. Dissolves in hot HClAq or H_2SO_4Aq. Unchanged by molten KOH or KNO_3; but dissolves in boiling KOHAq giving an aluminate and H. Atom of aluminium is trivalent in the gaseous molecule $AlCl_3$.	Oxidised superficially by heating in oxygen to full redness. Combines directly with Cl, Br, and I. Dissolves in acids, also in KOHAq, with evolution of H. Atom of gallium is trivalent in the gaseous molecule $GaCl_3$; existence of gaseous Ga_2Cl_6 is doubtful.	Oxidised at moderately high temperature in air. Soluble in most dilute acids. Does not decompose water at 100°. Atom of indium is trivalent in the gaseous molecule $InCl_3$.	Oxidises in air at ordinary temperatures. Combines directly with Cl, Br, I, P, and S. Dissolves in H_2SO_4Aq or HNO_3Aq with evolution of H. Decomposes water at red-heat with evolution of H. Atom is monovalent in the only known gaseous molecule TlCl.

461 **General formulae and characters of compounds.** The compounds $AlCl_3$, ($? Ga_2Cl_6$), $GaCl_3$, $InCl_3$, and $TlCl$, have been gasified; these formulae are molecular. Al_2Br_6 and Al_2I_6 also appear to exist as gases; but from the results of recent work it is probable that at higher temperatures the molecules $AlBr_3$ and AlI_3 are formed. The formulae given to the other compounds are generally the simplest formulae which express the compositions of the different bodies; they are not necessarily molecular.

The compounds of gallium and indium have not been very fully investigated.

Oxides: M_2O_3; also InO and probably GaO; also Tl_2O.

Sulphides: M_2S_3; also TlS.

Haloid compounds: MCl_3, in one case M_2Cl_6; also $GaCl_2$; also $TlCl$.

Salts: M_23X; also TlX ($X = SO_4$, $2NO_3$, $\frac{2}{3}PO_4$, &c.).

462 The oxides M_2O_3 are obtained by adding ammonia to solutions of salts of the several metals, and drying and heating the hydrated oxides so formed. The hydrated oxides obtained are $Al_2O_3.3H_2O$, ($? Ga_2O_3.3H_2O$), $In_2O_3.3H_2O$, and $Tl_2O_3.H_2O$. The oxides, and the hydrated oxides, are insoluble in water; but it is possible, by dialysing a solution of $Al_2O_3.xH_2O$ in Al_2Cl_6Aq, to obtain an aqueous solution of hydrated aluminium oxide. The hydrates $M_2O_3.xH_2O$, except $Tl_2O_3.H_2O$, are soluble in solutions of caustic potash; potassium aluminate, $K_2Al_2O_4$, is obtained by evaporating the solution of aluminium hydrate; the solution of indium hydrate gives a precipitate of the oxide when boiled; and the solution of gallium hydrate is decomposed by carbon dioxide with precipitation of gallic oxide.

The oxides M_2O_3 all dissolve in acids forming salts $M_2.3X$; the thallic salts are very unstable and are easily reduced to thallous salts, Tl_2X.

When thallic hydrate $Tl_2O_3.H_2O$ is suspended in concentrated potash solution and chlorine is passed into the liquid, the hydrate partially dissolves, forming a violet coloured liquid, which possibly contains the potassium salt of a thallium acid; no salts of this hypothetical acid have however been isolated.

Thallous oxide, Tl_2O, is strongly basic and alkaline. When a solution of thallous sulphate, Tl_2SO_4, is mixed with a solution of baryta in the proportion $Tl_2SO_4 : BaO_2H_2$, the liquid is filtered from precipitated barium sulphate, evaporated and allowed to crystallise, the hydrated oxide Tl_2O . $2H_2O$ is obtained. When this hydrate is heated to $100°$ in absence of air Tl_2O is formed. Thallous oxide is very soluble in water and the solution closely resembles KOHAq or NaOHAq in its properties; it has a corrosive action on the skin, a hot and burning taste, turns red litmus blue, absorbs and combines with carbon dioxide, and neutralises acids forming thallous salts, Tl_xX. There can be little doubt that the solution contains the hydroxide TlOH, and that the composition of hydrated thallous oxide is better expressed by the formula 2TlOH . H_2O than by the formula Tl_2O . $2H_2O$.

The **haloid compounds**, MCl_3, are obtained; (1) by **463** heating the elements in chlorine, $M = Ga$ and In; (2) by heating an intimate mixture of the oxides M_2O_3 and carbon in chlorine, $M = Al$ and In; or (3) by adding HClAq to the hydrated oxides M_2O_3, $M = Tl$. The other haloid compounds are obtained by similar methods.

The vapour density of aluminium chloride agrees with the formula $AlCl_3$; that of gallic chloride shews that at moderate temperatures the gaseous compound probably consists of molecules of Ga_2Cl_6, but at higher temperatures it consists of molecules of $GaCl_3$; gaseous indium chloride has the molecular composition $InCl_3$; thallic chloride has not been gasified, at $100°$ it separates into thallous chloride, TlCl, and chlorine. The chlorides, MCl_3, are all deliquescent solids; they are all, except probably $TlCl_3$, partially decomposed by hot water with formation of various oxychlorides of more or less complex compositions. These chlorides all combine with many other chlorides to form double compounds; e.g. $AlCl_3 . PCl_5$; $AlCl_3 . POCl_3$; $AlCl_3 . KCl$; $InCl_3 . 3KCl$; $TlCl_3 . 3KCl$.

Thallous chloride, TlCl, is formed as a white precipitate when HClAq is added to a solution of thallous oxide. This chloride is only slightly soluble in water; it has been gasified without decomposition; it combines with various chlorides to form double compounds; the compound $2TlCl . PtCl_4$ resembles, and is isomorphous with, $2KCl . PtCl_4$.

The **sulphides** M_2S_3 are obtained by the direct union **464** of their elements at high temperatures. A compound $K_2S . Tl_2S_3$ (possibly a thio-thallate of potassium) is produced by heating thallous sulphide (Tl_2S) with sulphur and potassium carbonate. A similar compound of potassium and aluminium sulphides seems to exist.

Thallous sulphide, Tl_2S, is obtained by the direct union of its elements, or by adding sulphuretted hydrogen, or ammonium sulphide, to a neutral solution of a thallous salt. A compound $Tl_2S.As_2S_3$, analogous to $K_2S.As_2S_3$, is known.

465 Salts of the form M_23X ($X = SO_4$, $2NO_3$, CO_3, $\frac{2}{3}PO_4$, &c.) are obtained by dissolving the different hydrated oxides, $M_2O_3.xH_2O$, in acids and evaporating. The salts are generally soluble in water; the thallic salts are easily decomposed, sometimes even by solution in water, and are readily reduced to thallous salts, Tl_2X. The sulphates, with the exception of thallic sulphate, combine with sulphates of the alkali metals to form alums, $M_2.3SO_4.X_2SO_4.24H_2O$ ($X = $ alkali metal except Li.) Thallous sulphate (Tl_2SO_4) forms an alum with aluminium sulphate, $Al_2.3SO_4.Tl_2SO_4.24H_2O$.

Several basic salts of aluminium, and a few of the other elements, are known.

The thallous salts, Tl_2X, are generally stable bodies; they are similar to, and usually isomorphous with, salts of potassium.

466 Of the four odd-series elements of Group III., three, viz. aluminium, gallium, and indium, are evidently very closely related; the fourth, thallium, is to a great extent separated from the others.

Thallium shews marked similarities with the alkali metals which belong to Group I.; at the same time the properties of some of its compounds—e.g. the acidic character of Tl_2S_3, and the probable existence of an unstable potassium thallate—suggest relations with the most negative element of Group III., viz. boron.

Boron, which is the first member of the group, to some extent summarises the properties of the other members. Oxide of boron is acidic; aluminium oxide is acidic towards strong alkalis; the solubility of $Ga_2O_3.xH_2O$ and $In_2O_3.xH_2O$ in concentrated potash solution shews that these oxides have feebly marked acidic functions; an acidic oxide of thallium probably exists. Oxide of boron also shews basic functions towards some acidic oxides, e.g. SO_3 and P_2O_5; all the other oxides of the group are basic towards most acids. The positive character of boron is shewn in its interaction with steam at high temperatures.

Neither the group-character nor the family-character distinctly preponderates in Group III. The even-series elements from *scandium to ytterbium* form a closely related class; the odd-series elements from *aluminium to indium* are also very similar; *boron* at one end of the group, and *thallium* at the other, are separated from the other members of the group; although differing widely in most of their properties, boron and thallium approach each other in some respects.

CHAPTER XXV.

THE ELEMENTS OF GROUP IV.

		2	4	6	8	10	12
	Even series.	C=11·97	Ti=48	Zr= 90	Ce=139·9	—	Th=231·8
Group IV. {		3	5	7	9	11	
	Odd series.	Si=28·3	Ge=72·3	Sn=117·8	—	Pb=206·4

Even-series elements	CARBON.	TITANIUM.	ZIRCONIUM.	CERIUM.	THORIUM.
Atomic weights	11·97	48	90	139·9	231·8
		The *molecular weights* of these elements are unknown.			
Sp. grs. (approx.)	3·3 (diamond)	?	4·15	6·7	7·7
Atom. weights spec. gravs.	3·6	?	21·7	20·9	39·4
Sp. heats	·463 (?)	·148	·0666	·0448	·0276

The *melting points* of these elements have not been determined ; cerium is said to fuse considerably above 500° but under 900°, and zirconium at a higher temperature than cerium. Carbon, titanium, and thorium, have not been melted.

Appearance, and general physical properties.	Colourless, transparent, highly refractive, crystals (diamond); also black, amorphous, powder; also black, lustrous, crystalline, solid (graphite). S. G. of *graphite* 2·25; of *amorphous carbon* abt. 1·9. Graphite is a fair conductor of electricity. Amorphous carbon is very porous and absorbs gases freely.	Iron-grey, lustrous, powder.	Black powder, resembling amorphous carbon; also hard, brittle, very lustrous, grey, crystals. Very porous, and absorbs large volumes of gases.	Grey, lustrous, solid; very ductile; fairly malleable.	Dark-grey, lustrous, metal-like, powder.
Occurrence and preparation.	Diamond and graphite occur native; enormous quantities of CO_2 occur in the air; carbonates of Ca, Mg, Fe, &c. are common minerals. CO_2 found in all waters. Carbon compounds form chief parts of all living organisms. Diamond has not been prepared artificially. Amorphous carbon prepared by heating oil, fat, &c. in absence of air.	Titanium oxide, silicate, and a few other compounds, occur in certain rare minerals. Many iron-ores contain small quantities of Ti compounds. Prepared by reducing vapour of K_2TiF_6 by K or Na.	Zirconium oxide, ZrO_2, occurs in a few rare minerals. Prepared by reducing K_2ZrF_6 vapour by K or Na; or by aluminium, when the Zr crystallises out on cooling.	Occurs as silicate in the rare mineral *cerite;* also in very small quantities in various minerals, and in some clays. Prepared by electrolysing Ce_2Cl_6 mixed with NaCl.	Thorium oxide, ThO_2, is found in a few rare minerals. Prepared by reducing $ThCl_4$ by Na or K.

Group IV. *continued.*

Even-series elements	CARBON.	TITANIUM.	ZIRCONIUM.	CERIUM.	THORIUM.
General chemical properties.	Heated in air or oxygen, burns to CO and CO₂; diamond burns only at very high temperatures in oxygen. Combines directly with H when electric sparks are passed from carbon poles in atmosphere of H. Combines directly with S at high temperatures. Combines with many metals to form bodies resembling alloys. Graphite is oxidised by heating with KClO₃ and HNO₃Aq to *graphitic acid* C₁₁H₄O₅; no other form of carbon gives this acid. Exhibits allotropy very markedly. Atom is tetravalent in gaseous molecules.	Burns brilliantly when heated in air or oxygen, forming TiO₂. Combines directly with Cl. No hydride known. Combines directly with N at high temperatures. Decomposes water at 100°, giving TiO₂ and H. Atom is tetravalent in gaseous molecule TiCl₄.	Amorphous Zr burns when heated in air or oxygen; crystalline is superficially oxidised. Insoluble in most acids, but easily dissolved by HFAq. Oxidised by molten KOH, KNO₃, or KClO₃. Atom is tetravalent in gaseous molecule ZrCl₄.	Burns to CeO₂ when heated in air or oxygen. Combines directly with Cl, Br, and I; also with S, and P. Decomposes warm water slowly with formation of Ce₂O₃ and H.	Burns when heated in air to ThO₂. Soluble in HClAq, but HNO₃Aq. Properties have been little examined. Atom is tetravalent in gaseous molecule ThCl₄.

General formulae and characters of compounds. Carbon is **468** characterised by the enormous number of compounds which it forms by combining with some or all of the elements hydrogen, oxygen, and nitrogen.

The chlorides MCl_4, except $CeCl_4$, and the fluorides MF_4, have been gasified and their vapour-densities determined; the oxides CO and CO_2, and the sulphide CS_2, have also been gasified; the formulae of these compounds are therefore molecular; the formulae of the other compounds are the simplest that express their compositions.

None of the elements of the family except carbon forms any compound with hydrogen: the hydrides of carbon are exceedingly numerous.

Oxides: CO, $(? TiO)$; $(? C_2O_3)$, Ti_2O_3, Ce_2O_3; MO_2, $M =$ any element of the family.

Sulphides: CS, CS_2, $(? C_2S_3)$, TiS_2, Ce_2S_3, ThS_2.

Haloid compounds: MX_4, $M =$ any element of the family; C_2Cl_6, &c., Ti_2Cl_6, Ce_2Cl_6; $TiCl_2$.

Acids: H_2CO_3Aq; salts of the form M_2TiO_3 and M_2ZrO_3 are known ($M=$ an alkali metal); the hydrates $TiO_2.xH_2O$ and $ZrO_2.xH_2O$ are acidic.

Salts: carbon does not form salts by replacing the hydrogen of acids; $M.2X$ ($X=SO_4$, $2NO_3$, &c.) $M=Ti$, Zr, Ce, Th; $Ce_2 3X$, and $Ti_2 3X$.

469 Oxides. The *dioxides* MO_2, are produced by heating the elements in oxygen; the dioxides of titanium, zirconium, cerium, and thorium, are also obtained by precipitating solutions of salts of these elements by ammonia, and drying and heating the hydrated oxides so obtained. Carbon dioxide is most easily obtained by decomposing a metallic carbonate by an acid, e.g. $CaCO_3 + 2HClAq = CO_2 + H_2O + CaCl_2Aq$. The dioxides, with the exception of CO_2, are solids, insoluble in water, some of them insoluble also in most acids; carbon dioxide is a colourless, odourless, gas which can be condensed to a liquid, and, at a very low temperature, to a snow-like solid.

Carbon dioxide dissolves freely in water; the solution reddens blue litmus and interacts with alkalis to form salts M_2CO_3: from these alkali carbonates, carbonates of most metals can be obtained. An aqueous solution of carbon dioxide probably contains *carbonic acid*, H_2CO_3; but this compound has not been isolated; the sulphur compound *thiocarbonic acid*, H_2CS_3, is known as a solid.

By precipitating solutions of salts ($M2X$) of titanium, zirconium, cerium, and thorium, *hydrated dioxides* $MO_2.xH_2O$ are obtained. Many of these hydrated oxides seem to exist; the following, obtained by drying under different conditions, are among the more important, $TiO_2.H_2O$, $TiO_2.2H_2O$; $ZrO_2.H_2O$, $ZrO_2.2H_2O$; $ThO_2.2H_2O$; $2CeO_2.3H_2O$. These hydrated dioxides are soluble in acids, and from these solutions salts of the form $M.2X$ ($X=SO_4$, $2NO_3$, &c.) are obtained. The solution of $CeO_2.xH_2O$ in acids seems to contain cerous salts $Ce_2.3X$, as well as ceric salts $Ce2X$.

Most of the dioxides $MO_2.xH_2O$ exhibit acidic properties. None of them dissolves in solutions of alkalis to form salts. When however *titanium dioxide*, TiO_2, is fused with sodium carbonate, a quantity of carbon dioxide is evolved corresponding with that calculated on the assumption that *sodium titanate* Na_2TiO_3 is produced; the fused mass is separated by water into $NaOHAq$ and $NaHTiO_3$. *Zirconium dioxide*, ZrO_2, fused with Na_2CO_3 behaves similarly to TiO_2; the *zirconates* Na_2ZrO_3 and Na_4ZrO_4 are said to have been obtained. *Thorium dioxide*, ThO_2, does not decompose Na_2CO_3 when heated with this salt, nor does it form salts by fusion with solid potash or soda. The fact that when *hydrated cerium dioxide*, $CeO_2.xH_2O$, is precipitated by adding potash to a solution of ceric sulphate, $Ce(SO_4)_2$, the whole of the potash cannot be removed from the precipitate by washing with water indicates the possible formation of an unstable potassium cerate.

Carbon monoxide, CO, is obtained by heating the dioxide with carbon ($CO_2 + C = 2CO$), or by the interaction of (1) formic acid, or (2) oxalic acid, with concentrated hot sulphuric acid;

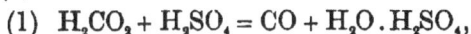

(1) $H_2CO_2 + H_2SO_4 = CO + H_2O.H_2SO_4$,

(2) $H_2C_2O_4 + H_2SO_4 = CO + CO_2 + H_2O.H_2SO_4$.

This oxide is a poisonous gas; salts of formic acid H_2CO_2 can be obtained from it, and as the oxide is produced from the acid, the oxide is sometimes called *formic anhydride*.

The *sesquioxides* Ti_2O_3 and Ce_2O_3 are obtained by reducing the higher oxides MO_2; in the case of titanium this is done by heating TiO_2 to redness in dry hydrogen, in the case of cerium it is better to heat cerous oxalate, $Ce_2(C_2O_4)_3$, in hydrogen. Ti_2O_3 is oxidised to TiO_2 by continued heating in air; Ce_2O_3 is very easily oxidised by mere exposure to air. Both oxides dissolve in acids forming salts; a large series of cerous salts $Ce_2.3X$ is known; only titanous sulphate, $Ti_2(SO_4)_3.8H_2O$, seems to have been obtained as a definite solid.

The sulphides, with the exception of CS_2, have not **470** been much examined. *Titanium disulphide*, TiS_2, is a dark coloured stable solid, obtained by the interaction of titanium tetrachloride and sulphuretted hydrogen at high temperatures. *Thorium disulphide*, ThS_2, resembles TiS_2; it is obtained by passing hydrogen mixed with hydrogen sulphide over heated ThO_2. *Cerium sesquisulphide*, Ce_2S_3, is a golden coloured solid formed by passing sulphuretted hydrogen over hot CeO_2. None of these sulphides exhibits any tendency to form thio-salts by interacting with alkali sulphides, or by fusion with alkalis.

Carbon disulphide is a mobile, very refractive, liquid, boiling at $47°$; its vapour is very inflammable. It is obtained by the direct union of its elements at high temperatures. This sulphide is the anhydride of *thio-carbonic acid* H_2CS_3; when CS_2 interacts with concentrated Na_2SAq sodium thio-carbonate Na_2CS_3 is formed; the acid H_2CS_3 is obtained as a dark

yellow oily liquid by decomposing one of its alkaline salts by
dilute HClAq. Thio-carbonic acid and its salts are very
easily decomposed, to CS_2 and H_2S in the case of the acid, and
to H_2S and carbonates in the case of the salts.

471   The haloid compounds of the carbon family of the form
MX_4 where $X = Cl$ or Br, are prepared, except in the case of
the carbon compounds, by strongly heating an intimate
mixture of the oxides MO_2 and carbon in a stream of chlorine,
or bromine; tetrachloride of cerium has not yet been prepared.
$TiCl_4$ is a liquid boiling at $136°$, the other tetrachlorides are
solids; they have all been gasified without decomposition.
These tetrachlorides all combine with various other chlorides,
and in some cases with other compounds, to form double
compounds; e.g. $TiCl_4.PCl_5$; $3TiCl_4.4NOCl$; $TiCl_4.4NH_3$;
$ZrCl_4.2NaCl$; $2ThCl_4.8NH_4Cl$.

The *tetrafluorides* MF_4, $M = Ti$, Zr, Ce, or Th, are obtained
by dissolving the hydrated dioxides $MO_2.xH_2O$ in aqueous
hydrofluoric acid and evaporating. The tetrafluorides of
titanium, zirconium, and thorium form compounds with potassium fluoride of the form $2KF.MF_4$; as in some cases the
corresponding hydrogen compounds $2HF.MF_4$ have been
isolated and the potassium compounds have been obtained by
neutralising aqueous solutions of these hydrogen compounds,
it seems better to regard the compounds in question as the
potassium salts of *fluotitanic*, *fluozirconic*, and *fluothoric acids*,
H_2MF_6 ($M = Ti$, Zr, Th). A double fluoride of cerium and
potassium is known, but its composition is different from
that of the K_2MF_6 salts; it is represented by the formula
$2CeF_4.3KF.2H_2O$.

Cerous chloride, or *cerium sesquichloride*, Ce_2Cl_6, is obtained
by heating in chlorine a mixture of the corresponding oxide,
Ce_2O_3, and carbon.

Carbon forms several haloid compounds; CX_4, C_2X_6, C_2X_4,
where $X = Cl$ or Br; and CI_4. *Carbon tetrachloride* CCl_4,
trichloride C_2Cl_6, and *dichloride* C_2Cl_4, have been gasified without decomposition. The first of these compounds is obtained
by the interaction of chlorine and chloroform, $CHCl_3$; the
second by the interaction of chlorine and ethylene dichloride,
$C_2H_4Cl_2$; and the third by reducing C_2Cl_6 by means of hydrogen
evolved in contact with the carbon trichloride.

472   Salts. Salts of carbon, i.e. compounds obtained by replacing the hydrogen of acids by carbon, are unknown. Titanium forms two series of salts; *titanous salts* represented by

the sulphate $Ti_2(SO_4)_3$ which is obtained by dissolving the corresponding oxide Ti_2O_3 in concentrated sulphuric acid and evaporating; and *titanic salts* obtained by dissolving the dioxide, TiO_2, in acids. Most of the titanic salts are basic salts, e.g. $TiO.SO_4$; $5TiO_2.N_2O_5.xH_2O$; $2TiO_2.P_2O_5.xH_2O$; &c. ; a few normal salts are known, e.g. $Ti(SO_4)_2.3H_2O$.

Zirconium forms one series of salts, the *zirconic salts*, represented by the sulphate $Zr(SO_4)_2$, and one or two others. Most of the zirconium salts are basic salts, e.g. $3ZrO_2.SO_3$; $3ZrO_2.2N_2O_5$; $5ZrO_2.4P_2O_5$, &c.

The salts of thorium are all *thoric salts*, e.g. $Th(SO_4)_2$; $Th(NO_3)_4.xH_2O$; $Th_3(PO_4)_4.xH_2O$, &c. The thoric salts are usually normal; a few basic salts are known, e.g.

$$2ThO_2.7SeO_2.xH_2O.$$

Cerium resembles titanium in that it forms two classes of salts. The *cerous salts* $Ce_2.3X$ are numerous ; e.g.

$$Ce_2(SO_4)_3.xH_2O ; Ce(NO_3)_3.xH_2O ; CePO_4, \&c.$$

The chief representatives of the *ceric salts* are

$$Ce(SO_4)_2.xH_2O, \text{ and } Ce(NO_3)_4.$$

The salts of all the metals of the family form many double salts generally by combination with salts of the alkali, and alkaline earth, metals.

Carbon is evidently separated from the other even series **473** members of Group IV. by its distinctly non-metallic character. The other elements are all metals in their physical properties ; their oxides are basic, but most of them shew acidic functions. As the atomic weight increases the elements become more distinctly metallic in their chemical properties. In a few respects, e.g. existence of M_2O_3 and corresponding salts, *cerium* is more closely related to *titanium* than to any other member of the family.

The three elements *titanium, zirconium, and thorium* exhibit very marked similarities ; the existence of the compounds K_2MF is characteristic.

474

Odd-series elements	SILICON.	GERMANIUM.	TIN.	LEAD.
Atomic weights	28·3	72·3	117·8	206·4
		The *molecular weights* of these elements are unknown.		
Sp. grs. (approx.)	2·5	5·5	7·3	11·4
Atom. weights / *spec. gravs.*	11·2	13·1	16·1	18·1
Melting points (approx.)	abt. 1000°	900°	230°	330°
Sp. heats	·203	·077	·055	·031
Appearance, and general physical properties.	Brown, amorphous powder; also as greyish-black, needle-shaped, very hard, metal-like, lustrous, crystals; also as crystalline plates resembling graphite in appearance. Graphitic Si conducts electricity.	Grey-white; lustrous; crystalline; very brittle.	White; crystalline; lustrous; not hard; rather brittle, but ductile and malleable at certain temperatures. Also as a grey powder, S. G. 5·8, produced by keeping ordinary tin at very low temperatures for some time.	White with greyis tinge; soft; crystal line; very malleable ductile, but not t nacious.
Occurrence and preparation.	Enormous quantities of silicates occur as clays, felspars, &c. Prepared by reducing vapour of $SiCl_4$ by K or Na; or by reducing K_2SiF_6 by K or Al.	The sulphide is found in *argyrodite* a rare mineral (chiefly Ag_2S) from Freiberg. Prepared by reducing GeO_2 in hydrogen, or by strongly heating the same oxide with carbon.	Found native; but chiefly as oxide SnO_2; very widely distributed, but not in very large quantities. Prepared by reducing SnO_2 by carbon, or ,by electrolysing aqueous solutions of salts.	Found native in small quantities; chief ore is PbS, widely distributed in fairly large quantities. Prepared by redu ing PbO by carbo or potassium cya nide; also by ele trolysis of solutior of salts.
General chemical properties.	Amorphous Si burns to SiO_2 when heated in air; crystalline Si does not oxidise even when heated in oxygen. Soluble in HFAq; not in HClAq or HNO_3Aq. Amorphous Si, but not crystalline Si, dissolves in KOHAq forming K_2SiO_3 and H. Oxidised by molten KOH. Combines directly at high temperatures with Cl, also with S, and N. Combines with several metals to form bodies resembling alloys. Exhibits allotropy. Atom is tetravalent in gaseous molecules.	Oxidised to GeO_2 by nitric acid; insoluble in HClAq; dissolves in H_2SO_4Aq. Atom is tetravalent in gaseous molecules $GeCl_4$ and GeI_4.	Burns to SnO_2 when strongly heated in air. Dissolved by dilute HNO_3Aq; oxidised to SnO_2 by conc. nitric acid. Soluble in KOHAq with evolution of H, and formation of K_2SnO_3. Combines directly with Cl, and with S. Atom is divalent and tetravalent in gaseous molecules.	Oxidised super-ficially in moist air heated in air burn to PbO and highe oxides. Dissolved by sul phuric acid and nitri acids. Combines directly with Cl and S. Atom is divalent an tetravalent in gas eous molecules.

475 General formulae and chemical characters of compounds.

Silicon is the only member of the family which forms a compound with hydrogen. Numerous compounds of silicon with carbon, hydrogen, and oxygen, the silico-organic compounds, are known.   The compounds $SiCl_4$, $SiBr_4$, SiI_4; $GeCl_4$, GeI_4, GeS; $SnCl_2$, $SnCl_4$; $PbCl_2$, $Pb(CH_3)_4$, have been gasified and their vapour densities have been determined.

Oxides: MO_2, M = any element of the family; MO, M = Ge, Sn, Pb; also Sn_2O_3, Pb_2O_3, and Pb_3O_4.   Hydrates of most of these oxides are known : some of them are acidic.

Sulphides: MS_2, M = any element of the family except Pb ; GeS, SnS, PbS.

Haloid compounds: MX_4, M = any element of the family except Pb ; MCl_2, M = any element of the family except Si ; Si_2X_6.

Salts: salts of Si are unknown; MX, X = SO_4, $2NO_3$, $\frac{2}{3}PO_4$, &c.; a few stannic salts $Sn.2X$ are known.

Salts derived from acidic hydroxides; M_2XO_3, X = Si, Sn, Pb ; also many complex silicates, some complex stannates, and probably one or two plumbites M_2PbO_2; (M = K, Na, &c.).

Silicon hydride, SiH_4, is obtained mixed with hydrogen **476** by decomposing any alloy of magnesium and silicon by hydrochloric acid.   The hydride is obtained pure by the interaction between sodium and the compound $SiH(OC_2H_5)_3$.   Silicon hydride is a colourless gas, condensing under great pressures to a colourless liquid ; it is decomposed by heating to about 400^0; it is very inflammable when mixed with air ; the gas interacts with an aqueous solution of potash to form potassium silicate and hydrogen ; $SiH_4 + 2KOHAq + H_2O = K_2SiO_3Aq + 4H_2$.

Oxides.   The *dioxides* SiO_2, GeO_2, *and* SnO_2 are obtained **477** by strongly heating the elements in air ; GeO_2 and SnO_2 are more readily obtained by oxidising the elements by concentrated nitric acid.   *Silica*, SiO_2, is usually prepared by decomposing an alkali silicate by an acid, evaporating to dryness, heating, and removing the alkali salt of the acid by solution in water. *Lead dioxide*, PbO_2, is obtained by evolving oxygen in contact with a lower oxide of lead in presence of an alkali ; the usual method is to suspend PbO in concentrated KOHAq and pass in chlorine, or to boil PbO with KOHAq and KClOAq.   The *oxides GeO, SnO, and PbO*, are obtained by adding an alkali to solutions of the corresponding chlorides, MCl_2, and drying the hydrated oxides so produced.

The oxides are all white, or nearly white, solids, insoluble in water.   Silicon dioxide, SiO_2, when strongly heated is insoluble in ordinary acids except hydrofluoric ; GeO_2 probably forms salts by interacting with acids, but these salts have not yet been isolated; hydrated stannic oxide, $SnO_2.xH_2O$, dissolves in most acids to form salts, but few have been isolated, and most of these are basic salts; plumbic oxide, PbO_2, dissolves in HClAq, HNO_3Aq, &c. to form plumbous salts, $PbCl_2$, $Pb.2NO_3$, &c. but it appears to dissolve without change in acetic and phosphoric acids, and these solutions possibly contain plumbic salts, $Pb.2X$. All the dioxides, MO_2, are more or less distinctly acidic. Hydrated silicon dioxide, $SiO_2.xH_2O$, dissolves in solutions of caustic alkalis, and silicates are obtained from these solutions; of the vast number of silicates which are known, many occur in minerals; those which may be called normal silicates belong to the forms M_2SiO_3 and M_4SiO_4, $M = K$, Na, $\dfrac{Ba}{2}$, $\dfrac{Mg}{2}$, &c. The other hydrated dioxides $MO_2.xH_2O$, where $M = Sn$ or Pb, dissolve in concentrated aqueous potash, or in molten potash containing a very little water, and salts are obtained on evaporating *in vacuo;* these salts belong to the form X_2MO_3 where $X = K$ or Na.

The *monoxides SnO and PbO* dissolve in acids to form salts; GeO also dissolves in acids, but no salts have yet been isolated; a few stannous salts and a considerable number of plumbous salts, MX, have been isolated. *Plumbous oxide,* PbO, dissolves in conc. potash or soda solution; a few salts of the form M_2PbO_2 have been obtained, $M = K$, Na, or Ag.

The *sesquioxides Sn_2O_3 and Pb_2O_3* are obtained by the action of weak oxidisers on solutions of stannous or plumbous salts, MX, in presence of an alkali; Sn_2O_3 easily oxidises in air to SnO_2; Pb_2O_3 interacts with dilute acids as if it were $PbO.PbO_2$, a plumbous salt (PbX) is formed and PbO_2 remains.

478   Sulphides. The only *sulphide of silicon* is SiS_2; it is prepared by passing carbon disulphide vapour over a heated mixture of silica and carbon. The *sulphides GeS_2 and SnS_2* are obtained by passing sulphuretted hydrogen into acidified solutions of germanic and stannic chlorides, respectively. GeS is obtained by heating GeS_2 in hydrogen; and SnS by passing sulphuretted hydrogen into an acidified solution of stannous chloride. When sulphuretted hydrogen is passed into an acidified solution of a lead salt, PbS is pptd. There are indications of the existence of a higher sulphide of lead than PbS, but none has been isolated. Both *stannous and stannic sulphide,* SnS and SnS_2, dissolve in solutions of alkali sulphides to form alkali thio-stannates, M_2SnS_3; PbS fused with an alkali carbonate probably forms an alkali thio-plumbate;

GeS dissolves in potash solution, probably forming a thio-germanate.

Haloid compounds. The *tetrachlorides* MCl_4, M = Si, **479** Ge, Sn, are obtained by heating the elements in a rapid stream of chlorine; $SiCl_4$ is better prepared by heating a mixture of silica and carbon in chlorine; $GeCl_4$ by heating a mixture of germanium and mercuric chloride ($HgCl_2$); $SnCl_4$ by passing chlorine into stannous chloride, $SnCl_2$. When chlorine is passed into a solution of $PbCl_2$ in HClAq, the gas is absorbed; on heating this solution, chlorine is evolved; addition of water to the solution ppts. lead dioxide PbO_2; the solution probably contains $PbCl_4$, but this compound has not yet been certainly isolated.

The *tetrachlorides*, $SiCl_4$, $GeCl_4$, and $SnCl_4$, are liquids boiling at moderate temps.; $SiCl_4$ boils at $57°$, $GeCl_4$ at $86°$, $SnCl_4$ at $115°$. $SiCl_4$ is decomposed by water to silica (SiO_2) and hydrochloric acid; $SnCl_4$ dissolves in water, and on evaporation various hydrates, $SnCl_4.xH_2O$, are obtained. $SnCl_4$ combines with many chlorides and other compounds to form double compounds; e.g.

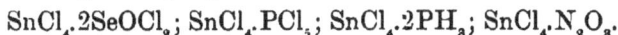

$$SnCl_4.2SeOCl_2; \; SnCl_4.PCl_5; \; SnCl_4.2PH_3; \; SnCl_4.N_2O_3.$$

The *tetrafluorides* MF_4, M=Si, Ge, Sn, are obtained by the interaction of hydrofluoric acid with the oxides MO_2. SiF_4 is gaseous, the others are solids. They all combine with potassium fluoride to form characteristic salts K_2MF_6. When SiF_4 is passed into water, silicic acid and fluosilicic acid H_2SiF_6, are formed ($3SiF_4 + 3H_2O + Aq = 2H_2SiF_6Aq + H_2SiO_3$). *Potassium fluosilicate*, K_2SiF_6, is obtained by neutralising this acid by potash. Acids of the form H_2MF_6 when M = Ge or Sn are not known.

Silicon trichloride, Si_2Cl_6, is obtained by heating the tetrachloride with silicon: no corresponding haloid compounds of the other members of the family are known.

The *dichlorides* MCl_2, where M = Ge, Sn, or Pb are formed when the metals are heated in hydrochloric acid gas; $PbCl_2$ is usually obtained by adding HClAq to the solution of a lead salt, and crystallising from hot water. $GeCl_2$ is a liquid; the others are solids. $SnCl_2$ is decomposed by much water giving *oxychlorides*, $xSnO.ySnCl_2.zH_2O$. $PbCl_2$ dissolves in hot water without change; *oxychlorides of lead* are obtained by heating this solution with lead oxide, PbO. No fluoride of lead has yet been isolated.

480   Salts.   Salts which have been definitely isolated belong to the forms MX and M2X, X = SO_4, $2NO_3$, $\frac{2}{3}PO_4$, &c. Silicon oxide does not shew any basic functions.   The other dioxides, MO_2, dissolve in acids.   From the solutions of GeO_2 in acids no salts have yet been obtained. SnO_2 forms *stannic salts* by interacting with acids; only a few of these salts have been obtained as definite solids, and most of these are basic salts, e.g. $2SnO_2.P_2O_5.xH_2O$; $2SnO_2.As_2O_5$. PbO_2 dissolves in concentrated acetic and phosphoric acids without evolving oxygen, but definite salts have not been obtained from these solutions; with HClAq, PbO_2 forms $PbCl_2$ and evolves chlorine, with H_2SO_4Aq it forms $PbSO_4$ and evolves oxygen.

The monoxides SnO and PbO interact with acids to form salts, MX; salts of germanium GeX have not yet been isolated.

The chief *stannous salts* which have been obtained as solids are the sulphate $SnSO_4$, and the arsenate $SnHAsO_4.xH_2O$; a few basic salts are known, e.g.

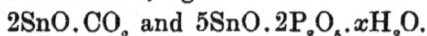

$$2SnO.CO_2 \text{ and } 5SnO.2P_2O_5.xH_2O.$$

The *plumbous salts* form a stable and very definite series, e.g. $PbSO_4$, $PbCO_3$, $Pb2NO_3$, $Pb_3(AsO_4)_2$, PbC_2O_4, $PbSiF_6$. Many basic carbonates, e.g.

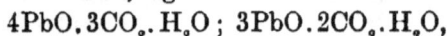

$$4PbO.3CO_2.H_2O \; ; \; 3PbO.2CO_2.H_2O,$$

are known; basic lead nitrates are also very numerous. Many lead salts form double salts by combination with salts of alkali and alkaline earth metals.

481   Carbon and silicon, which are, respectively, the first even-series, and the first odd-series, member of Group IV., shew most marked similarities; both are to some extent separated by their distinctly non-metallic characters from the other members of the group.   The other seven elements are fairly closely allied.   The properties of several elements in Group IV. seem to be distinctly conditioned by the properties of the elements coming before and after them in their respective series; this is very marked in the cases of tin, placed in Series 7 between indium and antimony, cerium, placed in Series 8 between lanthanum and didymium, and lead, placed in Series 11 between thallium and bismuth.

482   The history of the element *germanium*, in Group IV., and of the elements *scandium* and *gallium* in Group III., is peculiarly interesting.   When Mendelejeff published his memoir on the periodic law, these elements had not been discovered. Mendelejeff predicted the properties of the three elements; he stated the atomic weight, spec. grav., general physical properties, and the formulae and chemical characters of the chief compounds, of each element. The descriptions given by Mendelejeff of the elements in question, several years before these elements were discovered, might almost be adopted now as descriptions of germanium, scandium, and gallium, so exactly in nearly every particular have they been realised.

There were two gaps in Group III., in Series 4 and 5, respectively. The differences between the values of the atomic weights of the elements in Series 2 and 4, in the various groups beginning with Group I., and of course omitting Group III., are 32, 31, 36, 37, 36, 36·5; hence, it was argued, in Group III. the difference will probably be about 33. The differences between the values of the atomic weights of the elements in Series 3 and 4 are 16, 16, 20, 20, 20, 19·5; hence, in Group III. the difference will probably be about 17. Boron, 11, occupies the position III.—1; now $11 + 33 = 44$. Aluminium, 27, occupies the position III.—3; now $27 + 17 = 44$. Therefore, it was concluded that the atomic weight of the element which is to occupy the position III.—4 would be about 44. Similar reasoning led to the value 69 for the atomic weight of the element in III.—5.

The elements in Group III., when Mendelejeff's prediction was made, shewed a gradation of properties from the non-metallic boron to the distinctly metallic thallium; boron was succeeded by the metal aluminium; the elements of the group did not fall very distinctly into two families. One of the unknown elements would find a place in Series 4 succeeding the positive metals potassium and calcium, and followed by the elements titanium, vanadium, chromium, and manganese, all of which are metals but several shew decidedly negative functions: the other unknown element would find a place in Series 5, following the decidedly metallic elements copper and zinc, and followed by the metal-like non-metal arsenic, which is again followed by the non-metals selenion and bromine. The relations of the unknown element in III.— 4 to aluminium should, it was argued, be fairly similar to those of titanium to silicon, or of vanadium to phosphorus; the unknown element would probably less closely resemble aluminium than calcium resembles magnesium, or potassium resembles sodium; but it would more closely resemble aluminium than vanadium resembles phosphorus, or chromium resembles sulphur; because when members of Series 3 and 4 are compared it is found that the resemblance is most marked in the lower members of the series.

The unknown element to be placed in III.—4 would shew analogies with boron; therefore although it must be similar to aluminium, it probably would not form an alum. But the unknown element to be placed in III.—5 must resemble aluminium more distinctly than the other unknown element does; therefore it probably would form an alum.

As calcium, which occupies in Group II. a position similar to that to be occupied by one of the unknown elements in Group III., is distinctly more positive than the first member of its own family (beryllium), but is very similar to the other members of its own family (strontium and barium), so probably would the unknown element to be placed in III.—4 closely resemble the succeeding even-series members of its group (yttrium, lanthanum, ytterbium).

The elements coming in Series 3, 5, and 7, are unlike each other in Group I., are similar but not very closely and intimately related in Group II., are very similar in Group V., and yet more similar in Groups VI. and VII.; therefore it was concluded that the unknown element in III.—5 would be distinctly similar to, but yet would shew differences from, both aluminium and indium.

Reasoning such as this guided Mendelejeff when he tabulated the properties of the elements scandium and gallium in Group III., and the element germanium in Group IV., while yet these elements were unknown.

CHAPTER XXVI.

THE ELEMENTS OF GROUP VIII.: AND RECAPITULATORY.

THE elements of Group VIII. are divided into three **483**
sections. These elements are not placed in any of the
ordinary series; but they find their places, one section, *iron,*
nickel, and *cobalt,* between Series 4 and 5; another section,
rhodium, ruthenium, and *palladium,* between Series 6 and 7;
and the third section, *iridium, osmium,* and *platinum,* between
Series 10 and 11.

It is probable that Group VIII. will some day be completed
by the discovery of three elements to come between Series 8
and 9.

Copper, which is the first element of Series 5; silver, which
is the first element of Series 7, and gold, which is the first
element of Series 11, that is, the three elements which in
order of atomic weights immediately succeed the respective
sections of Group VIII., are sometimes placed in Group VIII.

Group VIII.	Section 1.	Fe= 55·9	Ni= 58·6	Co= 59	
	Section 2.	Rh=104	Ru=104·4	Pd=106·2	**484**
	Section 3.	Ir=192·5	Os=193 (?)	Pt=194·3	

Section 1.	IRON.	NICKEL.	COBALT.
Atomic weights	55·9	58·6	59
	The *molecular weights* of these elements are unknown.		
Sp. grs. (approx.)	7·8	8·9	8·6
Atom. weights / spec. grave.	7·2	6·6	6·8
Sp. heats	·114	·108	·107
Melting points (approx.)	1500°—1600°	1400°—1500°	1400°—1500°

Section 1.	IRON.	NICKEL.	COBALT.
Appearance, and general physical characters.	Greyish-white; lustrous; crystalline; malleable; ductile; fair conductor of electricity; hard; magnetic. Iron obtained by electrolysis of FeCl₂Aq is said to be silver-white and very soft.	White; lustrous; malleable; ductile; tenacious; hard; slightly magnetic.	
Occurrence, and preparation.	Found native but not in large quantities; oxides, sulphides, carbonates, &c. occur in enormous quantities, and very widely distributed. Prepared by reducing Fe₂O₃ by C at very high temperatures; or by reducing Fe₂O₃ or FeCl₃ by H; or by electrolysis of FeCl₂Aq.	Metal is found in meteorites. Chief ore contains arsenide of Ni; sulphides, silicates, &c. also occur, not widely distributed, but in considerable quantities. Prepared by reducing NiO by C or H.	Closely resembles nickel.
General chemical properties.	Oxidised, chiefly to Fe₃O₄, by strongly heating in oxygen. Slowly oxidised by exposure to ordinary moist air. Combines directly with Cl, Br, and I, also with S. Dissolved by most acids. Forms compounds resembling alloys with C and Si. Decomposes steam at red-heat.	Slowly oxidised in moist air; oxidised by strongly heating in oxygen. Dissolved by most acids. Decomposes steam at red-heat.	

485   General formulae and characters of compounds. Nickel and cobalt very closely resemble each other in their chemical properties. The salts of cobalt are generally pink when hydrated and blue when anhydrous; the hydrated salts of nickel are usually green, and the anhydrous salts, yellow. Cobaltic chloride Co_2Cl_6 form a large series of compounds with ammonia, e.g.

$$Co_2Cl_6.10NH_3.2H_2O; \quad Co_2Cl_6'.10NH_3; \quad Co_2Cl_6.12NH_3.$$

These compounds resemble the chromium-ammonia compounds; corresponding nickel compounds are not known.

The cyanides of iron and cobalt form compounds with potassium cyanide of the forms $K_4M(CN)_6$, and $K_3M(CN)_6$, (M = Fe or Co); the acids of which these compounds are salts, viz. $H_4M(CN)_6$ and $H_3M(CN)_6$ have been obtained. Nickel cyanide does not form a corresponding salt; the compound $K_2Ni(CN)_4$ is known.

The only compound of the three metals the vapour density of which has been determined is Fe_2Cl_6; the valency of the atom of iron cannot be decisively determined from the composition of this molecule; the atom is probably tetravalent.

The formulae for the compounds of the three metals, with the exception of the haloid compounds, are the simplest by which their compositions can be expressed.

Oxides. MO, M_3O_4, M_2O_3 : hydrates of all are known.

Sulphides. MS, MS_2.

Haloid compounds. M_2X_4, M_2X_6.

Salts. MX, M_3X ; $X = SO_4$, $2NO_3$, $\frac{2}{3}PO_4$, &c.

Oxides. *Ferrous oxide* FeO is difficult to prepare free from ferric oxide, as it combines very rapidly with oxygen. The hydrated oxide $FeO.H_2O$ is obtained by pptg. ferrous sulphate dissolved in air-free water with potash in absence of oxygen.

Nickelous oxide NiO, and *cobaltous oxide* CoO, are obtained by ppg. solutions of the corresponding salts by alkalis and heating the ppts. out of contact with air. These oxides combine with oxygen when carefully heated in air, forming the oxides M_2O_3 which at a higher temperature are decomposed to MO and oxygen.

The protoxides, MO, dissolve in acids forming salts MX.

The *oxides* M_3O_4 are formed by heating the oxides MO in air ; Fe_3O_4 is also obtained by adding an alkali to a hot mixture of ferrous and ferric sulphates (or other salts) in the ratio $FeSO_4 : Fe_2(SO_4)_3$. *Ferroso-ferric oxide*, Fe_3O_4, interacts with acids to form both ferrous and ferric salts ; e.g.

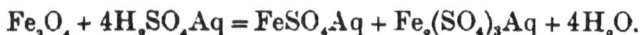

$$Fe_3O_4 + 4H_2SO_4Aq = FeSO_4Aq + Fe_2(SO_4)_3Aq + 4H_2O.$$

The corresponding oxides of nickel and cobalt form nickelous salts only, and evolve oxygen, or chlorine if hydrochloric acid is used.

Ferric oxide Fe_2O_3 is obtained by adding an alkali to a solution of a ferric salt, e.g. to $Fe_2(SO_4)_3Aq$, and drying and heating the hydrated oxide, $Fe_2O_3.3H_2O$, so obtained. This oxide interacts with acids to form ferric salts.

Nickelic oxide Ni_2O_3, and *cobaltic oxide* Co_2O_3, are obtained by oxidising solutions of nickel or cobalt salts in presence of an alkali ; e.g. by passing chlorine into potash containing $NiO.xH_2O$ or $CoO.xH_2O$ in suspension. These oxides dissolve in acids to form salts MX and evolve oxygen, or chlorine if hydrochloric acid is used ; they are decomposed to MO and oxygen when heated in air.

When ferric oxide is heated with potash and a little bromine, or when very finely divided iron is heated with potassium nitrate, and the product is poured into water, a reddish

solution is obtained which is decomposed by addition of a little
nitric acid giving a pp. of ferric hydrate and evolving oxygen.
From the quantities of ferric hydrate and oxygen thus obtained,
the existence, in the red solution, of a salt K_2FeO_4 *potassium
ferrate*, is inferred. The corresponding barium salt, $BaFeO_4$,
is said to have been obtained as a solid. Corresponding
nickelates or cobaltates are unknown.

487 Sulphides. The sulphides MS are obtained by adding
hydrogen sulphide or ammonium sulphide to aqueous solutions of salts of the three metals. None of these sulphides
shews any acidic functions. When *ferrous sulphide*, FeS, is
heated with sulphur, *iron disulphide*, FeS_2, is obtained.

488 Haloid compounds. The metals dissolve in hydrochloric acid to form solutions of the chlorides MCl_2; crystals
of the hydrated chlorides $MCl_2.xH_2O$ are formed on evaporation.

Ferrous chloride is easily oxidised to a basic ferric chloride
by evaporating its solutions in air; *nickelous* and *cobaltous
chlorides* are stable in air. Ferrous chloride has been gasified
but the vapour density has not been finally determined; the
numbers obtained seem to point to the existence of gaseous
molecules having the composition Fe_2Cl_4 at moderate temperatures, and the composition $FeCl_2$ at higher temperatures.

Ferric chloride, Fe_2Cl_6, is obtained by heating iron in a
stream of chlorine; crystals of the hydrate $Fe_2Cl_6.12H_2O$ are
obtained by dissolving iron in *aqua regia*, or by passing
chlorine into a solution of ferrous chloride, and evaporating.
Nickelic and *cobaltic chlorides*, M_2Cl_6, are very unstable and
are easily decomposed to the chlorides MCl_2 and chlorine.

489 Salts. Iron forms two series of salts; the *ferrous salts* FeX,
and the *ferric salts* $Fe_2.3X$. Nickel and cobalt form only one
series of definite stable salts MX. Ferrous salts are very
numerous; they are more or less easily oxidised to basic ferric
salts. Several normal ferric salts exist, but the greater number
are basic salts. Both series of salts form numerous double
salts. Ferric sulphate forms alums, $Fe_2 3SO_4.M_2SO_4.24H_2O$,
where $M = K$, Na, or NH_4.

490 Iron is distinctly related to manganese, the last element of
Series 4. The relation is shewn in the composition of the
oxides MO, M_3O_4, and M_2O_3, of the salts MX and $M_2.3X$,
and in the existence of ferrates analogous to the manganates,
K_2MO_4.

Iron is distinctly metallic, but the formation of ferrates
shews that it has negative tendencies.

Nickel and cobalt are less like manganese than iron is; they are decidedly metallic in their chemical characters. The three elements of Section 1 of Group VIII. form a link connecting the negative metals chromium and manganese, which are the highest members of Series 4, with the positive metal copper which forms the first member of Series 5.

491

Section 2.	RHODIUM.	RUTHENIUM.	PALLADIUM.
Atomic weights	104	104·4	106·2
	The *molecular weights* of these elements are unknown.		
Sp. grs. (approx.)	11·8	12·1	11·3
$\frac{Atom. weights}{spec. gravs.}$	8·8	8·6	9·4
Sp. heats	·058	·061	·06
Melting points	abt. 2000°	above m. p. of rhodium.	abt. 1500°—1600°
Occurrence, and preparation.	These metals occur in small quantities in many platinum ores. They are usually separated in the form of ammonio-chlorides, $2NH_4Cl.MCl_4$; when these are strongly heated the metals are obtained.		
Appearance, and general physical properties.	Greyish-white; very hard; much less ductile than I'd; scarcely softened in oxyhydrogen flame.	White; lustrous; less ductile and malleable than I'd.	White; hard; lustrous; ductile, and malleable; most fusible of the metals of this section or of Section 3 of Group VIII.
General chemical properties.	Oxidises at a red-heat, when in powder; also combines with Cl at red-heat. Unacted on by any acid when pure; when alloyed with Pt, I'b, Cu, &c. dissolves in *aqua regia*. Oxidised by fusion with KNO_3 or BaO_2.	Oxidised by heating powdered metal in air. Oxidised by heating with KOH, or KNO_3, forming K_2RuO_4 which is soluble in water. Combines directly with Cl when heated. Slowly dissolved by *aqua regia*.	Oxidised superficially in ordinary air. Slowly dissolved by hot HClAq or H_2SO_4Aq, readily by HNO_3Aq.

Section 3.	IRIDIUM.	OSMIUM.	PLATINUM.
Atomic weights	192·5	193* (?)	194·3
	The *molecular weights* of these elements are unknown.		
Sp. grs. (approx.)	21·1	21·4	21·2
$\frac{Atom. weights}{spec. gravs.}$	9·1	9·0	9·1
Sp. heats	·0326	·0311	·0324
Melting points	abt. 2500°	infusible at full white-heat.	abt. 2000°
Occurrence, and preparation.	These metals occur in small quantities associated (?alloyed) with each other and frequently with rhodium, ruthenium, and palladium. They are usually separated as $2NH_4Cl.MCl_4$, and are obtained by strongly heating these compounds.		
Appearance, and general physical properties.	White; lustrous; brittle, but fairly malleable at red-heat.	White with tinge of blue; hard; crystalline; also a black, amorphous, powder.	Silver-white; very lustrous; fairly hard; very malleable and ductile; expands by heat less than any other metal.
General chemical properties.	When finely divided oxidises slowly when heated in air, and dissolves in *aqua regia*; in compact form is insoluble in all acids. Oxidised by fusion with potash and potassium nitrate. Combines directly with Cl.	Oxidises readily to OsO_4, when heated in air in state of finely divided powder: in this state is also oxidised to OsO_4 by nitric acid. Combines directly with Cl.	Not oxidised by heating in air or oxygen. Oxidised by heating with solid potash. Combines directly with Cl. Dissolved by *aqua regia*.

* Exact value of atomic weight of osmium is doubtful; numbers vary from 193 to 199.

492 General formulae and chemical characters of compounds of metals of Sections 2 and 3. The platinum metals—i.e. rhodium, ruthenium, palladium, iridium, osmium, and platinum—are characterised by their insolubility in most acids : the three metals placed in Section 3 are also characterised by their high specific gravities. Gold resembles the platinum metals both in being insoluble in most acids and in being very heavy. Gold and the platinum metals are often named the *noble metals*.

The double chlorides $2KCl.MCl_4$ are characteristic of the platinum metals. The chlorides of these metals combine with ammonia and form several series of more or less complex ammonia-compounds. The higher oxides of these metals are acidic in their reactions with alkalis. Osmium is characterised by the easily gasified oxide OsO_4; this is the only compound of the platinum metals the vapour density of which has been determined. The compounds of the platinum metals have not been fully investigated.

493 Oxides. All the metals form *protoxides* MO, and *dioxides* MO_2. The protoxides are usually unstable. The dioxides usually dissolve in acids, but few definite salts have been isolated. The *hydrated dioxides*, which are not obtained by direct interactions between the oxides and water, generally dissolve in alkali solutions; in some cases salts of the form K_2MO_4, or basic salts $xK_2O.yMO_2$, have been separated. *Trioxides*, MO_3, are known where M = Rh, Ir, or Os : these oxides dissolve in alkali solutions. *Tetroxides of ruthenium and osmium* are known, MO_4 : they are solids with low melting points and boiling about $100°$; they form salts by interacting with alkalis.

494 Chlorides. All the metals form *dichlorides* MCl_2, and *tetrachlorides* MCl_4. *Trichlorides* MCl_3, are known where M = Ru, Rh, or Ir. The di- and tetra-chlorides combine with alkali chlorides forming compounds of which $2KCl.MCl_2$ and $2KCl.MCl_4$ are representatives; these double compounds are usually known as *chloro-platinites* (rhodites, ruthenites, &c.), and *chloro-platinates* (rhodates, &c.), respectively.

495 Salts. Very few salts of the platinum metals have been isolated. They all seem to be unstable and ill-defined compounds. Most of the salts are either basic, e.g.

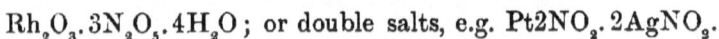

$Rh_2O_3.3N_2O_5.4H_2O$; or double salts, e.g. $Pt2NO_2.2AgNO_2$.

496 The platinum metals are evidently possessed of most of the physical properties which we are accustomed to associate with the metallic elements; but chemically considered they are both metallic and non-metallic. They differ considerably from all the other elements.

The properties of these elements are in keeping with the position assigned them by the periodic law: each section follows a series of elements the lower members of which are metallic, and the higher members are physically metallic but chemically both negative and positive.

We have now learned something of the classification of **497** elements and compounds based on the periodic law.

The elements fall into series and groups: each series, with the exception of the first, is composed of seven elements; Series 1 contains a single element, hydrogen; each group is divided into two families, those elements which occur in even series, and those which occur in odd series; a complete family consists of six elements. The nine elements, iron, nickel, cobalt, rhodium, ruthenium, palladium, iridium, osmium, and platinum, are placed to some extent apart from the others; the first section of this group (iron to cobalt) belongs both to Series 4 and Series 5, the second section (rhodium to palladium) belongs to Series 6 and 7, the third section (iridium to platinum) belongs to Series 10 and 11.

The properties of the elements in any series vary from that with the smallest atomic weight to that with the largest, so that the first and last members of the series are the most unlike one another. When the gradation of properties has been completed through a series, the properties, as it were, swing back to the starting point, and exhibit a similar gradation in the next series. The properties of the first members of Groups II. to VII., and of the first and second members of Group I., are to a certain extent typical of the properties of all the other members of these groups.

The properties of an element and its compounds are connected with the general properties of the group to which the element belongs and also with the gradation of properties in that group; they are likewise connected with the general properties of the series and with the gradation of properties in the series to which the element belongs. Inasmuch as the properties of each group and series are connected with the properties, and the gradation of properties, of the other groups and series, it follows that the relations of any element to other similar elements are to be elucidated only by studying the position of the specified element in the complete scheme of classification.

The periodic law seeks to connect the changes in the properties of the elements, and in the compositions and properties of compounds, with the changes in the atomic weights of the elements; and it endeavours to make this connexion definite and to present it in accurate terms.

If the atomic weight of an element is known, the place of the element in the classificatory scheme is determined, and therefore the properties of the element and its compounds can be stated in a general way. If the properties of an element and its compounds are determined, the position of this element in the orderly sequence of elements can be found, and thereby an approximately correct value can be deduced for the atomic weight of the element. We have had examples of the application of the periodic law both to the classification of elements and to the determination of the best values of the atomic weights of elements.

498 The valencies of the atoms of the elements are undoubtedly important factors determining the compositions of compounds. These valencies probably vary periodically with variations in the atomic weights of the elements; but the valencies of so few elements have been definitely determined that we are not at present able to state the connexions between changes of atomic valencies and of atomic weights.

The maximum valencies of all the elements in Series 2, except lithium, have been determined by considering the compositions of gaseous molecules of compounds of these elements with monovalent atoms. The valencies are as follows;—

Groups.

	I.	II.	III.	IV.	V.	VI.	VII.
Series 2.	Li	Be	B	C	N	O	F
Valency.	(? one)	two	three	four	three	two	one.

The atom of lithium is probably monovalent; assuming this atom to be monovalent, we see that the maximum valency of the atoms of the elements of Series 2 increases from the first to the middle member, and then decreases to the last member.

We cannot assert that maximum atomic valency varies in all the series exactly as it does in Series 2; nevertheless, the assumption that it does thus vary would certainly in some cases lead to results which are confirmed by observation.

Thus, the assumption requires the atoms of the elements in Group IV. to be tetravalent; the maximum valencies of the atoms of all the members of this group, except cerium, have been determined and the atoms have been found to be tetravalent. On the other hand, the assumption requires the atoms of the elements in Group VI. to be divalent; but the existence of the gaseous molecules $MoCl_3$, $TeCl_4$, WCl_6, UCl_4, proves that the maximum valency of some at least of these atoms is greater than two.

The conception of atomic valency—the conception, that is, that every atom in a molecule directly interacts with a limited number of other atoms—has been deduced from the study of gaseous molecules, and is strictly applicable to gaseous molecules only. But the greater number of the compounds of inorganic chemistry are non-gasifiable bodies; the conception of atomic valency cannot therefore, at present, be made use of, otherwise than in a broad and general way, in questions regarding the conditions which determine the compositions of compound molecules.

The periodic law asserts that the compositions of compounds vary periodically with variations in the atomic weights **499** of the elements. If we are content to use the expression *composition of a non-gasifiable compound* as meaning the ratio of the numbers of atoms forming the chemically reacting weight of the compound, then we can express the compositions of classes of similar compounds by general formulae, and we can trace connexions between these formulae and the atomic weights of the elements.

Thus, let $X = F$, Cl, Br, I, OH, NO_3, ClO_3, $\dfrac{O}{2}$, $\dfrac{S}{2}$, $\dfrac{SO_4}{2}$, $\dfrac{CO_3}{2}$, &c., then the compositions of a great many important compounds of the elements of the different groups may be thus expressed;—

Group I.

General formula	RX

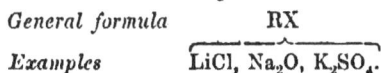

Examples	LiCl, Na₂O, K₂SO₄.

Group II.

General formula	RX₂

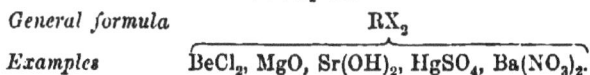

Examples	BeCl₂, MgO, Sr(OH)₂, HgSO₄, Ba(NO₃)₂.

Group III.

General formula RX_3

Examples BF_3, Sc_2O_3, $Al_2(SO_4)_3$.

Group IV.

General formulae RX_2 RX_4

Examples SnO, $SnCl_2$, $PbSO_4$; CCl_4, CeO_2, $Sn(SO_4)_2$.

Group V.

General formulae RX_3 RX_5

Examples $BiCl_3$, $NO_2(OH)$, $Bi(NO_3)_3$; PF_5, $NbCl_5$, $PO(OH)_3$.

Group VI.

Gen. formulae RX_3 RX_4 RX_6

Examples Cr_2O_3; $TeCl_4$, UCl_4, $U(SO_4)_2$; CrO_3, $SO_2(OH)_2$, WCl_6, $UO_2(SO_4)$.

Group VII.

General formulae RX RX_2 RX_4 RX_7

Examples ICl; MnO, $MnSO_4$; MnO_2; $MnO_3(OH)$, $ClO_3(OH)$.

Group VIII.

General formulae RX_2 RX_3 RX_4

Examples $FeSO_4$, $Ni(NO_3)_2$; $Fe_2(SO_4)_3$, $AuO(OK)$; $PtCl_4$;

                                                   RX_8

                                    $PtPCl_6$, $PtClSO_2(OH)$.

The compositions of the well marked compounds of the elements of a group become more varied as we pass from the lower to the higher groups.

It would however be out of place in an elementary book to attempt to generalise the connexion between the changes in the forms of compounds and the variations in the atomic weights of the elements. Let it suffice to note that the arrangement of the elements in accordance with the periodic law indicates the existence of such a connexion, and points the way by which the nature of this connexion may be elucidated.

500   If the properties of the elements and their compounds vary periodically with variations in the atomic weights of the elements, accurate determinations of the atomic weights of all the known elements must be demanded in chemistry. The values given for these constants in the tables on pp. 57, 59, and 60, are given in round numbers only. The atomic weights of many elements have been determined with great accuracy, those of other elements only with a fair degree of accuracy, and those of a few elements with but little accuracy. The following table presents the most trustworthy results of the various determinations.

Atomic weights of the elements.

Element	At. Wt.	Element	At. Wt.	Element	At. Wt.
Aluminium	27·02	Hydrogen	1	Ruthenium	104·4
Antimony	120	Indium	113·4	Scandium	44
Arsenic	74·9	Iodine	126·53	Selenion	78·8
Barium	136·8	Iridium	192·5	Silicon	28·3
Beryllium	9·08	Iron	55·9	Silver	107·66
Bismuth	208	Lanthanum	138·5	Sodium	23
Boron	10·9	Lead	206·4	Strontium	87·3
Bromine	79·75	Lithium	7·01	Sulphur	31·98
Cadmium	112	Magnesium	24	Tantalum	182
Caesium	132·7	Manganese	55	Tellurium	125
Calcium	39·9	Mercury	199·8	Thallium	203·64
Carbon	11·97	Molybdenum	95·8	Thorium	231·8
Cerium	139·9	Nickel	58·6	Tin	117·8
Chlorine	35·37	Niobium	94	Titanium	48
Chromium	52·4	Nitrogen	14·01	Tungsten	183·6
Cobalt	59	Osmium	193 (?)	Uranium	240
Copper	63·2	Oxygen	15·96	Vanadium	51·2
Didymium	144	Palladium	106·2	Yttrium	89·6
Erbium	166	Phosphorus	30·96	Ytterbium	173
Fluorine	19·1	Platinum	194·3	Zinc	64·9
Gallium	69	Potassium	39·04	Zirconium	90
Germanium	72·3	Rhodium	104		
Gold	197	Rubidium	85·2		

CHAPTER XXVII.

WE have been endeavouring to connect the changes of composition with the changes of properties exhibited by certain kinds of matter called elements and compounds. We have divided the elements and their compounds into classes, each of which is more or less distinctly marked off from the others. The compositions of compounds have been represented by formulae which exhibit the number of combining weights of each element combined in one reacting weight of a compound ; or, in the language of the only theory of the structure of matter which has been found capable of explaining observed facts, the formulae of compounds exhibit the number of atoms of each element combined in one molecule of a compound. Some of the formulae of compounds also suggest reactions by which these compounds are formed, or reactions which occur between them and other substances ; such formulae not only state the compositions of the compounds but also indicate certain of their properties.

Our study of changes of properties and changes of composition has shewn that the connexions between these cannot be wholly perceived or understood so long as we look only to the compositions and properties of the substances forming a chemical system at the beginning of a reaction, and at the compositions and properties of the substances forming the system at the close of the reaction. It became necessary for us to pay some regard to the changing systems during the process of change. By doing this we were led to picture to ourselves many apparently simple chemical occurrences as consisting of two or more parts, and to regard the state of equilibrium finally attained by a system of chemically interacting substances as frequently the result of direct and reverse changes occurring between different members of the system.

The examination of the distribution of the chemically interacting substances in a system free to settle down into equilibrium led us to give a definite meaning to the term affinity as applied to acids and bases. We found it possible to obtain measurements of the relative affinities of various acids and bases, and thus to attach to each acid and base a constant number which conveys much quantitatively accurate information regarding the amounts of various chemical changes in which the acids and bases play important parts. We were able to see that there are definite connexions between the compositions of acids and bases and the values of the affinity-constants of these compounds. The accurate development of this connexion is in the future.

Although we have endeavoured to separate the chemical from the physical parts of the events we have studied, we have found the two classes of phenomena sometimes so inextricably interwoven that it was impossible wholly to ignore the physical aspects of certain chemical occurrences. Chemical changes, we found, are accompanied by changes of energy, and, as the net result, part of the energy of the initial system is always degraded when the system has attained to chemical equilibrium.

That we might form clear mental pictures of the mechanism of chemical changes, we found it almost necessary to adopt the conceptions and the language of the molecular and atomic theory. This theory put before us two definite portions of each kind of matter, the atom and the molecule; it enabled us to form well-defined conceptions of each of these extremely minute masses. Of the two conceptions, we found the atom the more definite; we were obliged to confine ourselves to gases when attempting to reason accurately concerning molecules, and even then we found it necessary to allow some latitude to our notion of the molecule, such latitude as is implied in the physical definition of the gaseous molecule as "that minute portion of a gas which moves about as a whole, so that its parts, if it has any, do not part company *during the motion of agitation of the gas.*"

As the molecular theory has been developed from the study of gaseous phenomena, and is as yet strictly applicable only to gases, we found it advisable to speak of chemical changes occurring between solid or liquid substances as being inter-

actions between reacting weights,—meaning thereby aggregates
or collocations of atoms—rather than between molecules, of
the bodies taking part in the changes.

The study of the interactions of gases led to the conception
of the gaseous molecule as a structure built up of definite
numbers of atoms arranged in a definite manner; chemical
facts obliged us to connect the properties of gaseous molecules
not only with the nature, and the number, but also with the
arrangement, of their parts. It was sometimes necessary to
count two or more atoms in a molecule as a single atom, so
far as certain chemical changes were concerned; we thus
gained the conception of the compound radicle.

As a guide in our attempts to learn something about the
arrangement of the parts of molecules, we made use of the
hypothesis of atomic valency, which asserts that each atom
forming part of a gaseous molecule is capable of directly interacting with a limited number of other atoms. We agreed to
measure the maximum number of atoms between which and
any specified atom there could be direct intramolecular action
by the maximum number of atoms of hydrogen, fluorine,
chlorine, bromine, or iodine, with which the specified atom
combines to form a gaseous molecule.

The more our study of chemistry advanced the more
importance were we led to attach to those constants, the
atomic weights of the elements, until at last we arrived at a
system of chemical classification, based on the atomic weights
of the elements, which as it is developed seems to include in
itself all other classificatory schemes.

Chemistry is the daughter of alchemy. The object of
both has always been to find the changeless foundation of
changing phenomena.

Alchemists dreamt of the philosopher's stone, and worked
hard to find it. Chemists have found the elements, and beneath the elements they have found the atoms, and beneath
the atoms they sometimes think they perceive the atoms of
the one element, of which all the known elements and compounds, it may be, are developed forms.

INDEX.

The numbers refer to paragraphs.